# HANDBOOK of POWER CYLINDERS, VALVES and CONTROLS

**1st Edition**

**Compiled by Michael J. Barber C.Eng. M.I.Mech.E.and the Editors of The Trade & Technical Press Limited**

**ISBN 85461 100 2**



Published by
The Trade & Technical Press Limited
Crown House, Morden, Surrey SM4 5EW
England

*Printed in Great Britain by TTP (Printers) Limited, East Molesey, Surrey, England.*

# PREFACE

The need to increase efficiency and exploit the benefits of hydraulics and pneumatics to the full has lead to dramatic changes in fluid power systems design.

Where great force and torque are required, hydraulic cylinders and valves have that unique capability. Pneumatics provides for fast control and now with the introduction of micro-electronics, the advantages of torque and speed are married with precision and 'brains'.

This Handbook takes a positive and practical approach to the subject and combines in one volume, essential technical information on the principles and control of hydraulics and pneumatics, together with detailed coverage of all types of hydraulic and pneumatic cylinders, valves and their electronic controls. Hundreds of helpful illustrations and line drawings are included to emphasize component construction and function.

The Handbook of POWER CYLINDERS, VALVES & CONTROLS will be useful to all engineers, design engineers and maintenance personnel in their understanding and selection of hydraulic and pneumatic system components.

**The Publishers**

## ACKNOWLEDGEMENTS

Abex – Denison Ltd
Association of Hydraulic Equipment Manufacturers
Atlas Copco Ltd
Henry Berry & Co Ltd
Robert Bosch Ltd
British Compressed Air Society
British Standards Institute
Brown & Sharpe Fluid Power Ltd
Compair Automation Ltd
Crouzet Ltd
Enerpac Ltd
Festo Ltd
IMI Norgren Enots Ltd
Kay Pneumatics Ltd
IMI Martonair Ltd
Journal of Applied Pneumatics
Mecman Ltd
Monsun-Tison Ltd
Moog Controls Ltd
Origa Ltd
Parker Hannifin Ltd
RGS Electro-Pneumatics Ltd
Ross Valve UK Ltd
Sacol Powerline Ltd
Schrader Bellows/Division of Parker Hannifin
TI Reynolds Ltd
Vickers Systems Ltd Lang Pneumatic
Voith Engineering Ltd

# CONTENTS

# ACKNOWLEDGMENTS – ILLUSTRATIONS AND TABLES


| Page Number | | Company |
|---|---|---|
| 11 | | Bosch Hydraulics |
| 12 | | Bosch Hydraulics |
| 13 | | Bosch Hydraulics |
| 27 | | Ciba-Geigy |
| 30 | | Parker Hannifin |
| 46 | | Vickers Systems Ltd/Libby Owens Ford |
| 47, | bottom | G.L. Rexroth Ltd |
| 48, | bottom | Bosch Hydraulics |
| 53 | | Abex Denison Ltd |
| 60 | | Vickers Systems Ltd/Libby Owens Ford |
| 61 | | Monsun-Tison Ltd |
| 62 | | Hydraulic, Pneumatic, Mechanical Power Ltd (Power International) |
| 71 | | Vickers Systems Ltd/Libby Owens Ford |
| 73, | bottom | Double 'A' Hydraulics Ltd |
| 74 | | G.L. Rexroth Ltd |
| 79 | | Commercial Hydraulics |
| 80, | top | Commercial Hydraulics |
| | bottom | Vickers Systems Ltd/Libby Owens Ford |
| 84 | | V.A. Hawe |
| 88 | | Pratt Precision Hydraulics |
| 91, | top | Sacol Powerline Ltd |
| 93 | | V.A. Hawe |
| 94 | | Mecman Ltd |
| 98 | | T I Reynolds Ltd |
| 102 | | Double 'A' Hydraulics Ltd |
| 106, | bottom | Pratt Precision Hydraulics |
| 109 | | Sacol Powerline Ltd |
| 110, | top | Sacol Powerline Ltd |
| | bottom | Double 'A' Hydraulics |
| 113 | | Dowty Hydraulic Units Ltd |
| 114 | | Enerpac Ltd |
| 118 | | G.L. Rexroth Ltd |
| 120 | | M.A.N. |
| 121, | top | Kalmar |
| | bottom | Clark |
| 123 | | Air Power and Hydraulics Ltd |
| 125 | | Sacol Powerline Ltd |
| 126 | | Sacol Powerline Ltd |
| 127, | bottom | Joseph Young & Sons Ltd |
| 128 | | Enerpac Ltd |
| 132 | | Abex Denison |
| 133, | top | Abex Denison |
| | bottom | Double 'A' Hydraulics Ltd |
| 134, | top | Double 'A' Hydraulics Ltd |
| | bottom | Double 'A' Hydraulics Ltd |
| 135 | | Abex Denison |
| 136, | top | Legris |
| 136, | bottom | Vickers Systems Ltd/Libby Owens Ford |
| 138 | | G.L.Rexroth Ltd |
| 139, | top | G.L. Rexroth Ltd |
| | bottom | G.L. Rexroth Ltd |
| 141 | | G.L. Rexroth Ltd |
| 142 | | Double 'A' Hydraulics Ltd |
| 143 | | G.L. Rexroth Ltd |
| 144 | | G.L. Rexroth Ltd |
| 145 | | G.L. Rexroth Ltd |
| 146, | top | G.L. Rexroth Ltd |
| 147 | | Hytrol Hydraulic Valves |
| 149 | | Double 'A' Hydraulics Ltd |
| 150, | top | Monsun-Tison Ltd |
| | bottom | G.L. Rexroth Ltd |
| 151 | | Dowty Hydraulic Units Ltd |
| 152 | | Moog Controls Ltd |
| 153 | | Moog Controls Ltd |
| 156 | | G.L. Rexroth Ltd |
| 215, | top | Air Automation |
| | bottom | Compair Maxam |
| 218, | top | IMI Norgren Enots Ltd |
| | bottom | IMI Norgren Enots Ltd |
| 219, | top | Parker Hannifin |
| 222, | top | IMI Norgren Enots Ltd |
| | bottom | IMI Norgren Enots Ltd |
| 223 | | IMI Norgren Enots Ltd |
| 224, | top | IMI Norgren Enots Ltd |
| | bottom | IMI Norgren Enots Ltd |
| 225 | | IMI Norgren Enots Ltd |
| 227, | bottom | Compair Maxam |
| 231 | | Festo Pneumatics |
| 235 | | Asco (UK) Ltd |
| 237, | top | IMI Norgren Enots Ltd |
| | bottom | IMI Norgren Enots Ltd |
| 238, | top | Parker Hannifin |
| | bottom | Atlas Copco Ltd |
| 239 | | Festo Pneumatics |

| Page Number | Company |
|---|---|
| 240 | Festo Pneumatics |
| 241 | Festo Pneumatics |
| 242 | IMI Martonair Ltd |
| 243 | Parker Hannifin |
| 244 | Parker Hannifin |
| 248, top | |
| bottom | Mecman Ltd |
| 249 | Bosch Hydraulics |
| 250 | IMI Martonair Ltd |
| 251, top | Origa Ltd |
| 253, top | Mecman Ltd |
| 254 | IMI Martonair Ltd |
| 259, bottom | Vickers Systems Ltd/Libby Owens Ford |
| 260 | H. Kuhnke Ltd |
| 261 | Festo Pneumatics Ltd |
| 262, top | Vickers Systems Ltd/Libby Owens Ford |
| centre | IMI Martonair Ltd |
| bottom | Vickers Systems Ltd/Libby Owens Ford |
| 263, top | Baldwin |
| bottom | Vickers Systems Ltd/Libby Owens Ford |
| 264, top | Vickers Systems Ltd/Libby Owens Ford |
| bottom | Compair Maxam |
| 265, bottom | Atlas Copco |
| 266, top | 'Consultair' |
| bottom | Air Automation |
| 267, top | Air Automation |
| bottom | IMI Martonair Ltd |
| 268, top | Vickers Systems Ltd/Libby Owens Ford |
| bottom | Parker Hannifin |
| 269 | Air Automation |

| Page Number | Company |
|---|---|
| 271, top | Atlas Copco |
| centre | IMI Martonair Ltd |
| bottom | Atlas Copco |
| 272, top | Atlas Copco |
| 275 | Maxam Power Ltd |
| 276 | Atlas Copco |
| 277 | Maxam Power Ltd |
| 278 | Mecman Ltd |
| 281 | Compair Automation Ltd |
| 283 | Origa Ltd |
| 284 | IMI Martonair Ltd |
| 286 | IMIMartonair Ltd |
| 288 | IMI Martonair Ltd |
| 292, top | Ross Valve UK Ltd |
| 293 | IMI Nogren Enots Ltd |
| 294 | Mecman Ltd |
| 295, top | RGS Electro-Pneumatics Ltd |
| bottom | Drallim Controls |
| 296, top | IMI Norgren Enots Ltd |
| bottom | IMI Norgren Enots Ltd |
| 297 | IMI Norgren Enots Ltd |
| 298 | Atlas Copco |
| 299, top | Crouzet Ltd |
| 305 | IMI Norgren Enots Ltd |
| 306, top | IMI Norgren Enots Ltd |
| bottom | Kay Pneumatics Ltd |
| 307, top | IMI Norgren Enots Ltd |
| bottom | Compair Automation Ltd |
| 310 | Compair Automation Ltd |
| 311 | IMI Martonair Ltd |
| 312 | Atlas Copco |
| 314 | Mecman Ltd |
| 315, bottom | Kay Pneumatics Ltd |
| 316, top | Parker Hannifin |
| bottom | Parker Hannifin |
| 317 | Burkert Contromatic Ltd |
| 321 | IMI Norgren Enots Ltd |

# SECTION 1

## Hydraulic Principles

BASIC THEORY
PROPERTIES OF HYDRAULIC FLUIDS
PERFORMANCE CALCULATIONS
CIRCUIT SYMBOLS

# Basic Theory

THE FLUID power cylinder, or linear actuator, is one of the most useful machines to provide a direct method of converting pressure energy into an output force, capable of doing useful work with high efficiency of conversion.

The working medium can be a pressurized liquid (*hydraulic* cylinders) or pressurised air (*pneumatic* cylinders), both offering specific advantages and having attendant disadvantages which will be examined throughout this book. In a large number of applications, the two can be regarded as competitive, or possible alternatives. In others, experience, or the particular service requirements, favour one or other system, where the main advantages of that system can be fully exploited.

**Hydrostatics**

The majority of hydraulic systems are *hydrostatic* in character, *ie* transmittiing power by *static pressure* energy. However, most practical systems involve fluid flow through pipes and valves to give movement to the cylinder rod. This involves forces generated by motion and thus *hydrodynamic* phenomena. Hydraulic actuators, therefore, basically work on a hydrostatic principle, while flow through the pipelines and valves to the actuators conform to hydrodynamic laws.

The laws of static pressure, first expounded by Blaise Pascal (1623 to 1662), a French philosopher, state that:

a) A pressure transmitted through a fluid at rest, gaseous or liquid, neglecting its weight, will be the same at all points in the fluid.

b) The pressure at any point in a static fluid will be exerted equally in any direction.

c) The pressure will act normal to *ie* at right angles or perpendicular to) any surface in contact with the fluid, or at any section taken through the fluid.

The name of Pascal has been perpetuated, as the International System of Units (SI) now recognizes the *Pascal* (Pa) as the basic unit for pressure measurements.

If the weight of the fluid is considered, in a purely hydrostatic system, a static pressure is produced by virtue of a mass of fluid supported by a base area. This pressure (P) is dependent only on the height of the column of liquid (H) and its density (ρ).

$$P = H \times \rho \times g$$

where g is the acceleration due to gravity.

In engineering terms $P = H \times w$

where w is the specific weight of the fluid.

In the case of fluid in a container, *eg* a tank or reservoir, hydrostatic pressure exerts a force (F) on the base area:

$$F = A \times P \text{ (in consistent units)}$$

where A is the base area.

The pressure at any point within the fluid is determined by its by its depth below the surfaces of the liquid, or *head* (h) – Figure 1. A head, so determined, is called the *potential* head and may be less than the total head available. The total head, represented by h + z in Figure 1, is called the *piezometric* head. It follows that the piezometric head equals the potential head plus any additional head available or:

$$\text{piezometric head} = \frac{P}{w} + z$$

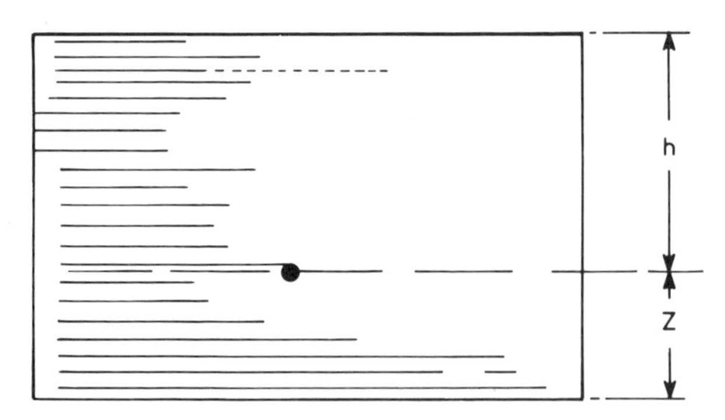


*Figure 1*

The piezometric head is constant for all fluids at rest. This is a special case of the *Bernoulli* equation where the velocity terms are zero. In fact, hydrodynamic equations are directly applicable to fluid statics by rendering all the terms involving fluid motion as zero. Conversely, the application of hydrostatic principles to the transmission force through a fluid must take velocity into account if fluid movement is involved, although in practice this may be small enough to be negligible. This applies generally when flow rates are small, the fluid performs a 'passive' role in the transmissions of pressure and the small amount of fluid movement involved has no significant effect on the performance of the system.

As stated in Pascal's laws, the pressure exerted on any surface immersed in the fluid is thus equal to the product of the fluid in a closed system is transmitted equally to all parts of the system and acts perpendicularly to all surfaces in contact with the fluid. For practical hydrostatic application the weight of fluid can be ignored if the potential head involved is very small compared with the applied pressure.

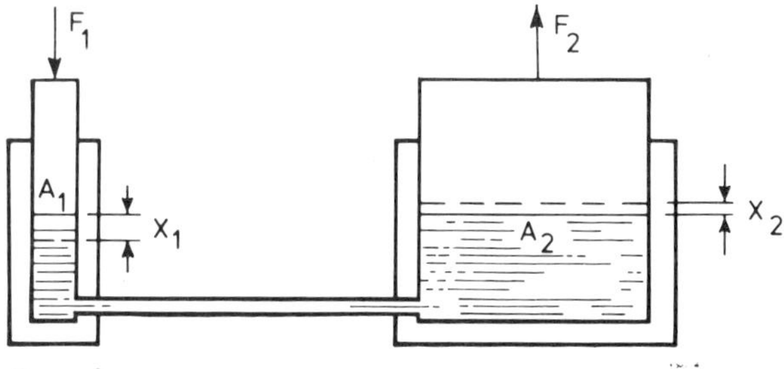


*Figure 2*

**Force Multiplication**

Forces exerted on surfaces by hydrostatic pressure are proportional to areas, so force multiplication is possible by using pistons of different areas as shown in Figure 2. Here $F_1$ applied to piston $A_1$ develops a pressure of $F_1/A_1$ which is transmitted throughout the fluid. Thus the resulting pressure on piston area $A_2$ is $(F_1/A_1)$ or:

$$\text{Output force } F_2 = \text{input force } F_1 \times \frac{A_2}{A_1}$$

This force multiplication is achieved in the ratio $A_2/A_1$.

This is a valid hydrostatic system since the displacement of fluid is small enough for velocity loss to be neglected.

In a practical system, the modifying factors are the weights of the respective pistons ($W_1$ and $W_2$) and the friction of the piston seals ($f_1$ and $f_2$) - see Figure 3. The true output can then be calculated as:

$$\text{Output force } F_2 = (F_1 + W_1 - f_1) \frac{D_2^2}{D_1^2} - (W_2 - f_2)$$

Respective piston travels are not modified in any way and are:

$$X_2 = X_1 \times \frac{D_1^2}{D_2^2}$$

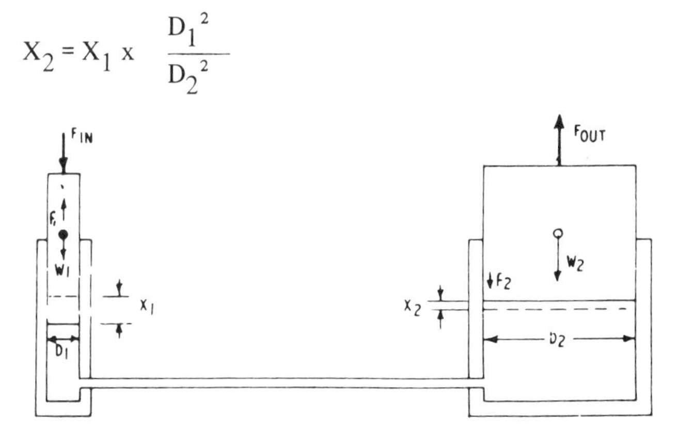


*Figure 3*

Similarly the work done on piston 1 is equal to the work done by the loaded piston 2.

$$\text{or } F_1 \times X_1 = F_2 \times X_2$$

These equations define the performance in a basic hydrostatic system of force multiplication over pressure ranges where the compressibility of the fluid is negligible. The first practical use of these principles was made by Joseph Bramah in 1975 when he patented the hydraulic press. A simple hand pump was used to provide the input force $F_1$ and the output force $F_2$ was generated by an early ram type hydraulic cylinder.

Not all such systems are necessarily force multipliers, however. They are equally suitable for force transfer with high efficiency. The typical automobile brake system may, in fact, employ slave cylinders of smaller diameter than the master cylinder. All force multiplication required is applied at the mechanical output, where a maximum pedal force of the order of 445N (100 lbf) with a master cylinder of 25 mm (1 in) diameter can yield a system of 42 to 56 bar (600 to 800 lbf/in$^2$), with virtually 100% force transfer for the slave cylinders.

**Pressure Generation**

In early water hydraulic systems pressure was generated by using a weight loaded hydraulic accumulator which consisted of a large hydraulic ram supporting a series of weights. This produced a constant static pressure throughout the system. The accumulator was recharged and the weights raised by using hand or steam driven pumps. In modern hydraulic systems gas charged accumulators are used where a static pressure has to be maintained or stored without the expenditure of power.

Modern hydraulic systems usually have high speed positive displacement pumps to provide the fluid flow and pressure required. It must be remembered that the pump only provides the fluid flow and a pressure is generated once a resistance to that flow is encountered.

Where a hydraulic pump acts as the transmitter in a force transfer system, the ram is 'sized' on the basis of the output force required and the fluid pressure available from the pump, *ie:*

$$\text{ram diameter } d = \sqrt{\frac{4 \times F}{\pi \times P}}$$

where F = output force required
P = pressure available from pump.

The delivery required from the pump is governed by the speed of working required:

$$\text{delivery } Q = \frac{\pi D^2}{4} \times \frac{S}{t}$$

where S = stroke
t = time to complete stroke.

Pump output required then follows as:

$$\frac{F}{k} \times \frac{S}{t}$$

where k is a constant depending on units employed

In engineering units:

$$\text{Power out (kW)} = \frac{F \times S}{10^5 \times t}$$

where F is in Newtons
S is in cm
t is in seconds

$$\text{Power out (hp)} = \frac{F \times S}{6600 \times t}$$

where F is in lbf
S is in inches
t is in seconds

These equations can also be written in terms of flow and pressure as:

$$\text{Power out (kW)} = \frac{Q \times P}{600}$$

where Q is in lit/min
P is in bar

$$\text{Power out (hp)} = \frac{Q \times P}{1430}$$

where Q is in gal/min
P is in $\text{lbf/in}^2$

These relationships assume that fluid velocity components are negligible, *ie* the system is truly hydrostatic. In practical systems it may be necessary to take into account velocity components and back-pressure effects; also seal friction and, in the case of large vertical rams, ram weight. Pressure drops in pipework and through valves and orifices are available in tabular of nomographic form for flow velocities of fluids of different viscosities. The *Reynolds* number is used to determine if the flow is laminar or turbulent.

**Cylinder performance**

For the purpose of determining the thrust developed by a hydraulic cylinder, frictionless motion is assumed, with no internal leakage. The thrust then follows directly from the piston area and fluid pressure, *ie*:

$$F = A \times P \quad \text{(in consistent units)}$$

where A = effective piston area
P = fluid pressure.

The effective piston area is the full piston area in the case of a single-rod cylinder extending, or the annulus area in the case of a single-rod cylinder retracting, or either stroke of a through-rod cylinder (see Figure 4):

$$F = 0.7854\ D^2 \times Pe \text{ (single-rod, extending)}$$
$$F = 0.7854\ (D^2 - d^2)\ Pe \text{ (other cases)}$$

where D = cylinder bore
d = rod diameter
Pe = effective fluid pressure
= actual fluid pressure entering the cylinder.

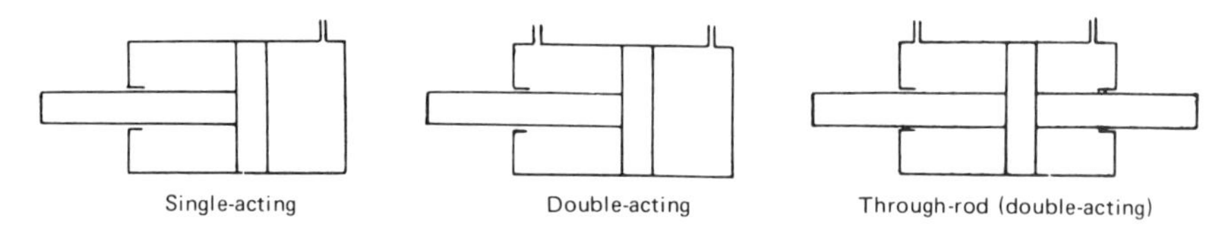


*Figure 4*

This is suitable for arbitrary cylinder sizing, with an allowance for frictional losses of the order of 2 to 5%. A further allowance may be made to correct for the fact that the effective pressure (Pe) will be lower than the nominal system pressure by virtue of *pressure drop* in the lines to the cylinder. Also the nominal system pressure may fluctuate between high and low values, which can be determined accurately if an accumulator is incorporated in the system. Using the nominal system pressure as a preliminary figure, a deduction of 10% is commonly taken to allow for pressure drop.

Thus a cylinder selected to work off a nominal 140 bar (2000 $lbf/in^2$) supply pressure would be 'sized' on an effective pressure of, say 126 bar (1800 $lbf/in^2$) and the thrust force required uprated by some 5% over the actual thrust requirements. On this approximate basis, a suitable cylinder bore size could be calculated directly from:

$$D = \sqrt{\frac{F}{0.7 \times P}} \quad \text{in consistent units}$$

In practice, the nearest standard size would be selected to the calculated value of bore (D); and preferably the next largest standard size. The method is inexact, ignores several other factors affecting the performance of the cylinder, and can result in excessively 'oversizing' and in some cases 'undersizing'. Nevertheless, it is a simple method of initially sizing a cylinder for the majority of applications for a single-rod cylinder extending.

Since the effective piston areas are different, it is obvious that the performance or thrust in one direction, is reduced in the case of a single-rod double acting cylinder by the amount: $0.7854 \times Pe \times d^2$.

This may or may not be significant, depending on if the cylinder is 'working' both ways, or only working against a load in the extending direction. If double-working, the cylinder size may have to be determined on the lower output force available on the retracting stroke. The appropriate working formula rendered in similar terms becomes:

$$D = \sqrt{\frac{F}{0.7 \times P} + d^2}$$

Pressure drop in the lines to the cylinder may be calculated directly to arrive at the correct value of Pe. To be exact, such calculations should take into account the pressure drop through the cylinder ports, although this is commonly ignored because of the relatively low flow velocities involved. The true pressure drop, however, is not given by simple calculations, but must take into account *back pressure* effects which govern the actual flow through the lines. This can be illustrated by reference to Figure 5.

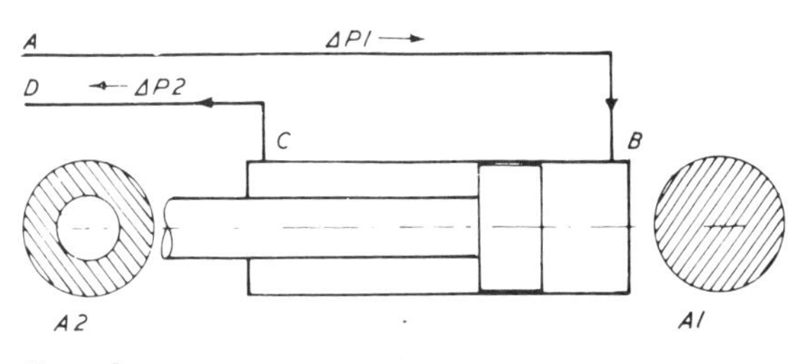


*Figure 5*

Assuming, for simplicity, that the lines to the cylinder are of equal length and size, the pressure drop in either line will be directly proportional to the flow rate in each line. With the cylinder extending, the flow rate in the inlet line AB will be proportional to the piston area $A_1$, and the flow rate in the outlet line CD will be proportional to the annulus area $A_2$. The pressure drop in line AB and CD will thus be in the ratio $A_1 : A_2$.

The pressure drop in line AB will represent initial pressure drop $\triangle P_1$ from the system pressure applied at the far end of the line. The pressure drop in line CD will represent a back pressure on the piston, opposing movement, which can only be overcome by further pressure on the other side of the piston. This additional pressure represents a further loss of effective pressure. However, since the piston area is greater than the annulus area (to which the back pressure is applied), this additional loss will be less than the actual back pressure ($P_2$) in the ratio $A_2 : A_1$.

In other words:

Pressure drop in inlet line = $\triangle P_1$
Additional loss of effective pressure due to overcoming back pressure in return line

$$= \frac{A_2}{A_1} \times \triangle P_2$$

Therefore true effective pressure (Pe) equals system pressure (P) less each of these losses:

$$Pe = P - \triangle P_1 \triangle \frac{A_2}{A_1} \times \triangle P_2$$

Also

$$P_2 = \frac{A_2}{A_1} \times \triangle P_1$$

Because of the flow rate relationship previously established, this can be rendered more conveniently in the form:

$$\text{Total pressure drop} = \Delta P \left(1 + \frac{1}{r}\right)$$

where $\triangle P$ is the calculated pressure drop in the inlet line to the cylinder
r is the ratio of effective piston areas $A_1 : A_2$

With the reverse direction of operation, line DC becomes the inlet and line BA, the outlet. Similar analysis will show that in this case

$$\text{Total pressure drop} = \Delta P \left(r + \frac{1}{r}\right)$$

Such a method of calculation enables an accurate estimate of pressure drop to be made, which may be necessary for critical application. It will also show how performance can be modified if a large differential exists between the two back pressures, as could occur with a large area ratio, *ie* larger than usual rod diameter. Such loss of performance will generally be apparent by one or other of the cylinder movements slowing up to adjust to the considerable difference in total pressure drops. In normal applications however,back pressure effects are usually ignored, and their possible effect on performance is usually negligible provided the ratio of areas is 4 : 3 or less (4 : 3 being the usual proportion for cylinders with standard rod sizes).

# Properties of Hydraulic Fluids

WATER WAS originally used as an hydraulic fluid in the first presses and in early hydraulic cylinders. Leather *U-seals* were used at the rod gland and minor leakages were not considered important in the dockside and marine applications of the early 19th century.

Although the disadvantages of using water were appreciated, such as the relatively high freezing point, low boiling point, and the corrosive affect on steel; oil was not universally adopted as an hydraulic fluid until the late 1920s.

Joseph Bramah had forseen this with a patent of 1802 which described an oil hydraulic cylinder used with a planing tool, but it was the development of machine tools giving fine surface finishes, together with the introduction of synthetic rubber seals, which made it practically possible.

The first oil hydraulic cylinders were developed for automobile brakes and power steering but were soon adopted by the machine tool industry, which appreciated the advantage of using a fluid able to lubricate and protect the hydraulic equipment used.

Today, due to the safety requirements of using fire-resistant fluids and the increasing cost of mineral oils, the trend is towards using water or water-based fluids as the hydraulic fluid.

Basic fluid parameters of main significance in hydraulics are *density* (or alternatively *specific gravity*), *viscosity*, *specific heat* and *compressibility*.

The density of fluid is defined as the mass per unit volume, *viz*

$$\text{density } (\rho) = \frac{\text{(mass) lb}}{\text{volume}} = \frac{\text{lbf}}{\text{volume x g}} = \frac{\text{kgf}}{\text{volume x g}}$$

in consistent units

Although rendered obsolete by modern unit definitions, specific weight remains a convenient engineering unit, defined as weight per unit volume, or

$$\text{specific weight (w)} = \frac{\text{lbf}}{\text{volume}} = \frac{\text{kgf}}{\text{volume}}$$

in consistent units

Specific gravity is a dimensionless quantity and is the ratio of the density (or specific weight) of a fluid to the density (or specific weight) of water. In the case of equations for engineering calculations it is often desirable to render density or specific weight, where it appears as a factor, in terms of specific gravity, thus avoiding any possible confusion between the true numerical values of density and specific weight which are to be employed in the formula.

The significance of specific gravity as a hydraulic fluid parameter is that it gives an indication of the weight of the fluid in the system, or more directly a comparison of fluid weights for a given system where different fluids may be considered. Also the higher the specific gravity of the fluid the more difficult to lift the fluid in the suction part of the system. The design of the suction side may therefore need particular care in order to avoid the possibility of cavitation and erratic pump operation.

**Viscosity**

The viscosity of a fluid is a measure of its internal resistance. *Dynamic* viscocity ($\mu$) is defined in terms of the force in dynes between two parallel laminae or layers of fluid each 1 cm$^2$ in area with a slip velocity of 1 cm/sec between them, the corresponding unit of viscosity being the poise. Because the dynamic viscosity of real fluids determined in poises is invariably a fractional quantity, the more usual unit employed for expressing dynamic viscosity is the *centipoise,* or one hundredth of a poise. The significance of dynamic viscosity is that it is effectively a *friction coefficient.*

For engineering calculations it is usually more convenient to employ *kinematic* viscosity *(v)* rather than dynamic viscosity, this being determined as the absolute dynamic viscosity divided by the mass density of the fluid. The standard unit is the *stoke*(St) but for the same reason as above the practical unit is invariably taken as a *centistoke* or one hundredth of a stoke (cSt).

Kinematic viscosity is used for the calculation of flow characteristics, and thus dynamic pressure.

**Practical Viscosity Values**

Logically all fluid viscosity values should now be quoted in centistokes. Actual measurement of fluid viscosity, however, continues to be made in *viscometers* yielding values in arbitrary scales: *Redwood No 1 seconds, Saybolt Universal Seconds (SUS),* or *Engler degrees.*

There are no exact conversions between these scales, nor for conversion of arbitrary viscosities in seconds or degrees to kinematic viscosities in centistokes, ft$^2$/sec, or in$^2$/sec units. Close approximate conversions can be made by reference to conversion scales or conversion tables. It should be noted that such conversions apply only at the same temperature as the original measurement.

In the case of a *non-Newtonian fluid* the instantaneous viscosity of the fluid is dependent on the shear stress in that fluid at that particular moment. If necessary, a specific viscosity figure can be obtained with a viscometer which ensures a uniform shear rate throughout measurement. Such a figure will, however, have limited practical value, unless the shear stability characteristics of the fluid are also known.

The variation of viscosity with temperature is one of the most significant parameters with hydraulic fluids, affecting both the performance and selection of a fluid. This can be fully expressed by plotting a characteristic curve for the fluid on an ASTM chart (Figure 1). The scales of an ASTM chart are so designed that the characteristic curve for most fluids is linear. Given a number of spot readings for viscosity and temperature, a close approximation to the viscosity-temperature characteristics of that fluid at intermediate temperatures can be obtained by joining these points with a straight line and extending this line in either direction for temperature outside the range covered by the spot values.

This chart (Figure 1) also shows equivalent kinematic viscosity values in four different scales and the viscosity ranges covered by SAE oil number ratings – these are still widely used for general classifications of oil viscosities. See also Figures 2 and 3 (pages 12 and 13).

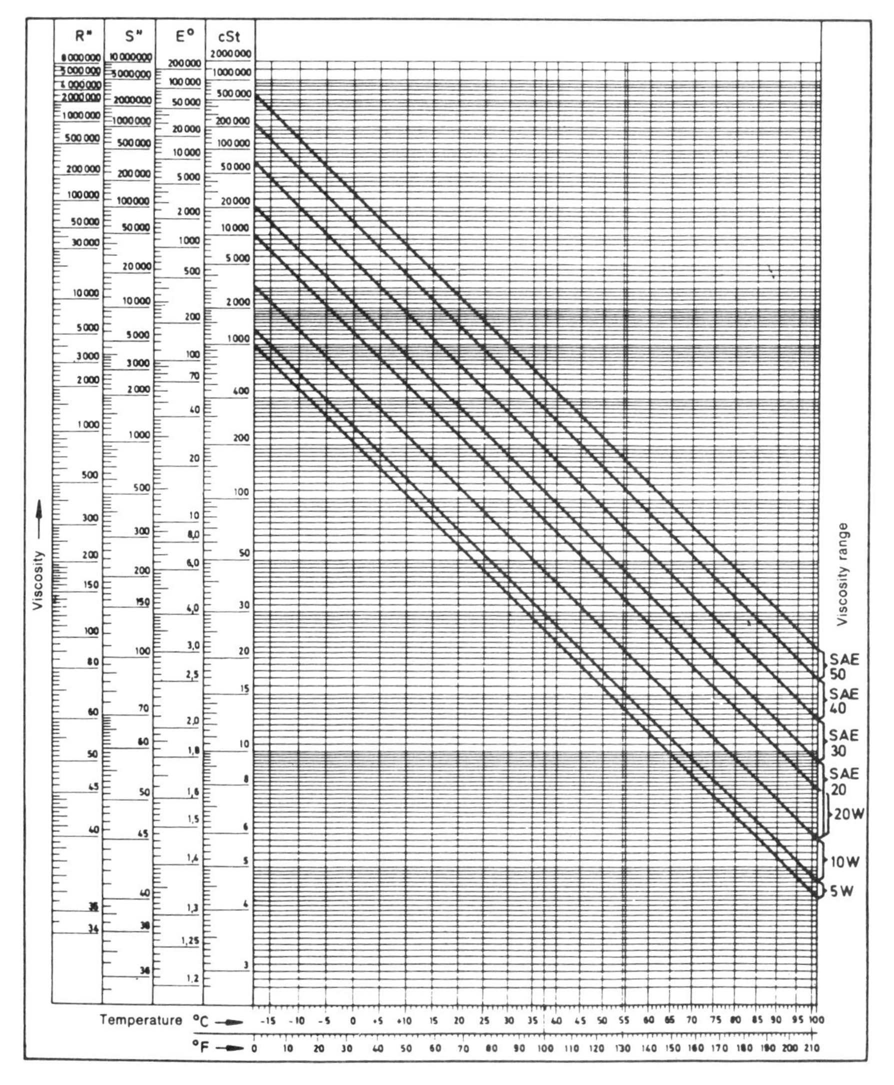


*Figure 1*
*Viscosity-temperature chart.*

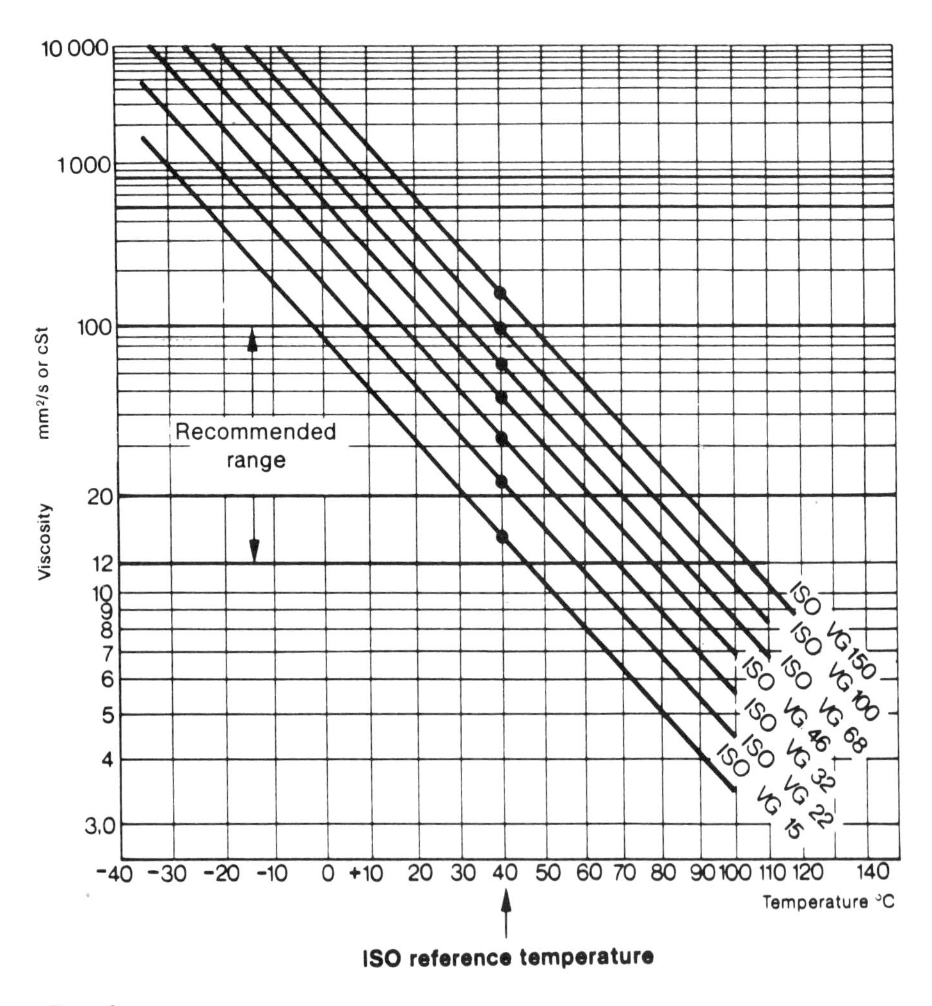
10 000
1000
100
20
12
10
9
8
7
6
5
4
3,0
Viscosity mm²/s or cSt
Recommended range
ISO VG 150
ISO VG 100
ISO VG 68
ISO VG 46
ISO VG 32
ISO VG 22
ISO VG 15
-40 -30 -20 -10 0 +10 20 30 40 50 60 70 80 90 100 110 120 140
Temperature °C
ISO reference temperature

*Figure 2*

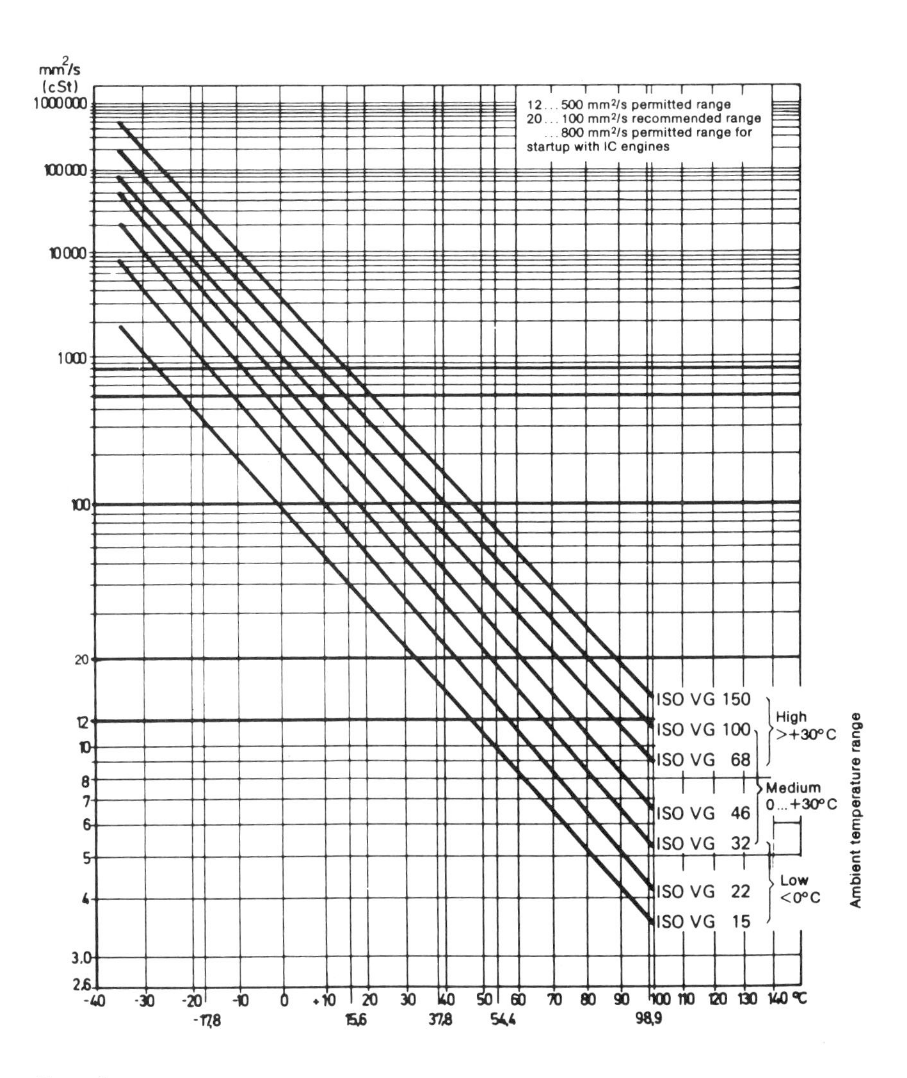

mm²/s
(cSt)
1000000
100000
10000
1000
100
20
12
10
8
7
6
5
4
3.0
2.6
12 ... 500 mm²/s permitted range
20 ... 100 mm²/s recommended range
... 800 mm²/s permitted range for startup with IC engines
ISO VG 150
ISO VG 100
ISO VG 68
ISO VG 46
ISO VG 32
ISO VG 22
ISO VG 15
High >+30°C
Medium 0 ... +30°C
Low <0°C
Ambient temperature range
-40 -30 -20 -10 0 +10 20 30 40 50 60 70 80 90 100 110 120 130 140 °C
-17,8
15,6
37,8
54,4
98,9


*Figure 3*

**ISO/BS Viscosity Classification**

Viscosity classification for industrial liquid lubricants defined by British Standard BS 4231 establishes 18 viscosity grades in the range 2 to 1500 centistokes covering approximately (in the case of mineral oils) the range from kerosene to cylinder oils and thus also embracing the normal range of mineral-oil-based hydraulic fluids. Classification is based on the principle that the mid-pont kinematic viscosity of each grade should be approximately 50% higher than that of the preceding one. Using this numbering system, oil viscosities are quoted as ISO viscosity grade (number), or ISO Viscosity Grade VG (number) – see Table 1. Table 2 gives corresponding kinematic viscosities at various temperatures (related also to Viscosity Index). VDMA classifcations for hydraulic fluids are given in Table 3.

**TABLE 1 – ISO VISCOSITY CLASSIFICATIONS**

| ISO viscosity grade | Mid-point kinematic viscosity cSt at 40.0 °C | Kinematic viscosity limits cSt at 40.0 °C | | ISO viscosity grade | Mid-point kinematic viscosity cSt at 40.0 °C | Kinematic viscosity limits cSt at 40.0 °C | |
|---|---|---|---|---|---|---|---|
| | | Minimum | Maximum | | | Minimum | Maximum |
| ISO VG 2 | 2.2 | 1.98 | 2.42 | ISO VG 68 | 68 | 61.2 | 74.8 |
| ISO VG 3 | 3.2 | 2.88 | 3.52 | ISO VG 100 | 100 | 90.0 | 110 |
| ISO VG 5 | 4.6 | 4.14 | 5.06 | ISO VG 150 | 150 | 135 | 165 |
| ISO VG 7 | 6.8 | 6.12 | 7.48 | ISO VG 220 | 220 | 198 | 242 |
| ISO VG 10 | 10 | 9.00 | 11.0 | ISO VG 320 | 320 | 288 | 352 |
| ISO VG 15 | 15 | 13.5 | 16.5 | ISO VG 460 | 460 | 414 | 506 |
| ISO VG 22 | 22 | 19.8 | 24.2 | ISO VG 680 | 680 | 612 | 748 |
| ISO VG 32 | 32 | 28.8 | 35.2 | ISO VG 1000 | 1000 | 900 | 1100 |
| ISO VG 46 | 46 | 41.4 | 50.6 | ISO VG 1500 | 1500 | 1350 | 1650 |

**Viscosity Index**

Viscosity Index (VI) is a single number representation of the viscosity-temperature characteristics of a fluid. The higher the viscosity index the smaller the change in viscosity unit temperature, and *vice versa,* although this is only a rough guide as to actual change. VI values may extend beyond 100, when the correct designation is $VI_E$ *(Extended Viscosity Index).*

**Specific Heat**

The specific heat of a fluid is defined as the ratio of the heat required to raise the temperature of a given volume of fluid by one degree to that required to raise the temperature of the same volume of water by the same amount. It can thus be specified either in Btu or calorie-gm-cm-units. The specific heat is not constant but varies with temperature, although for practical calculation a constant value is often assumed based on a nominal temperature range.

**TABLE 2 – ISO VISCOSITY CLASSIFICATION WITH CORRESPONDING KINEMATIC VISCOSITIES AT VARIOUS TEMPERATURES FOR DIFFERING VISCOSITY INDICES (BS 4231 : 1982)**

| ISO Viscosity grade | Kinematic Viscosity range | Approximate kinematic viscosity at other temperatures for differing values of viscosity index | | | | | |
|---|---|---|---|---|---|---|---|
| | | Viscosity Index = 50 | | | Viscosity Index = 95 | | |
| ISO VG | cSt at 40 °C | cSt at 20 °C | cSt at 37.8 °C | cSt at 50 °C | cSt at 20 °C | cSt at 37.8 °C | cSt at 50 °C |
| 2 | 1.98 to 2.42 | (2.87 to 3.69) | (2.05 to 2.52) | (1.69 to 2.03) | (2.92 to 3.71) | (2.06 to 2.52) | (1.69 to 2.03) |
| 3 | 2.88 to 3.52 | (4.59 to 5.92) | (3.02 to 3.70) | (2.38 to 2.84) | (4.58 to 5.83) | (3.01 to 3.69) | (2.39 to 2.86) |
| 5 | 4.14 to 5.06 | (7.25 to 9.35) | (4.37 to 5.37) | (3.29 to 3.95) | (7.09 to 9.03) | (4.36 to 5.35) | (3.32 to 3.99) |
| 7 | 6.12 to 7.48 | (11.9 to 15.3) | (6.52 to 8.01) | (4.68 to 5.61) | (11.4 to 14.4) | (6.50 to 7.98) | (4.76 to 5.72) |
| 10 | 9.00 to 11.0 | 19.1 to 24.5 | 9.68 to 11.9 | 6.65 to 7.99 | 18.1 to 23.1 | 9.64 to 11.8 | 6.78 to 8.14 |
| 15 | 13.5 to 16.5 | 31.6 to 40.6 | 14.7 to 18.0 | 9.62 to 11.5 | 29.8 to 38.3 | 14.6 to 17.9 | 9.80 to 11.8 |
| 22 | 19.8 to 24.2 | 51.0 to 65.8 | 21.7 to 26.6 | 13.6 to 16.3 | 48.0 to 61.7 | 21.6 to 26.5 | 13.9 to 16.6 |
| 32 | 28.8 to 35.2 | 82.6 to 108 | 31.9 to 39.2 | 19.0 to 22.6 | 76.9 to 98.7 | 31.7 to 38.9 | 19.4 to 23.3 |
| 46 | 41.4 to 50.6 | 133 to 172 | 46.3 to 56.9 | 26.1 to 31.3 | 120 to 153 | 45.9 to 56.3 | 27.0 to 32.5 |
| 68 | 61.2 to 74.8 | 219 to 283 | 69.2 to 85.0 | 37.1 to 44.4 | 193 to 244 | 68.4 to 83.9 | 38.7 to 46.6 |
| 100 | 90.0 to 110 | 356 to 454 | 103 to 126 | 52.4 to 63.0 | 303 to 383 | 101 to 124 | 55.3 to 66.6 |
| 150 | 135 to 165 | 583 to 743 | 155 to 191 | 75.9 to 91.2 | 486 to 614 | 153 to 188 | 80.6 to 97.1 |
| 220 | 198 to 242 | 927 to 1180 | 230 to 282 | 108 to 129 | 761 to 964 | 226 to 277 | 115 to 138 |
| 320 | 288 to 352 | 1460 to 1870 | 337 to 414 | 151 to 182 | 1180 to 1500 | 331 to 406 | 163 to 196 |
| 460 | 414 to 506 | 2290 to 2930 | 488 to 599 | 210 to 252 | 1810 to 2300 | 478 to 587 | 228 to 274 |
| 680 | 612 to 748 | 3700 to 4740 | 728 to 894 | 300 to 360 | 2880 to 3650 | 712 to 874 | 326 to 393 |
| 1000 | 900 to 1100 | 5960 to 7640 | 1080 to 1330 | 425 to 509 | 4550 to 5780 | 1050 to 1290 | 466 to 560 |
| 1500 | 1350 to 1650 | 9850 to 12600 | 1640 to 2010 | 613 to 734 | 8390 to 9400 | 1590 to 1960 | 676 to 812 |

**TABLE 3 – VDMA FLUID CLASSIFICATION**

| Hydraulic Fluids | | Approximate viscosity at 50 °C (122 °F) cSt |
|---|---|---|
| With additives for increasing the resistance to ageing and for improving the protection against corrosion. | HL 16<br>HL 25<br>HL 36<br>HL 49<br>HL 68 | 12 to 20<br>21 to 29<br>32 to 40<br>44 to 54<br>62 to 74 |
| With extra additives for improving the behaviour in the boundary lubrication range. | HLP 16<br>HLP 25<br>HLP 36<br>HLP 49<br>HLP 68 | 12 to 20<br>21 to 29<br>32 to 40<br>44 to 54<br>62 to 74 |

In the absence of specific figures the following semi-empirical formula can be used to calculate the specific heat of typical hydraulic oils (See also Table 4).

$$\text{specific heat} = \frac{0.388 \times 0.00045t}{\sqrt{sg}}$$

Where sg = specific gravity of the fluid at 60 °F
t = temperature
specific heat is then given in units of Btu/lb per F with temperature in °F; or cal/g per C with temperature in °C

The actual specific heat of a hydraulic oil may be appreciably modified by the presence of additives, and also of contaminants in the fluid.

**Compressibility of Fluids**

Specifically, compressibility (usually designated by the symbol ß) is the reciprocal of the *fluid bulk modulus* (K). However, the bulk modulus is not constant, tending to increase with temperature and decrease non-linearly with pressure. The instantaneous value of the bulk modulus at any pressure is called the *tangent bulk*

**TABLE 4 – SPECIFIC HEAT OF TYPICAL FLUIDS cal/g °C**

| Fluid | Temperature – °C | | | | |
|---|---|---|---|---|---|
| | 15 | 20 | 25 | 60 | 80 |
| Mineral oil (typical) | – | 1.0 | – | – | – |
| Water-glycol | – | – | 0.718 | 0.774 | 0.785 |
| Phosphate ester | 0.43 | – | – | – | – |
| Silicone | 0.39 | – | – | – | – |
| Water | 1.0011 | 1.000 | 0.9992 | 1.0000 | 1.0033 |

*modulus* (Kt), and the mean value of the bulk modulus from atmospheric pressure to any pressure (P) is called the *secant bulk modulus* (Ks):

$$K_t = - V \frac{dP}{dV}$$

where V is the volume at pressure P

$$K_s = - \frac{V_o P}{V_o - V}$$

where $V_o$ is the volume at atmospheric pressure.

The values of the tangent and secant moduli will tend to coincide at lower pressures (*ie* as P approaches atmospheric pressure), and for pressures up to about 70 bar (1000 lbf/in$^2$) the difference can usually be ignored. For general engineering calculations a 'typical' bulk modulus value may be quoted and used for pressures up to 700 bar (10000 lbf/in$^2$). Logically this should be the secant modulus. For working at specific high pressures, however, the tangent modulus should be used, if known. (See also Table 5).

**TABLE 5 – COMPRESSIBILITY OF TYPICAL FLUIDS**
**(Compressibility expressed as percentage reduction in volume)**

| Fluid | Pressure | |
|---|---|---|
| | 70 bar (1000 lbf/in$^2$) | 700 bar (10000 lbf/in$^2$) |
| Water | 0.34 | 3.3 |
| Water-in-oil emulsions | 0.35 | 3.5 |
| Water-glycol | 0.26 | 2.6 |
| Mineral oils (typical) | 0.35 | 3.4 |
| Phosphate ester | 0.25 | 2.5 |
| Chlorinated hydro carbon | 0.24 | 2.4 |

Typical working figures are a reduction in initial volume of 0.5% per 1000 lbf/in$^2$ (0.00735% per atmosphere) for hydraulic oils and 0.4% per 1000lbf/in$^2$ (0.0059% per atmosphere) for water. Such figures are reasonably valid for pressures up to 700 bar (10000 lbf/in$^2$) and over a temperature range of 10–100 °C (50–200 °F).

The compressibility of a normal hydraulic oil at 20 °C (68 °) and 70 bar (1000 lbf/in$^2$) can also be estimated quite accurately from its kinematic viscosity at 22 °C (72 °F), using the empirical formula

'compressibility' $= 3.5 - 0.2 \log \nu$
where $\nu$ = viscosity in centistokes and compressibility is given in in$^3$/lb x $10^6$ units.

Alternatively, 'compressibility' may be expressed in terms of relative density, *viz:*

$$\text{'compressibility'} = \left( \frac{1 - \triangle\ 1000}{\triangle_0} \right) \times 100\% \text{ per } 1000 \text{ lbf/in}^2$$

where $\triangle_0$ = fluid density at atmospheric pressure
$\triangle_{1000}$ = fluid density at 70 bar (1000 lbf/in$^2$)

As a direct result of compressibility, the density of any real fluid will increase with pressure. In very high pressure systems this may be more significant than the volumetric change.

**Vapour Pressure**
The vapour pressure of a fluid is the pressure exerted by the saturated vapour in contact with the surface of the fluid at a specified temperature. The higher the vapour pressure the more volatile the fluid, and/or the nearer it is to boiling point. Fluids with a high vapour pressure, either due to their volatile nature or because of a high operating temperature, are therefore prone to 'flash' into vapour under suction conditions, thus setting a specific limit to the net positive suction head a pump can accommodate without cavitating.

Additionally, as the boiling point of a fluid approaches, the more volatile fractions will come off first, progressively changing the nature of the fluid. This is seldom significant in oil fluids at normal working temperatures, but with water fluids progressive loss of water may be experienced at quite moderate working temperatures.

**Aniline Point**
The aniline point of a mineral oil is the lowest temperature at which the oil is completely miscible with an equal volume of freshly distilled aniline. It is a general indication of the aromatic content of the oil (paraffinic oils having a high aniline point and aromatic oils a low aniline point), and because of this is sometimes used as a form of compatibility index.

**Surface tension**
Surface tension may be significant as affecting:

(1) Foaming characteristics at the free liquid surface or interface between two non-miscible fluids.

(2) The ability of the fluid to 'wet' a metal surface.

(3) Inherent leakage past seals and at joints, *etc.*

Characteristics (1) and (2) can be adequately controlled by additives if necessary. Item (3) is not normally significant in the case of oil fluids which have adequate surface tension to make sealing relatively easy. Fluids which have a surface tension of less than 30 dynes/cm$^2$, however, are troublesome to seal, needing particular attention to joints, and one must often accept that some leakage will be inevitable with practical designs of seals.

**Cloud Point**
The cloud point of a mineral oil is that temperature at which waxes or other solids normally present in solution tend to crystallize out, or come out of the solution. This can lead to clogging or partial choking of the system.

**Pour Point**
This is the temperature at which the thickening action of separation of the waxy constituents is so marked that the fluid ceases to flow. On the ASTM curve this would be marked by an an abrupt rise in viscosity. The pour point (temperature) is determined with regard to specific flow conditions.

**Thermal Expansion**
The coefficient of volumetric expansion of an oil remains practically constant over the usual range of working temperatures encountered in hydraulic systems. Specific values are largely related to the specific gravity of the oil. The coefficient of volumetric expansion, however, decreases rapidly with increasing pressure, *eg* of the order of 0.000025 per C per 70 bar (1000 lbf/in$^2$). The relative volume of a pressurized fluid is thus less than that predicted on the basis of volume correction coefficients alone.

**Thermal Conductivity** (See also Table 6)
The thermal conductivity is a measure of the ability of a fluid to dissipate heat. In a practical system heat dissipation may be hindered by the formation of boundary layer films and thus fluids which do not 'wet' the internal surfaces tend to have lower thermal conductivities. The lower the thermal conductivity of the fluid the higher its working temperature will tend to be, under similar operating conditions. Thus mineral oils, having a generally low thermal conductivity, will tend to run at higher working temperatures than water-based or water-glycol fluids used in a similar system.

**TABLE 6 – THERMAL CONDUCTIVITY OF FLUIDS**
**Cal/cm sec °C x $10^{-3}$**

| Fluid | | Temperature | | | |
|---|---|---|---|---|---|
| | Nominal | 10 °C | 50 °C | 70 °C | 80 °C |
| Mineral oil | 3.24 | – | – | – | – |
| Water-glycol | 9.61 | – | – | – | – |
| Phosphate ester | 5.30 | – | – | – | – |
| Glycerine | – | – | – | 6.8 | – |
| Water | – | 14.7 | 15.4 | – | 16.0 |

The thermal conductivity of a hydraulic oil can be estimated from the following semi-empirical formula:

$$\text{thermal conductivity} = \frac{0.821 - 0.000244t}{sg} \text{ Btu in/ft}^2\text{h}$$

$$= \frac{(1.2 - 0.00035t) \times 10^{-3}}{sg} \text{ cal gm/cm}^2 \text{ h}$$

where t = temperature in °C or °F, respectively
sg = specific gravity of fluid.

By comparison the thermal conductivity of water is some 4.5 times greater, and that of water-glycol solutions some 3 times greater.

**Flash Point**

The flash point broadly defines the relative fire hazard of a fluid. The flash point may also serve as an indication of the type of an oil or blend because the more volatile the oil the lower the *flash point* and *vice versa*. The flash point may be determined by various standard 'open' or 'closed' cup tests. These tests measure the tendency of the fluid to ignite when brought into contact with a hot surface. They are particularly relevant where spills may leak on to hot exhaust pipes, *etc*, but give similar results to auto-ignition temperature values Auto-Ignition Temperature (AIT).

**Spontaneous Ignition Temperature**

A steel block, with a thermometer set into it to allow its temperature to be read, is heated strongly from below. Droplets of the fluid under test are allowed to fall into a cavity in the block.

Conventional hydraulic oils vaporize at relatively low temperatures and ignite with a sharp report. The water containing types of water-in-oil emulsions and water-glycol mixtures form a steam blanket which must be dispersed so that extremely high temperatures are needed for spontaneous ignition.

Phosphate ester fluids, on the other hand, with their innate resistance to fire, cannot support combustion but will ignite spontaneously at very high temperatures. The behaviour of the various fluids in this test differs; interpretation of the results is difficult and not amenable to tabulation.

**Auto-Ignition Temperature (AIT) (ASTM D 2155–66)**

This is the temperature at which a fluid will ignite spontaneously. It is a measure of the tendency of the fluid to withstand overheating and contact with hot surfaces. This is considered to be one of the relevent tests for hydraulic fluids.

**Hydraulic Fluids**

In hydrostatic systems the prime requirement of the fluid is to transmit pressure, hence low viscosity and low compressibility are important features. Other factors which must be considered are: oxidation stability, load carrying ability, corrosion protection, cleanliness, seal compatibility air entrainment and foaming, viscosity index, pour point and filtration.

The viscosity level of an hydraulic fluid is normally selected on the basis of the

**TABLE 7 – GENERAL CHARACTERISTICS OF HYDRAULIC FLUIDS**

| | Mineral Oil | Phosphate Ester | Water–Glycol | Water–in-oil Emulsion | Oil–in-water Emulsion | Chlorinated Aromatics |
|---|---|---|---|---|---|---|
| Specific gravity (typical) | 0.864 | 1.275 | 1.060 | 0.916–0.94 | Less than 1.0 | 1.43 |
| Maximum service temperature °C | 110 | 150 | 65 | 65 | 65 | 150 |
| Water content (%) | None | None | 45 | 40 | 95 | None |
| Fire-resistance | Poor | Good | Excellent | Good | Good | Good |
| Viscosity index | High | Low | Very high | High | Low | Low |
| Lubricating properties | Excellent | Very good | Good | Fair | Fair | Fair-good |
| Special seals | No | Yes | No | No | No | Yes |
| Special Paints | No | Yes | Yes | No | No | Yes |
| Rust prevention | Very good | Fair | Fair | Good | Fair | Fair |
| Toxicity | None | Slight | None | None | None | Slight |

speed and discharge pressure of the pump which must be satisfactorily lubricated for optimum performance. Too low viscosity will cause pump slip, leakage and inadequate lubrication, while too high a viscosity will cause overheating and cavitation.

**Types of Fluids**
Hydraulic fluids can be grouped into four distict types, *ie*:

(1) *Mineral oils* – either uninhibited or treated with additives.

(2) *Emulsions* – water-in-oil or oil-in-water.

(3) *water-based glycols.*

(4) *Synthetic fluids:*

- (a) Phosphate esters;
- (b) Carboxylate esters;
- (c) Chlorinated hydrocarbons;
- (d) Poly alpha olefines;
- (e) Polyglycols;
- (f) Silicone fluids.

Of these groups (2) and (3), and (a), (b) and (c) from group (4) are classified as fire-resistant fluids.

Table 7 compares the general characteristics of the main types of fluids.

**Mineral Oils**
Mineral oils are the normal choice for industrial hydraulic systems, with the advantage of offering nearly all the requirements of an hydraulic fluid except for fire-resistance. Straight mineral oil lubricants can be regarded as suitable for hydraulic systems working under ideal conditions, particularly with low fluid temperatures and in perfectly clean systems.

As a general rule, however, such oils are compounded with special additives to produce hydraulic oils specifically intended for use in all practical hydraulic systems. Specialized oil hydraulic equipment is invariably designed around an oil of a specific viscosity, primarily to suit the requirements of the pump, and a hydraulic oil is always implied in such cases.

**Additives for Mineral Oils**
Most modern hydraulic oils are compounded with additives, notably oxidation inhibitors, corrosion inhibitors and anti-foam agents. Some oils may have less and others more (*eg* film strength improvers or anti-wear additives can be advantageous where high bearing loads are involved, and pour point depressants for fluids used in very low working temperatures, or starting up from cold in very low ambient temperatures). A separate additive is also commonly employed to improve the viscosity index of the oil.

The main cause of deterioration with a straight mineral oil is oxidation. The rate of oxidation is enhanced by heating (*eg* high working temperatures for the oil), agitation, which is present in most hydraulic systems, and the presence of contaminants which can act as catalyst (notably metal particles).

Apart from the loss of lubricating properties, the onset of oxidation is accompanied by the formation of soluble and insoluble degradation products, the latter being deposited in the system in the form of sludge. The oil also loses its ability to separate from water and air, both of which invariably have contaminants present, and will tend to become increasingly acid, which can lead to corrosion.

*Oxidation inhibitors* work by showing a preferential absorption for oxygen and thus remain effective as long as there is active additive remaining. Some, such as the phosphorous and sulphur compounds, also possess marked anti-wear and anti-corrosive properties and are thus multi-purpose additives. Oxidation additives are usually added in concentrations of up to 5%, this being the maximum figure for which such additives are fully effective.

*Corrosion inhibitors* are essentially rust inhibitors capable of adhering strongly to metallic surfaces and isolating them from contact with air and moisture. The selection of a suitable additive is quite critical, however, both to meet the service conditions concerned and avoid interaction with other additives. In particular, certain types of rust inhibitors have a degrading effect on oxidation inhibitors, while others may have a secondary effect of working as an emulsifying agent tending to emulsify any free water present in the oil.

*Anti-foam agents* are added to ensure effective release of entrained air from the oil surface in the tank without excessive foaming developing at the surface. Basically they are 'foam-breakers', causing an early disruption of the air bubbles as they appear. By this means, air normally dissolved in mineral oil and released at lower pressures, or any entrained air, is released with no adverse effects on the working of the system.

*Anti-wear additives* are basically film-strength improvers, greatly assisting in maintaining the full lubrication properties of the oil. A typical anti-wear additive is zinc dithiophosphate in proportions up to about 1%.

**VI Improvers**

The VI of an oil can be raised with additives. The additives used are normally polymerized methylacrylate esters, or Butane or Styrene oleofines, in proportions from 4% to 8%. All such additives are susceptible to shear break-down and so the initial VI achieved is seldom maintained in practice, the extent of the break-down being dependent on the rate of shear experienced by the oil. In general, an initial loss may be expected during the first few hours of working in the system, after which the VI should remain appreciably constant through out the useful life of the oil, unless continually subjected to shear stresses in a particular part of the system.

Modern polymeric VI improvers show very much better viscosity retention at even high shear rates than their earlier counterparts but still have limitations at extremely high pressure, *eg* 350 bar (5000 lbf/in$^2$). For very high pressure applications where no shear loss can be tolerated, synthesized hydraulic oils are to be preferred because they do not use polymeric thickeners and show no shear loss – see Figure 4

**Emulsions**

Water-in-oil and oil-in-water emulsions are similar in general behaviour, the particular difference being that in the former water is distributed in droplet form through an oil medium and in the latter oil droplets are distributed through a water medium. As a consequence, oil-in-water emulsions (with a water medium) are rather more fire-resistant than water-in-oil emulsions. On the other hand, the latter (with an oil medium) generally have better lubricating properties. (Water-in-oil

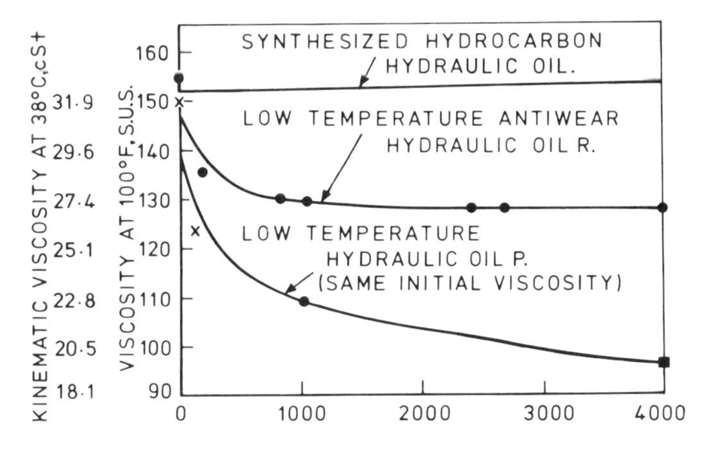


*Figure 4*
*Vane pump shear stability test results 70 bar (1000 lbf/in²).*

emulsions are more widely used and may have a water content of up to 60%, fire-resistance being directly related to water content).

Similar additives may be included as for mineral oils, notably oxidation inhibitors, anti-wear and anti-corrosion additives and also emulsifying agents to maintain the emulsion in stable form. VI improvers are not used because water-in-oil emulsions are non-Newtonian fluids with no VI as such. Their actual viscosity is dependent on the rate of shear and at very high shear rates reverts to that of the oil content itself. This generally limits their application to systems or components which do not produce high localized rates of shear, *eg* such emulsions would generally be unsuitable for use with high-speed vane pumps or with rolling bearings.

Properly formulated, water-in-oil emulsions can be quite stable, although some separation may occur if the fluid is allowed to stagnate. This will tend to result in an oil-rich layer or emulsion, or even a pure oil layer forming at the top of the tank level. This need not be significant, for there is usually sufficient agitation on re-starting to reform a consistant emulsion as the fluid is circulated through the system. A more likely cause of trouble is where separation is caused by the presence of contaminants, as such separation may be permanent, largely because the emulsifying agent has probably been exhausted by the presence of the contaminants. Therefore, water-in-oil emulsions are most reliable in clean systems; they also have strictly limited working temperatures, in common with other water-based fluids, and the possibility of water loss through evaporation and subsequent modification of the fluid make-up.

Oil-in-water emulsions are essentially water, containing approximately 2 to 5% of an emulsifying system to provide limited lubrication and anti-corrosion properties. Their use is limited to massive systems, *ie* which discharge to waste and in pressure transmitters such as pit-props where lubrication is not very demanding. Their main limitation has been their inability to lubricate pump bearings satisfactorily.

**Water-Based Glycols**

Water-glycol fluids originated as straight water-glycerine mixtures, with the glycerine content adjusted to give the required degree of protection against freezing in water-hydraulic systems. The glycerine content employed can range up to 50%. A secondary advantage offered by such mixtures is a raising of the viscosity of the fluid and an improvement in VI. Cost, however is relatively high, nullyfying one of the basic advantages of using water as hydraulic fluid.

Water-polyglycol mixtures have, however, been further developed as industrial fire-resistant fluids, offering superior protection to water-in-oil emulsions, and lower cost and minimal compatibility problems compared with phosphate esters.

The lubricating properties of these mixtures are greatly improved by the incorporation of anti-wear and load carrying additives to provide satisfactory lubrication under boundary film conditions; and their viscosity is increased by the addition of polymer type thickness, which also provide a high VI. They are reasonably stable, although they do need close control and regular checking of water and alkaline content.

The water content controls the fire-resistance of the fluid (increasing with increasing water content). Evaporation and loss of water are likely during service, more especially at higher system temperatures so the system design should aim to minimize such losses. Topping up is normally done with a pre-mixed water-glycol solution, following specific recommendations given by the fluid manufacturer. Water alone, or glycol alone, should not be used to top up a water-glycol mixture to compensate for volumetric loss.

**High Water Base Fluids (HWBF)**

The category of high-water-base hydraulic fluids falls into two groups under ISO classification – see Tables 8 and 9.

HFAE emulsions of oil in more than 80% water.

HFAS chemical solutions in more than 80% water.

HFAS class fluids, also known as HWBF or 5/95 fluids, normally contain a nominal 95% of water, usually mixed at the point of use with 5% of concentrate (commonly polymers). The water used must be within the hardness limits given by the manufacturer of the concentrate.

Because of the low viscosity of HFAS fluids – one centistoke, or the same as water-hydrodynamic bearing films in pumps and motors are much thinner than with mineral oils. Metal-to-metal contact can thus be more frequent, aggravating wear and also generating contaminants. Modification of the detail design of the pump may be necessary to accommodate the former and filtration down to 10 $\mu$m in pressure lines and 25 $\mu$m in return lines is recommended to remove wear products.

Filters and elements must be of a suitable type. Corrosion and erosion can also be a problem and in general HFAS fluids are not compatible with zinc and cadmium plating, aluminium (unless anodized), cork gaskets and normal paints and sealants. Stainless steel or GRP is recommended for reservoir construction. For seals, *Nitrile*, *Neoprene* and *Viton* are suitable elastomers.

HFAS fluids can produce problems with valves, notably leakage due to the low fluid viscosity. In this respect, *Poppet valves* are better than *sliding spool* valves but all types of standard valves can give acceptable service (except those with wetted components in aluminium).

Theoretically, an increase in leakage should also result in an increase in heat

## TABLE 8 – ISO CLASSIFICATION OF HYDRAULIC FLUIDS

| Particular Application | Specific Application | Composition and Special Properties | Symbol 150 – L | Typical Application | Remarks |
|---|---|---|---|---|---|
| Hydrostatic | General | Non-inhibited refined mineral oils | HH | | |
| | | With improved anti-rust anti-oxidation properties | HL | | |
| | | HL with improved anti-wear properties | HM | Highly loaded components | |
| | | HL with improved viscosity temperature properties | HR | | |
| | | HM with improved viscosity temperature properties | HV | Construction and marine equipment | |
| | | Synthetic fluids with no specific fire-resistant properties | HS | | Special properties |
| | Hydraulic slideway systems | HM with anti-stick slip properties | HG | Machines with combined hydraulic and oil-way lube systems | |
| | Applica-tions where fire-resistant properties required | Oil-in-water emulsions | HFAE | | Typically 80% + water |
| | | Chemical solutions in water | HFAS | | Typically 80% + water |
| | | Water-in-oil emulsions | HFB | | |
| | | Water polymer solutions | HFC | | Typically 80% – water |
| | | Phosphate esters containing no water | HFDR | | |
| | | Chlorinated hydrocarbons no water | HFDS | | Possible environment or health hazards |
| | | Mixtures of HFDR and HFDS with no water | HFDT | | |
| | | Other synthetics with no water | HFDW | | |
| Hydro-kinetic | Automatic trans-mission | | HA | | |
| | Couplers and converters | | HN | | |

**TABLE 9 – MINERAL OILS AND WATER-BASED FLUIDS COMPARED**

| **Physical properties of mixed fluid** | **Petroleum oils** | **HFC Water-glycol fluid** | **HFB Water-oil emulsions** | **HFA 95/5 Fluids** |
|---|---|---|---|---|
| Density | 0.85 to 0.9 | 1.05 | 0.95 | 1.00 |
| Typical viscosity cSt at 40 °C | Very low to very high | (40% $H_20$) 43 cSt 40 °C | (40% $H_20$) 65 cSt 40 °C | (95% $H_20$) 1 cSt 40 °C |
| Vapour pressure | Low | High | High | High |
| **Corrosion resistance** | | | | |
| Liquid phase | Good | Good | Good | Good |
| Vapour phase | Fair | Fair to poor | Fair to poor | Fair to poor |
| **Storage stability** | | | | |
| Room temperature | Excellent | Good | Requires care | Requires care |
| Low temperature | Excellent | Good | Poor | Poor |
| Lubricity | Excellent | Good | Good to limited | Limited |
| Bulk fluid temperature limits °C | −30 to +80 | −32 to +65 | −10 to +65 | +5 to +50 |
| Monitoring requirements | Viscosity, neutralization number | Viscosity, water content, pH | Viscosity, water content | Water content, pH, bacteria check |
| Fire resistance | Non-resistant | Good | Good | Excellent |

*Stability* – HFA fluids are stable from +5 °C to +50 °C. Below 0 °C freezing occurs which can separate the fluid. Above 50 °C evaporation is accelerated.

*Lubricity* – HFA have limited lubricating properties as indicated by the low viscosity. Additives can improve this so therefore varies with fluid manufacturer.

*Acidity* – To retard growth of bacteria HFA fluids should be alkaline at about 8 to 9.5 pH.

generation, although this is more than offset by the higher specific heat and thermal conductivity of the fluid. Thus HWBF systems tend to run cooler than mineral oil systems in actual practice.

**Synthetic Fluids**

Potential advantages of synthetic fluids are superior oxygen stability, high VI, lower viscosity, lower pour point and good lubricating properties (considerably better than water-based fluids, but not necessarily better than mineral oils).

*Phosphate esters* are the best-known and the most widely-used type. The performance of modern phosphate ester fluids is more or less directly comparable with that of mineral oils, particularly as they can be obtained in a wide range of viscosities. Their VI is lower than that of mineral oils, but can be enhanced by VI improvers. Bulk modulus is higher, however, which means that phosphate ester fluids are superior to mineral oils as regards compressibility effects at higher pressures – see Figure 5

The chief disadvantage of phosphate ester fluids is their very high cost, followed by their complete lack of compatibility with conventional elastomers and paint finishes. Until comparitively recently Butyl was the first choice material for elastomeric seals and packings, with possible alternatives in the more expensive

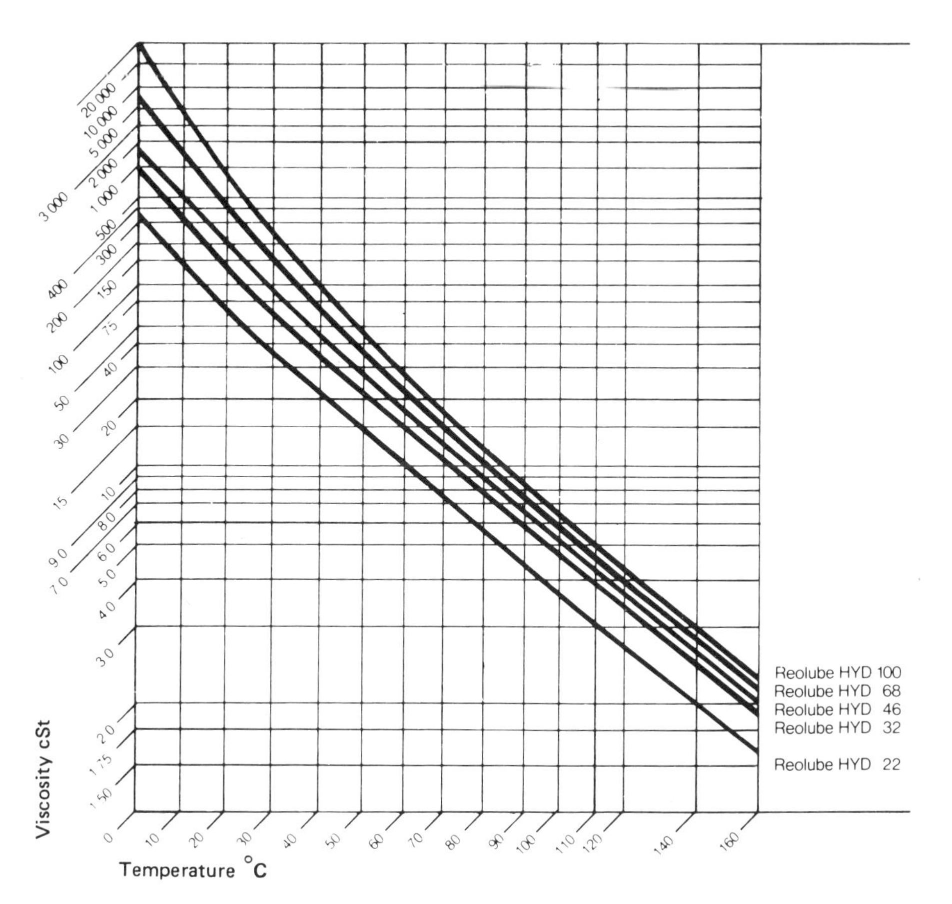


*Figure 5*
*Viscosity temperature relationship of a range of phosphate ester fluids.*

Viton* and silicone rubbers. Ethylenepropylene rubbers have replaced Butyl and are the standard choice elastomer for use with phosphate ester fluids.

About the only paints which are compatible with phosphate ester fluids are epoxy-based or polyurethanes. The latter have somewhat limited compatibility and so the former are preferred for painting reservoirs, *etc,* in which the fluid is used. It should be noted that the conventional paints used for external finishes arc readily stripped by split or leaking phosphate ester fluids.

A rather more minor disadvantage of phosphate ester fluids is their higher specific gravity compared with mineral oils. Maximum service temperature is generally higher and phosphate ester fluids can be worked at temperatures up to about 150 °C (300 °F) without degradation of the fluid.

**Chlorinated Hydro-Carbons**

*Chlorinated hydro-carbons* can be classed as *mild extreme-pressure lubricants,* with gravity at least comparable to, and probably better than, phosphate esters. Not all chlorinated compounds are good lubricants, however, polychlorinated diphenyls

*Dupont Registered Trade Name.

have a relatively poor performance in this respect. Specific gravity is again high (1.43) but such fluids can be produced with a wide range of viscosities. The cost is very high.

Chlorinated aromatics are not used on their own as commercial fire-resistant fluids, but normally mixed with phosphate ester fluids as phosphate ester 'blends', together with such additives as may be thought necessary (notably viscosity improvers).

**Silicone Fluids**

*Silicones* are another class of high cost fluids, prohibitively so for all but highly specialized applications. Their chief attraction is their suitability for working at high service temperatures up to the order of 360 to 370 °C (600 to 700 °F), with the added virtue of an extremely high VI so that reasonable viscosity values are maintained up to very high temperatures.

The performance of silicones as lubricants has been considerably enhanced by the introduction of improved silicone fluids, although these may show some slight loss of high temperature properties. They remain the sole commercial fluids available for working at temperatures in excess of 150 °C (300 °F).

All silicone fluids are, of course, fire-resistant, but would not normally be selected on this basis alone because of their high cost, compared even with phosphate esters.

# Performance Calculations

### Cylinder Speed of Operation

FLUID DEMAND of a cylinder is determined by the displacement volume, from which it follows that the speed of operation is related directly to the delivery available from the pump, *viz:*

$$\text{Time} = \frac{A_e \times L}{Q}$$

Where Ae is effective piston area (*ie* $0.7854 \times D^2$ or $0.7854\,(D^2 - d^2)$).
L is the length of stroke
and
Q is the pump delivery in consistent units.

For convenience, working formulae in various units are available, which can be used to determine the theoretical speed of operation of any hydraulic cylinder. Alternatively the demand required from the pump or system to produce the required speed of operation can be calculated. It does not necessarily follow that theoretical values are applicable in a practical system. Actual speed of operation can be modified by anything which could affect the actual delivery achieved at the inlet port, *eg* anything producing a throttling effect and also by back pressure effects.

In a system where constant pressure is maintained by either weight loaded or gas charged accumulator the flow rate (Q) depends on the flow rate into or out of the cylinder ports due to pressure drop through the valves and part of the system controlling the cylinder.

In practice it is generally recommended that flow velocities in inlet lines should be limited to a maximum of 4.5 m/s (15 ft/s) to minimize turbulence and pressure drop, and hydraulic shock. Table 1 is then a general guide to the piston speeds likely to be achieved with typical standard and oversize cylinder ports. Where higher piston speeds are required, inlet line size can be increased to increase delivery while maintaining a maximum flow velocity of 4.5 m/s (15 ft/s), with delivery to the cylinder through two or more ports.

Velocity limits may also be imposed by the construction of the cylinder. That is to say excessive velocities may be potentially damaging, or may affect cushioning

**TABLE 1 – PISTON SPEEDS WITH TYPICAL STANDARD AND OVERSIZE PORTS**

| Cylinder Bore in Inches | STANDARD CYLINDER PORT | | | | | OVERSIZE CYLINDER PORT | | | | |
|---|---|---|---|---|---|---|---|---|---|---|
| | Port Size BSP | Tube o.d. in Inches | Tube Bore in Inches | GPM Flow at 15 ft/sec | Piston Speed in Ft/Min | Port Size BSP | Tube o.d. in Inches | Tube Bore in Inches | GPM Flow at 15 ft/sec | Piston Speed in Ft/Min |
| 1½ | ½ | ⅝ | 0.529 | 8.50 | 112 | ¾ | ¾ | 0.654 | 12.98 | 170 |
| 2 | ½ | ⅝ | 0.529 | 8.50 | 63 | ¾ | ¾ | 0.654 | 12.98 | 96 |
| 2½ | ½ | ⅝ | 0.529 | 8.50 | 40 | ¾ | ¾ | 0.654 | 12.98 | 61 |
| 3¼ | ¾ | ¾ | 0.654 | 12.98 | 36 | 1 | 1 | 0.872 | 23.07 | 64 |
| 4 | ¾ | ¾ | 0.654 | 12.98 | 24 | 1 | 1 | 0.872 | 23.07 | 43 |
| 5 | ¾ | ¾ | 0.654 | 12.98 | 15 | 1 | 1 | 0.872 | 23.07 | 27 |
| 6 | 1 | 1 | 0.872 | 23.07 | 19 | 1¼ | 1¼ | 1.090 | 36.05 | 29 |
| 7 | 1¼ | 1¼ | 1.090 | 36.05 | 22 | 1½ | 1½ | 1.292 | 50.62 | 30 |
| 8 | 1½ | 1½ | 1.292 | 50.62 | 23 | 2 | 2 | 1.744 | 92.23 | 42 |
| 10 | 1½ | 1½ | 1.292 | 50.62 | 15 | 2 | 2 | 1.744 | 92.23 | 27 |
| 12 | 2 | 2 | 1.744 | 92.23 | 19 | | | | | |

suitability. Thus, in general, if the intended operating velocity of a hydraulic cylinder exceeds 35 m/min (120 ft/min), the supplier should be consulted regarding the suitability of the cylinder (and cushion, where applicable) for working at such piston speeds.

For particularly accurate determination of the speed of operation of a hydraulic cylinder the load travel curve should be plotted.

**Load-Travel Diagrams**

Cylinders are commonly sized on the basis of producing a required thrust to overcome the maximum force required. This can be far from an exact method, and potentially inefficient when the load varies with the distance travelled. It may, therefore, be useful to prepare a *load-travel diagram* in order to analyse the cylinder requirements and performance more completely – see Figure 1.

For a constant load throughout the travel, the load line will be horizontal, with an equal amount of work done over each part of the stroke. More usually the load-travel curve will depart considerably from this ideal. Thus, although the cylinder may be proportioned to produce a force equal to maximum load at all points along the travel, it may only be called upon to exert this force at a particular point of its stroke. The rectangular 'box' thus represents the potential work capacity of the

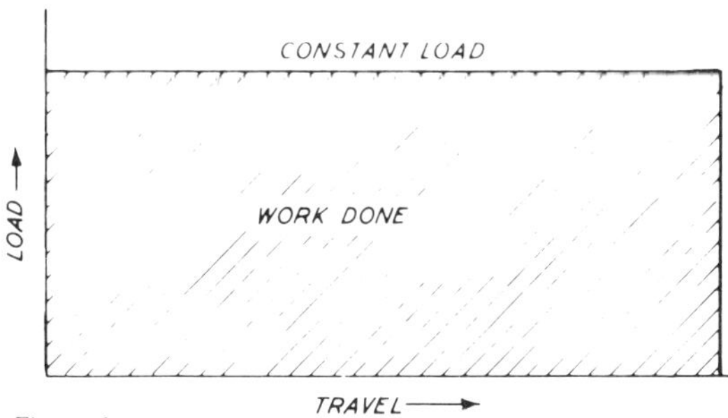


*Figure 1*

cylinder, while the area under the actual load-travel curve represents the actual work done. The ratio of the actual to the potential area is a measure of the efficiency of the cylinder as a working unit – see Figure 2.

While the actual efficiency achieved is not necessarily of primary importance, it is obviously desirable that it should be made as high as possible. Quite often an improvement may be found by re-thinking the linking geometry involved so that the load may be spread more evenly over the travel, increasing the area under the curve and enabling the maximum force requirement to be reduced. This means that the same duty can be performed by a smaller cylinder, which is working more efficiently.

The form of the load-travel curve may also assist in suggesting ways in which the efficiency can be improved. This can be applied particularly where the load-travel curve rises continuously to a peak, as in Figure 3. The efficiency in such a case may be substantially less than 25%. If a pre-load can be applied, however, so that the

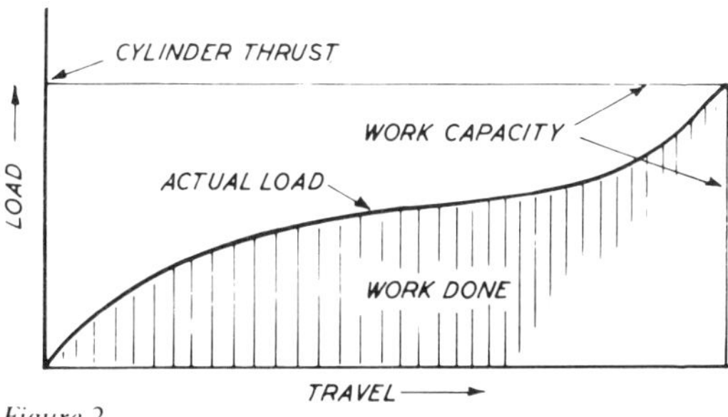


*Figure 2*

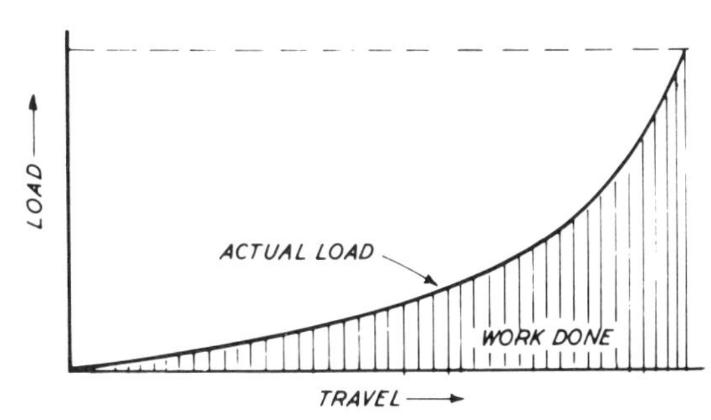


*Figure 3*

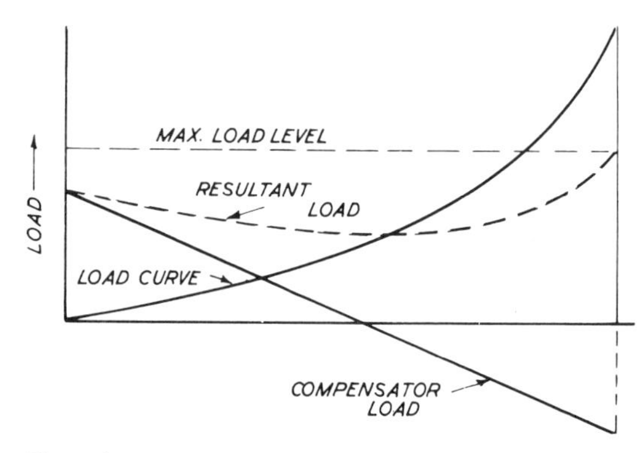


*Figure 4*

resultant of pre-load and normal load is closer to the horizontal, as in Figure 4, not only is the maximum force required substantially reduced, but the area under the load-travel curve now shows a very much higher efficiency. Despite the fact that the cylinder now has to work against an artificial load as well as the normal load, a smaller cylinder can be used for the duty required.

Pre-load devices normally take the form of springs or auxiliary cylinders. In the latter case, the auxiliary cylinder may be connected to the main circuit, or energised separately from an accumulator or even a gas bottle. They are generally referred to as compensators, which can, quite obviously, only be used in specific applications.

**Operating Time from the Load-Travel Curve**

The load-travel curve also provides a means of estimating the operating time of a cylinder with a good degree of accuracy, particularly where a variable delivery pump is being employed. Specific points can be taken on the load-travel curve, from which the pressure required at each point can be determined by dividing the load by the cylinder bore area. These values can be transferred to the pump characteristic curve to read delivery available at each pressure – see Figure 5.

In practice this will over-estimate performance because the pressure calculation is derived directly from the load only. For a true value, the total pressure drop in the lines, including back pressure effects, if significant, must be added to this value

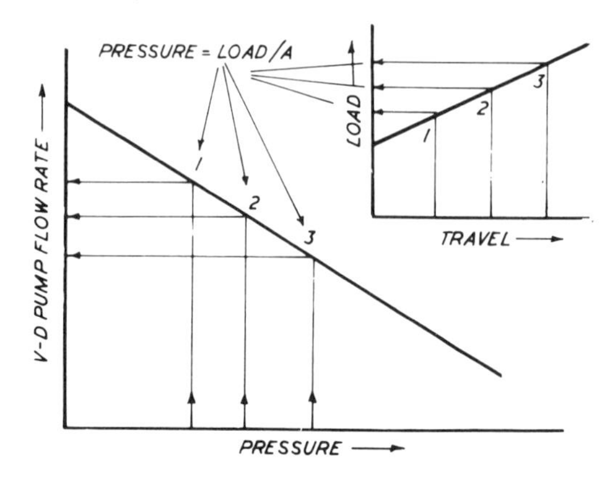


*Figure 5*

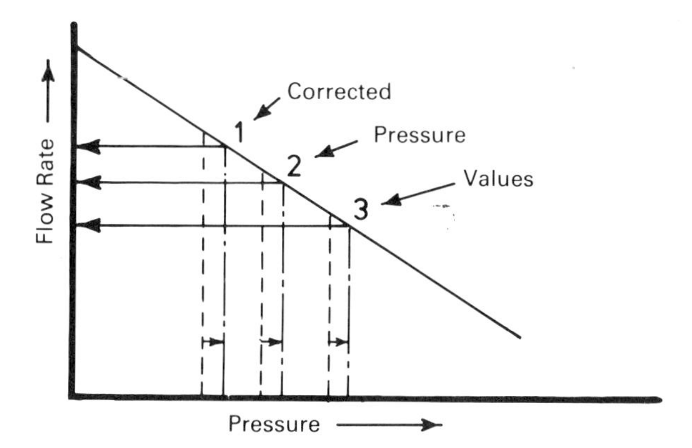


*Figure 6*

to obtain the true pump delivery – see Figure 6. For most purposes an estimate of pressure drop may be sufficient, unless the speed of operation is to be determined very accurately.

The actual velocity of the piston at each point then follows, by dividing the delivery available by the cylinder area, which can be plotted as a further curve of the *reciprocal* piston velocity against travel – see Figure 7. Enough points should be taken on the initial curve to enable a realistic *velocity-travel curve* to be plotted, paying particular attention to any irregular regions of the original load travel curve. The area under this curve will then give the full time of operation. If, for any reason; piston velocity is required to be known at particular points of the stroke it can be plotted separately as velocity against travel.

**Compressibility Effects**

Because of the compressibility of any real fluid, cylinder motion will be subject to a certain amount of springiness or elasticity, generally known as *compliance*. This effect is negligible at low to moderate pressures, but can become significant at pressures above 140 bar (2000lbf/in$^2$).

Cylinder compliance can be related to the fluid volumes on each side of the cylinder, and the bulk modulus of the fluid.

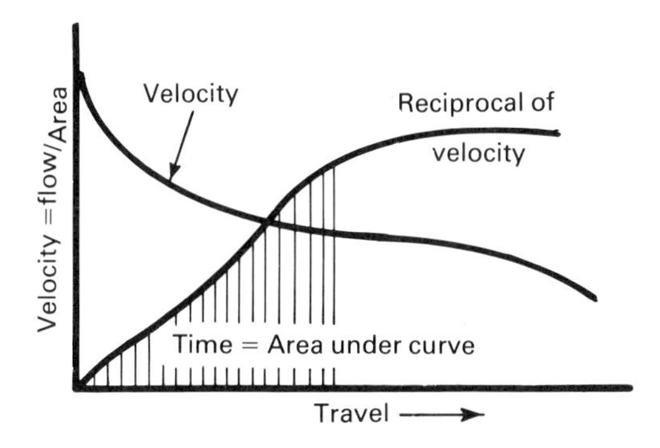


*Figure 7*

$$\text{Compliance } (\lambda) = \frac{1}{\beta \frac{1}{V_1} + \frac{1}{V_2}}$$

Where $\beta$ = bulk modulus of fluid
$V_1$, $V_2$ are the fluid volumes on each side of the piston – see Figure 8.

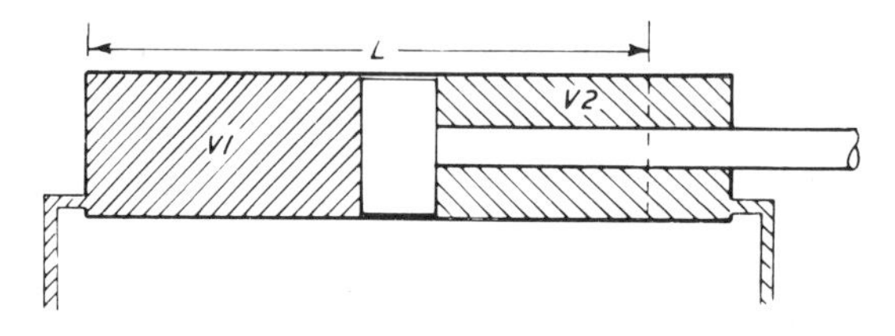


*Figure 8*

Compliance will reach a maximum value when $V_1 = V_2$ and the position of the piston giving maximum compliance is thus $L_1 \times A_1 = L_2 \times A_2$ where $L_1$ and $L_2$ are the effective cylinder port lengths and $A_1$ and $A_2$ are the effective piston areas.

In the case of a symmetrical cylinder (*eg a through-rod cylinder*) with equal volumes on either side of the piston, maximum compliance can be written directly in terms of the stroke (L)

$$\lambda \text{ maximum} = \frac{L}{4\ A\ \beta}$$

For critical applications, cylinder compliance can be determined over the full stroke range, as in Figure 9. The compliance ratio in this case is the ratio of the actual compliance at any particular stroke position to the maximum compliance.

For accurate analysis, the compliance of the fluid column lengths in the lines to the cylinder, should also be taken into account.

The significance of compliance is that cylinder is essentially a non-linear actuator, except for very small displacements at around the point of maximum compliance, and departs mostly from linear characteristics at each end of the stroke. In practice, however, the actual compliance present usually renders the non-linear characteristics negligible. Thus, for general applications, a cylinder can be considered to be a true linear actuator.

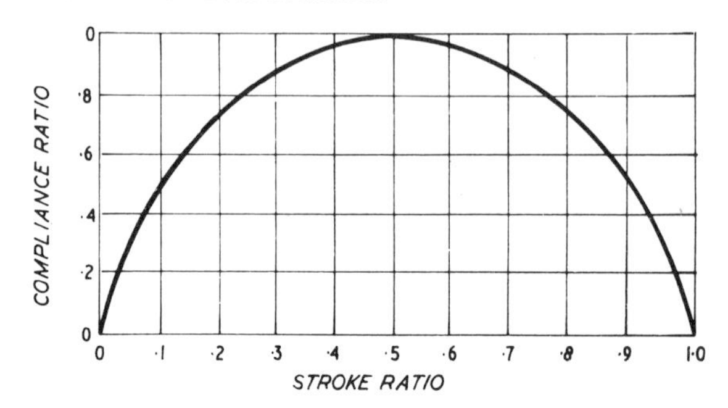


*Figure 9*

# Circuit Symbols

**Historical Development**

IT WAS realised, when the early hydraulic systems were constructed, that some form of diagram was required to assist the commissioning and operating engineers to understand the workings of the system.

The first diagrams were mainly pictorial, showing the relative positions and shapes of the various components, with lines drawn between connecting ports. They were useful for the installation engineers, but designers and 'trouble-shooters' found them difficult to use as the operation and flow paths through the components were not shown.

More complex cut-away diagrams were used showing sections through each component, to show the flow paths, with colour coding for different pressure levels. These were expensive to produce and had to be reproduced to show each different phase of circuit operation. They were ideal for instruction and are widely used for that purpose today. Designers still required some form of diagram which could be used initially as a 'design tool', and could later be used to show application and service engineers the correct operation of the system. This gave rise to the adoption of *graphical symbols* for fluid power diagrams.

The first standard set of symbols was introduced by the *Joint Industry Conference (JIC)* for hydraulic circuits in 1948 and for pneumatic circuits in 1950 in the USA. These were incorporated into a joint standard for hydraulic and pneumatic circuits by the *American Standards Association,* which was the predecessor of the *American National Standards Institute (ANSI),* in co-operation with the *American Society of Mechanical Engineers (ASME).*

Meanwhile on the other side of the Atlantic parallel activity was taking place at the *Comité Européan des Transmissions Oléohydrauliques et Pneumatiques (CETOP)*, which drew up a draft standard of graphical symbols for fluid power diagrams.

Inter-nation co-operation was carried out between the relative organisations to produce the ISO standard 1219, on which the British Standard BS 2917 is based, thus providing a truly international standard set of symbols which can be used for fluid power circuits throughout the world.

**Graphical Symbols**

Circuit diagrams are drawn to show the flow paths between the components used, by using single lines and arrow heads to indicate the direction of flow through the pipelines and the component symbols. Solid arrow-heads or triangles are used to indicate a liquid medium and hollow arrow-heads or triangles are for gaseous medium.

It is not the intention of this book to detail all the various symbols, which can be obtained from the relevant standards, but to outline the main principles involved so that they can be used in later chapters.

As well as showing the direction of flow, the graphical symbols indicate the type of control and the inter-connection between components in the system. The circuit diagram and symbols do not show the relative sizes or positions of the components nor their physical make-up.

Symbols of components having multi-stage functions, such as valves made up of two or more sections, can be shown in a simplified form on large circuit diagrams. The more detailed and complex form of the symbol is used to analyse the actual function of the component. For these symbols, or for symbols of more than one component joined together, a chain dotted enclosure line is used to form a rectangle around the symbol.

**Cylinders**

The symbols for the basic types of fluid power cylinders are shown in Figure 1 and these closely follow the physical make-up of the component.

In the case of a *single-acting cylinder,* fluid pressure is applied to one side only of the piston to produce the power stroke, return action on release of fluid pressure being by means of some external force, or a spring which is diagramatically shown. The spring may be postioned on the other side of the piston for a cylinder which retracts on the power stroke. Single-acting cylinders are commonly used in

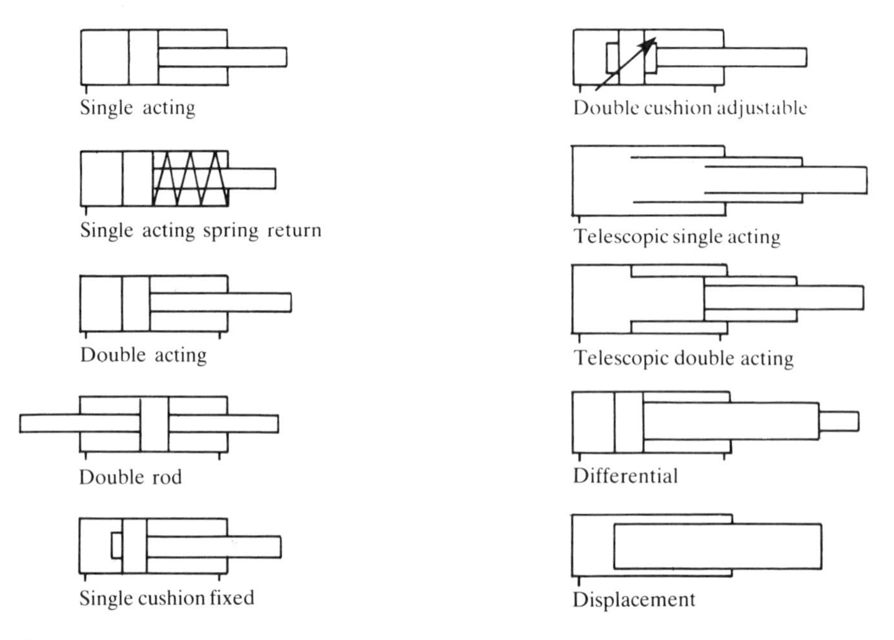


*Figure 1*

applications requiring *telescopic* or *displacement* cylinders as shown. The latter are also known as *plunger type* cylinders where the rod diameter is equal to the cylinder bore.

It is important that the symbol indicates where the diameter of the rod is significant compared to the diameter of the bore, as in the case of a differential cylinder. This enables allowance to be made for high flows from the cylinder when fluid is fed into a small annular volume.

Both *single-* and *double-acting* cylinders may be fitted with cushioning devices, to arrest the piston at the end of the stroke, which may be *single-* or *double-acting* and also either fixed or variable in performance. A diagonal arrow indicates that the cushioning is adjustable and this symbol is generally used to show that a component can be adjusted or varied.

**Pressure Controls**

Pressure control valves or devices are shown as a retangular box with an arrow indicating the flow path, the position of which is being controlled by a pilot pressure and opposed by a spring force as shown if Figure 2. Many basic graphical principles are used in this figure.

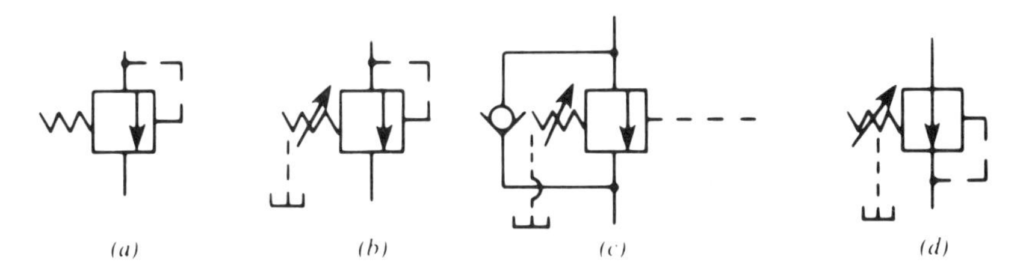


*Figure 2*

View (a) shows a normal *spring-loaded pressure relief valve* with a fixed pressure setting. The dashed line represents an internal pilot pressure connection. The arrow is always shown with a closed flow path, with the valve in the unrelieving position. This symbol is used as the simplified symbol for *single-* and *two-stage* relief valves.

View (b) shows a similar valve with an adjustable spring setting, shown by the diagonal arrow on the spring, and an external drain indicated by the dotted line draining to tank. This symbol, representing a reservoir, can be drawn adjacent to each valve, analogous to an earth symbol on electrical diagrams, or all the valve drains can be taken *via* dotted lines to a single reservoir.

The symbol in view (c) includes a reverse flow check valve, which allows flow through the valve in the reverse direction only, when the valve is closed. This is a symbolic representation of a *seated ball valve*. The pilot pressure connection is shown as coming from an external source which is usual with a sequence or unloading valve.

View (d) shows a *pressure-reducing* valve which is normally open, ash shown by the flow path arrow. The dashed line, indicating the internal pilot pressure, is taken from a position downstream of the valve and provides pressure to close the valve against the adjustable spring setting. An external drain, as shown, is always required with this type of valve, as when the valve is being closed to reduce downstream pressure, both the inlet and outlet lines are pressurized.

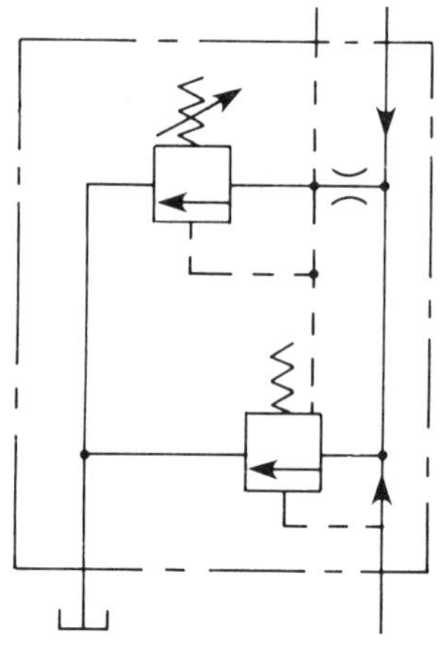

*Figure 3*

Figure 3 shows the detailed complete graphical symbol for a *two-stage balanced relief* valve with vent connection. The individual relief valve symbols for each stage are joined by the appropriate connecting lines and enclosed in a chain dotted box.

**Flow controls**

The symbol for the control of fluid flow is represented by and orifice or restriction as shown in Figure 4.

Symbol (a) represents a *fixed restriction* in a line while symbol (b) shows an *adjustable restriction* or *orifice*. In both of thses symbols the flow can be controlled in either direction.

The addition of a box in symbol (c) indicates that the *flow control* valve is pressure-compensated, *ie* a pressure control section is added to the valve to give a constant pressure drop across the valve and hence constant flow through the valve. A direction of flow arrow is added, as these valves can accurately control the flow rate in one direction only.

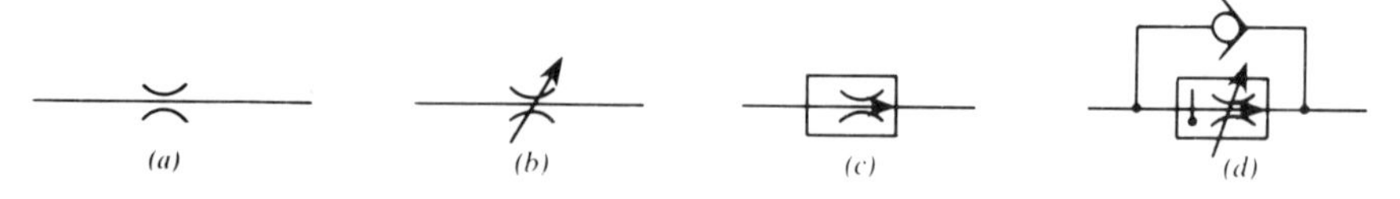


*Figure 4*

Symbol (d) shows a *pressure-* and *temperature-compensated* valve with adjustable flow control and a *reverse-free flow check* valve added. A detailed version of the simplified symbol shown in Figure 4 (c) is given in Figure 5 which shows the pressure compensator section of the valve.

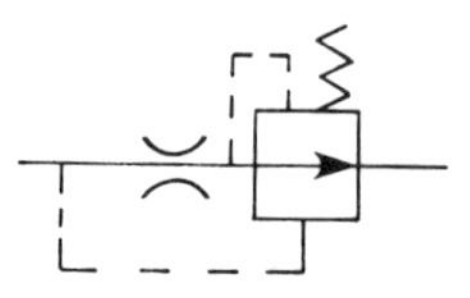

*Figure 5*

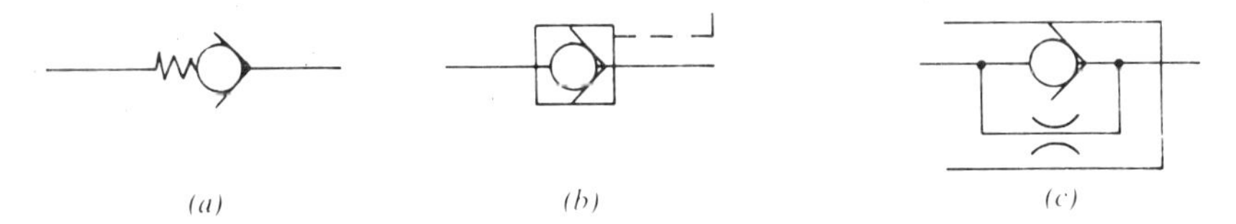

*Figure 6*

**Directional Controls**

The simplest form of *directional control* valve is the *non-return check* valve, as previously mentioned, which allows flow in one direction only. Various types of check valve symbols are shown in Figure 6.

Symbol (a) shows a *spring-loaded* check valve, where pressure has to overcome the spring force before the valve opens. This valve can therefore be used as a *pressure control* as well as a directional control valve.

The valve in symbol (b) is a *pilot-operated* check valve with the pilot pressure holding the valve open in the reverse flow condition. The pilot pressure line can also be shown on the other side of the box for a valve held closed by pilot pressure

Symbol (c) shows a check valve which allows free flow in one direction and restricted flow in the other direction, thus acting as a flow control valve.

The main type of directional control valve used in fluid power systems is the *multiple-position* valve and the symbol used shows a box for each position, containing the flow paths, for each position, through the valve. Figure 7 shows a *two-position four-way valve,* pilot-operated with a spring return.

*Figure 7*

The symbol is usually drawn with the valve in the spring return, *ie* un-pressurized or un-energized position as shown in view (a). If the transitional position, between the two main positions of the valve, is important to the circuit function, then this is shown between dashed lines as in view (b). The transitional position shown is one with all ports connected. By drawing the symbol in the un-energized or neutral position it is easy to see the circuit function, when the valve is moved to another position, from the flow path arrows on the diagram.

Figure 7 (b) should not be mistaken for a *three-position* valve as shown in Figure 8.

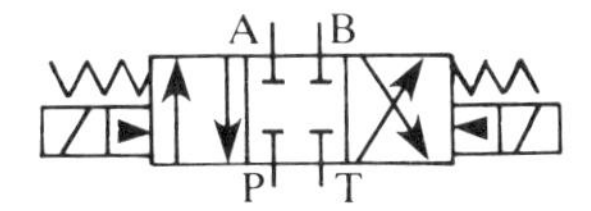


*Figure 8*

The symbol represents a three-postion four-way valve, spring-centred to a neutral position with all ports blocked, solenoid-operated and pilot-controlled, as indicated by the control symbols at each end of the diagram. These control symbols

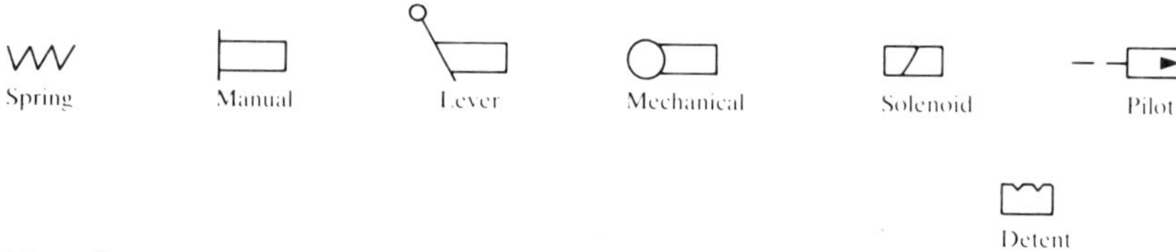


*Figure 9*

can be either a spring, manual, lever, mechanical, solenoid or pilot pressure or any combination used together as shown in Figure 9. The detent is a spring lock which holds the valve in the selected position until the next command is given. These are used for valves which are not spring-centred or spring-positioned.

The symbol shown in Figure 8 is a simplified symbol as the valve, in fact is a two-stage valve consisting of a solenoid valve mounted on a three-position spring-centred pilot-operated valve. For an analysis of the actual valve operation a detailed diagram is shown in Figure 10.

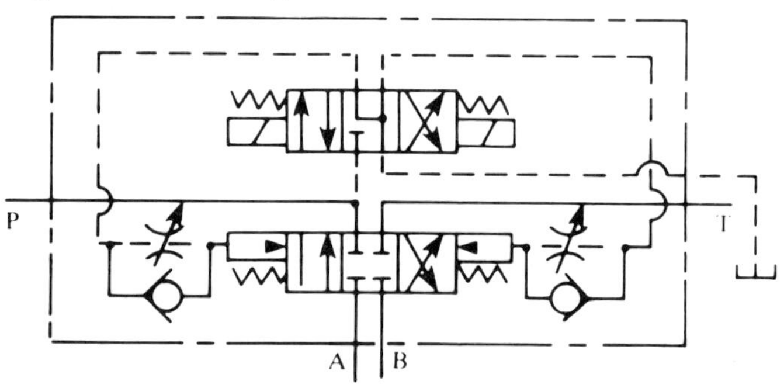


*Figure 10*

This diagram shows many of the principles already mentioned. The valve has an internal pilot pressure line, taken from the main pressure port (P), which is signalled *via* the three position solenoid valve to either side of the main three-position valve section. The flow control and reverse flow check valves, called 'pilot chokes', at each end of the main section are added to control the speed of operation of the valve.

Multiple position valves can be drawn with as many boxes as positions designed for the valve. One type of directional control valve, often used for mobile hydraulic applications, is the *multiple banked* valve. This consists of a number of multiple positions valves banked together, either in one casting or with a number of castings bolted together with long through bolts.

Each section often has a different centre or neutral position configuration to suit the requirements of the circuit, as shown in Figure 11. This valve has three banks of manually-operated three-positioned valves. It has an integral relief and check valve section and a through port to tank, to off load the pressure when none of the valve banks have been selected.

All the multiple position directional controls considered have a finite flow path position, as shown in each box, with transitional positions only shown within dashed lines when necessary. Multiple position valves capable of infinite positioning between certain limits are symbolized as in Figure 12.

The valve shown by the symbol is an *electrohydraulic servo* valve with a variable

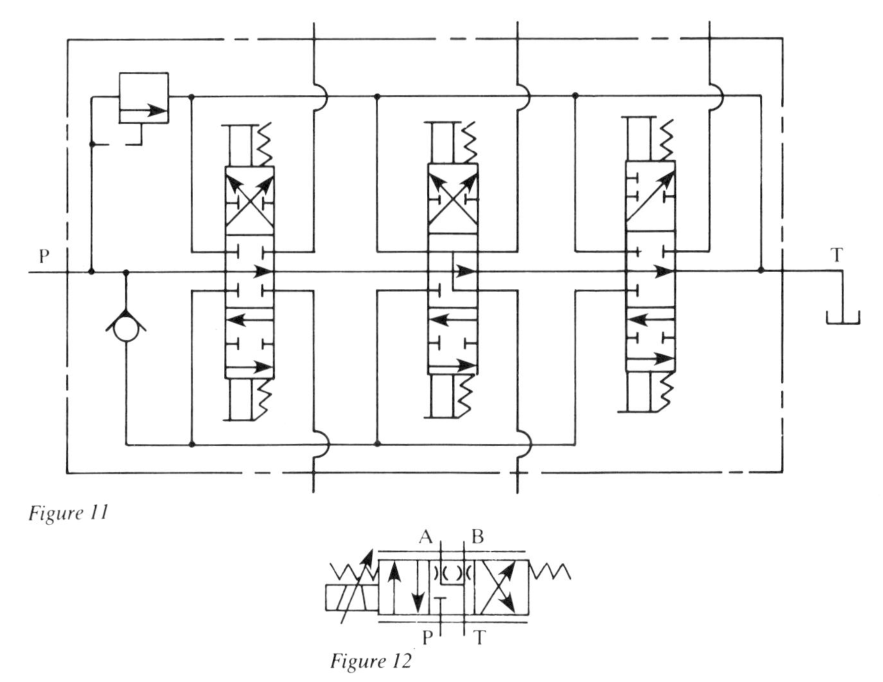


*Figure 11*

A B

P T

*Figure 12*

electric signal to give a proportional output from the valve, with the valve spring centred to the neutral position when the signal is removed. This type of directional control can also be used to control flow, by metering the flow with the valve in an intermediate position.

**Pumps and Motors**

Hydraulic pumps and motors are shown as circles with the direction of the flow and control functions added using the conventional symbols – see Figure 13.

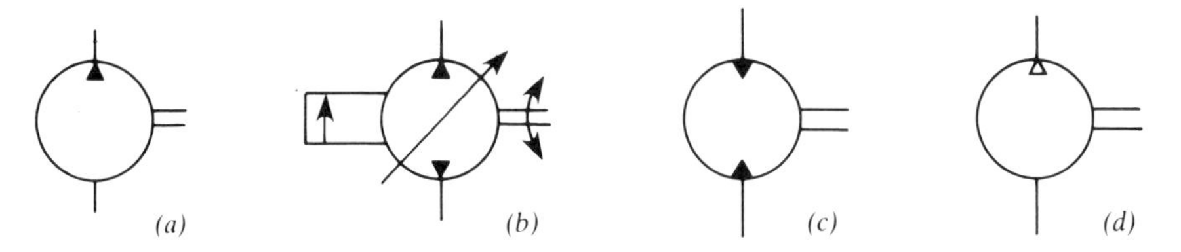


*Figure 13*

The pump shown in symbol (a) is a simple *unidirectional fixed displacement* pump, the mechanical input being shown as a solid shaft. In view (b) the symbol shows a *bidirectional variable* displacement pump with pressure-compensated control. The arrow on the input shaft shows that the rotation can also be bidirectional. View (c) shows a *fixed displacement bidirectional hydraulic motor* while a *fixed displacement air compressor* is shown by symbol (d).

There is a symbol designated by the relevant standards for all the components required in a hydraulic or pneumatic circuit, such as filters, heaters, coolers, shut-off valves, *etc;* and the reader is advised to refer to the latest issue of the standards when using a circuit diagram.

# SECTION 2

## Hydraulic Control

PRESSURE CONTROL
FLOW CONTROL
DIRECTIONAL CONTROL
SERVO CONTROL
CONTROL METHODS

# Pressure Control

IN AN hydraulic system, where the fluid is assumed to be virtually incompressible, pressure will rise sharply as soon as a resistance to fluid flow is encountered.

For this reason, every system must have some form of pressure-relieving device to protect the components from over-pressure and to limit the system pressure to the maximum design working pressure of the components. The device usually takes the form of a *pressure-relief valve* which opens to relieve the high pressure fluid to the low pressure side of the system or to the reservoir.

In systems where high pressures have to be maintained for long periods of time, using positive displacement pumps, the relief device is often built into the pump control, to reduce the pump flow when the maximum working pressure is reached. This conserves energy, as fluid flow, constantly being relieved at high pressure over a relief valve is a power loss, which is converted into heat in the fluid. The power loss can be calculated using the formula given in Section 1, *ie* Power loss = flow x pressure x a constant.

All pressure control devices have a finite operating time, which is usually greater than the rate of pressure increase in the system. A peak pressure is therefore developed before the device has fully operated, the value of which should be within the ultimate pressure which the components will withstand. Obviously the pressure peak is reduced in maximum value and band width by using a pressure relief device with a quick response rate.

Pressure control valves are used in hydraulic systems for may functions besides pure overload relief, and these are listed as follows: *sequence valves, unloading valves, counterbalance valves,* and *reducing valves,* each of which will be examined separately.

**Relief Valves**

The simplest type of pressure-relief valve consists of a spring-loaded ball valve or check valve. Its main limitation is that it has a tendency to 'chatter' which will ultimately wear a groove in the hardened ball and seat.

Important parameters in the design of such valves are (1) Pressure hysteresis – or the difference between 'cracking' or opening pressure and sealing pressure – and; (2) Proportional band-width – or the change in inlet pressure in response to increase in flow rate as the valve opens.

Poppet valves provide much greater control in design and manufacture over

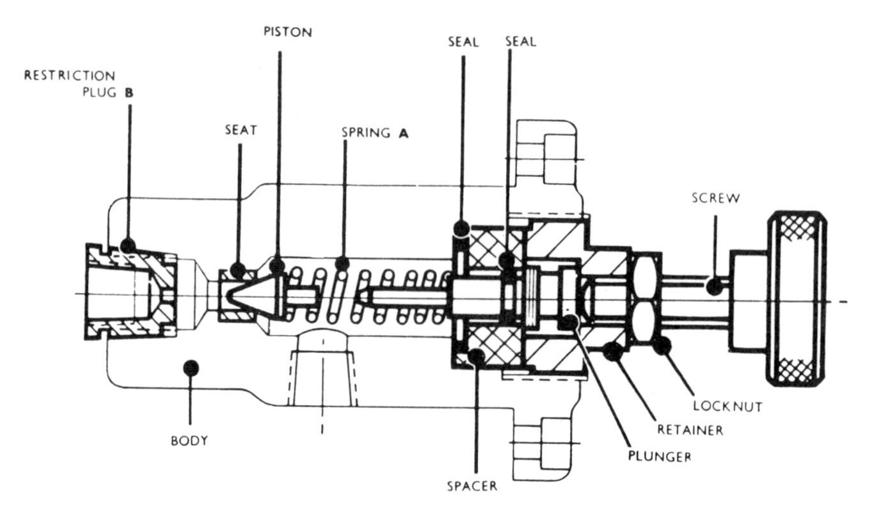


*Figure 1*
*Direct acting poppet relief valve.*

both pressure and proportional band-width, with throttling characteristics proportional to the square root of the valve stroke.

Sealing may, however, be a problem and an *O–ring* or similar *seat seal* is sometimes preferred to a metal-to-metal seat. The conical poppet is held down onto the seat by the action of a spring, the resistance of which can be adjusted by a screw (Figure 1).

The other configuration used for simple direct acting relief valves is the *plunger-type,* incorporating a hardened steel plunger sliding in the valve body. The degree of lift is dependent on flow rate as well as pressure so that flow variations can again initiate chatter, unless suitable damping is present.

One way of introducing self-damping is to use a two-diameter plunger to provide a bigger discharge area, with this end of the body sealed and discharge being directed through small holes in the body wall to restrict flow. The characteristics of this type of pressure relief valve can be adjusted by contouring the plunger, rendering the throttling characteristics proportional to the reciprocal of the stroke.

The main disadvantage of a direct acting relief valve is its high pressure over-ride at its maximum capacity. This type of valve is therefore most accurate at its rated flow rate only.

**Diferential–Relief Valves**

A two-diameter plunger is also used in the *differential-relief valve* (Figure 2). The main attraction of this configuration is that is reduces the size of spring required because the cracking pressure is applied to the annulus rather than the full plunger diameter. Port areas, on the other hand, are proportioned on the larger diameter.

The design may be further elaborated in the case of *pilot-operated* differential-relief valves to incorporate over-ride so that the valve will always open at a pre-determined maximum pressure regardless of adjustment of pilot pressure.

**Pilot-Operated Relief Valves**

Two stage *balanced piston type* relief valves are normally employed at higher power

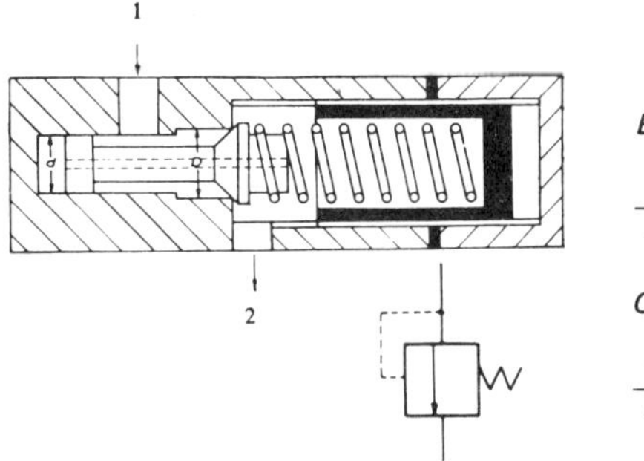


*Effective force on spring depends on*

$$\frac{\pi}{4}(D^2 - d^2).$$

*Capacity depends on*

$$\frac{\pi}{4}\ D^2$$

*Figure 2*
*Differential relief valve.*

levels, but can show advantages at all power levels. They are normally of integrated design with line and pilot connections.

A typical design for a pilot-operated relief valve is shown in Figure 3. The main relieving valve is arranged so that it is hydraulically-balanced and kept closed by a light spring.

Normally the pressure is equalized on either side of the spool by a small orifice. The pilot valve opens, when pressure overcomes the pilot spring setting and allows liquid to escape at a faster rate than it can be replenished through the orifice, so that the pressure falls above the spool. The excess pressure on the other side then lifts the valve and opens the exhaust port and liquid flows away to the tank in just sufficient quantity to keep the pressure constant.

In use the valve is set to the maximum working pressure or to some higher pressure if the working pressure is determined elsewhere. Valves of this type are very stable and their accuracy depends on that of the pilot valve and is not much affected by flow.

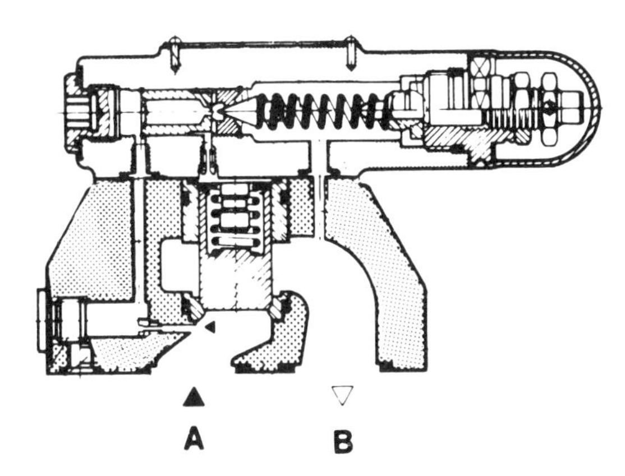


*Figure 3*
*Pilot-operated relief valve.*

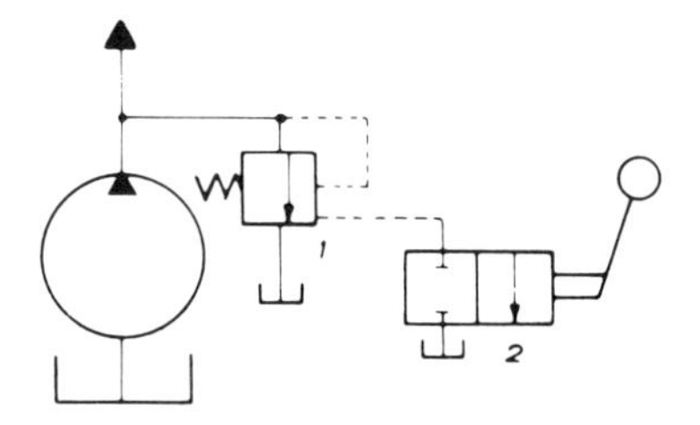


*Figure 4*
*Pilot-operated relief valve (1).*
*With unloading by two-way pilot valve (2).*

There is no need for the pilot-relief valve to be integral with the main valve and it is often convenient to connect it to another valve (Figure 4). A small two-way valve connected to the 'vent' or drain line will, when opened, cuase the pressure in the pilot system to fall to zero and the main valve will be opened, bypassing the pump output at a low pressure.

The actual pressure would be depend on the strength of the spring and the pipe friction. This method of 'venting' a relief valve is often used for unloading a pump, making a second unloading valve unnecessary.

It is also possible to work at more than one pressure, or from a remote position, by inserting a second pilot-relief valve between the two-way valve and drain, so that when the valve is open the maximum pressure setting is determined by this valve, providing the setting is less than the primary pilot valve.

**Sequence Valves**

A sequence valve can be used where it is necessary to ensure that sufficient pressure has built up in one circuit before fluid is admitted to another circuit. In its simplest form it comprises a spring-loaded *spool* valve with primary and secondary ports, the spool being normally positioned to shut off the secondary port. The primary pressure acts on the end of the spool, against spring pressure.

When sufficient pressure is present, the spool is moved against the spring, opening a connection through to the secondary port (Figure 5). The valve also has a throttling action, which prevents the primary pressure falling suddenly, but opens fully when the working pressure is reached. A non-return valve may be

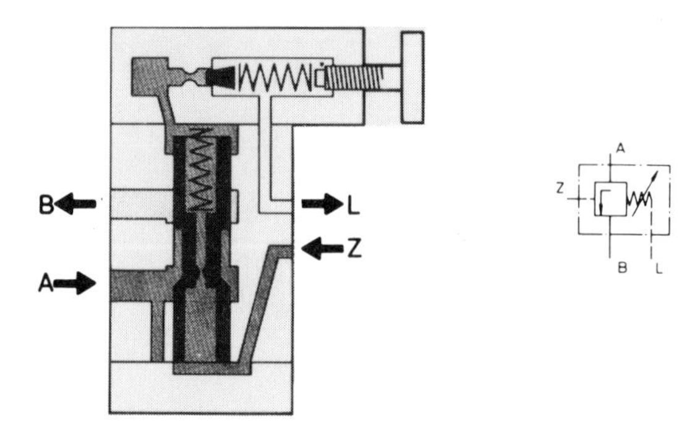


*Figure 5*
*Sequence valve.*

incorporated in the same body to provide free return flow of fluid in the non-controlled direction.

Basically, a simple sequence valve is similar to a pressure-relief valve, and a pilot-operated relief valve may, in fact, be adapted for this purpose merely by connecting the drain from the low-pressure side of the pilot section to the tank. Spool valves are, however, more normally used.

Sequencing valves have the particular limitation that they are reliant on system conditions remaining stable. Pressure variations can cause operation to be premature or delayed and also troublesome.

**Unloading Valves**

An unloading valve is very similar to a sequence valve and can be used in circuits where more than one pump or an accumulator maintains high pressure at low volume, while a large volume lower pressure pump is off loaded. The pilot pressure is usually from an external source, *ie* the high pressure pump line, and the valve is often internally drained.

Should any appreciable back pressure be present in the line from the unloading valve to tank, this will be additive to the spring setting, and under these conditions it may be advisable to use an external drain connection.

The main purpose of the valve is to off load the fluid from the large volume pump at minimum resistance to flow, but it must be remembered that there is still pressure in this part of the system even if it is only at 1 bar (14.5 $lbf/in^2$).

In an accumulator unloading valve, an integral check valve prevents return flow from the accumulator through the unloading valve, and the pilot pressure is sensed from a position between the accumulator and the check valve, to divert pump flow to tank when the accumulator is fully charged.

When the accumulator pressure drops to approximately 80 to 85% of the adjusted maximum, the valve closes and directs pump delivery to the accumulator and systems *via* the check valve. This is achieved by employing a small differential piston at the pilot relief section of the valve.

When pressure in the system drops 15%, the spring force acting on a smaller area is sufficient to start closing the valve, by which time the main spool will seat, bringing the pump back on load, while the system pressure may well have dropped a further 5%.

**Counterbalance Valves**

Counterbalance valves are usually spring-loaded spool valves, similar to sequence valves, but with integral reverse free flow check valves. They are used to control a load induced pressure to counterbalance and control the motion of a load.

They are often connected to the outlet of a cylinder to support a load against gravity and can be internally-piloted, externally-piloted, or both, depending on the circuit requirements.

The spring is usually internally drained as the outlet port of the valve is directed to tank *via* a directional control valve. An illustration of two piloting options is shown in Figure 6.

An internally piloted valve is shown in view (a), where the pressure setting of the valve must be higher than the pressure induced by the load, at the rod end of the cylinder, to allow for deceleration forces to brake the load. The load is easily raised by reverse flow through the bypass check valve. A disadvantage of this type of

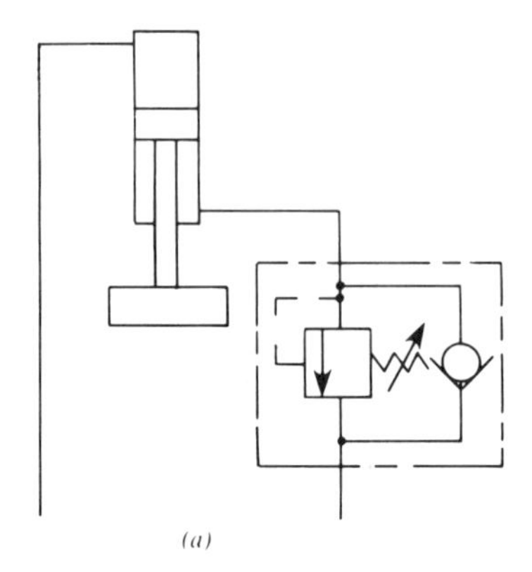


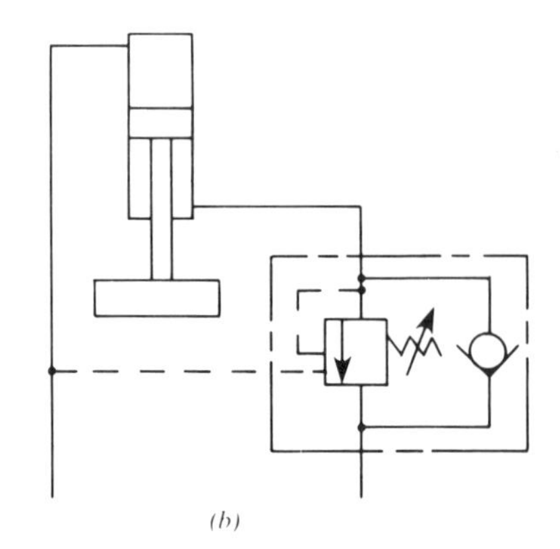


*Figure 6*
*Counterbalance valve circuit symbols.*

control is that the high pressure still has to be generated to extend the cylinder with the load removed.

It must be remembered that with spool valves, clearance between the spool and the body provides a leakage path. Therefore, this type of valve will not sustain a load for long periods of time.

With an externally-piloted valve, shown in view (b), the pilot is taken from the pressure input into the head end of the cylinder and will insure a positive control of the cylinder movement and restrain a load which tends to over-run the pump flow.

Counterbalance valves with external pilot connections are also known as *overcentre* valves as they can be used in applications where the load goes 'overcentre' and changes from a resistive to an over-running load. As soon as this happens pressure is lost from the head end of the cylinder and the valve, which is set at a low pressure setting, closes and a load induced pressure is set up in the rod end of the cylinder.

If the load has very high enertia, then an extra internal pilot section can be included in some valves, as shown in view (b), to act as a relief and braking valve. This internal pilot acts on a small pilot piston, of much smaller area than the external pilot connection, and will only over-ride the valve when the maximum permitted pressure in the rod end of the cylinder has been reached.

A disadvantage with the externally-piloted valve is that it is not quite as stable as the internally-piloted valve, in practice.

**Reducing Valves**

A reducing valve differs from the pressure control valves previously mentioned, in that it is a normally open valve used to maintain reduced pressures in certain parts of a system. It is actuated by pressure, sensed downstream of the valve, and tends to close as the pressure reaches the valve setting, thus preventing further pressure build-up. It can be obtained both as a direct-acting or pilot-operated valve, the latter having a wider range of adjustment and generally providing more accurate control.

The operating pressure is set by an adjustable spring in the pilot stage of the valve. If the main supply pressure is below the valve setting, fluid will flow freely from the inlet to the outlet.

An internal connection from the outlet port transmits the outlet pressure through the centre of the main spool to the pilot relief section. When outlet pressure rises to the valve setting, the main spool is unbalanced and moves to partly block the outlet port. Only enough flow is passed to the outlet to maintain the preset pressure.

If the valve closed completely, leakage past the spool could cause pressure to build up in the reduced pressure circuit. Instead a continuous bleed to tank is allowed to keep it slightly open and prevent downstream pressure rising above the valve setting. A seperate drain passage is provided to return this leakage to tank.

For this reason the use of reducing valves can affect the efficiency of a system, as pump flow is required to maintain the leakage flow, and this can cause heating of the fluid if high pressure differences are involved. It may be more efficient to employ a separate low pressure pump to provide fluid for the reduced pressure branch of the circuit.

# Flow Control

IN AN hydrostatic circuit, *flow control* is used to control the speed of hydraulic cylinders or actuators. For fluid to flow between two positions in a circuit there must be a *pressure drop,* or difference between the two positions. Conversely, where a pressure drop exists, flow will take place and this flow, and hence cylinder speed, will only remain constant while the pressure drop is constant.

The flow of a fluid in the circuit can be controlled from the output flow of a *variable displacement* pump or by the use of flow control valves.

**Flow Control Valves**

Simple flow control valves work on the basis of restricting the flow with either a fixed or variable orifice. In the latter case flow control valves fall into two categories:

(1) Non-compensated valves (*ie* simple throttle valves);

(2) Pressure-compensated, where the valve is designed to compensate for the effects of varying pressures and temperatures in the circuit and thus maintain a constant performance.

An inherent disadvantage of simple restrictions is that the flow through them is strictly dependent on pressure drop across the valve, and thus through-flow will vary with changes in load. As a consequence, their application is virtually limited to those systems where the load is constant, or where variations in flow rate, and thus operating speeds, are permissible.

**Pressure-Compensated Valves**

To provide a constant pressure drop across an orifice, and thus constant flow characteristics, a combination of two restrictors can be used: one fixed or adjustable and the other automatically variable. These two elements are normally combined in a single unit to produce a *pressure-compensated flow restrictor.*

A further refinement may be feedback or pressure from downstream to provide 'meter-in' control, when the variable throttle acts as a sensing orifice. One-way working can be provided by incorporating a non-return valve in the unit to provide free flow in one direction.

The main limitations of a pressure-compensated flow restrictor are that while constant flow characteristics are provided, independent of load, the controlled flow

*Pressure-compensated flow control valve.*

is throttled and surplus flow must be directed through another valve, resulting in the pump working at full relief setting continuously.

To overcome this loss of efficiency, a *pressure-compensated bypass regulator* valve can be used. Here the surplus flow is bypassed through the valve at working pressure and, at the same time, the controlled flow is not subjected to marked throttling losses. The use of a pressure-compensated bypass regulator therefore enables much higher circuit efficiences to be achieved.

Its main limitations are that, unlike the pressure-compensated restrictor, it cannot be used in parallel configurations. Also it is unsuitable for constant pressure systems, because it would bypass an unnecessarily high volume of fluid, resulting in lowered operating efficiency.

The difference in working of the two types can be seen in Figures 1 and 2, here drawn in the form of *two-port* and *three-port* valves, respectively, each with a fixed restrictor inserted in the outlet. In proprietary units the actual geometry may vary appreciably, but spring-loaded spool or piston types are normally used.

Temperature compensation may also be incorporated in flow control valves to adjust to change of fluid viscosity.

**Spill-Off Valves**

A spill-off flow control valve is shown in Figure 3. It is of the balanced-piston type and is based on the relief valve which embodies this principle. The main oil supply passes through the lower balance chamber, past the speed control orifice to the actuator. The upper balance chamber is connected to the actuator line.

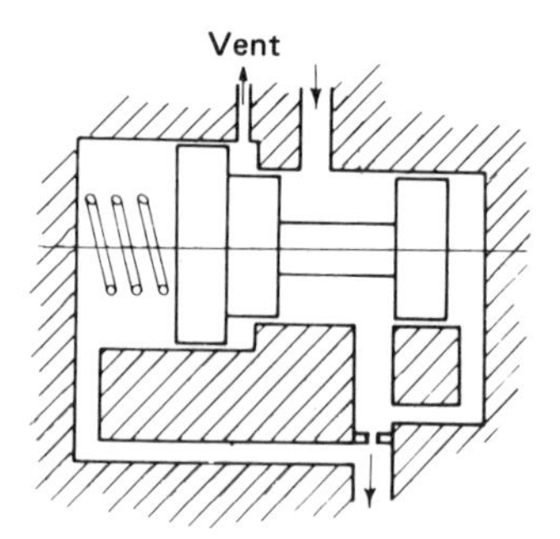


*Figure 1*

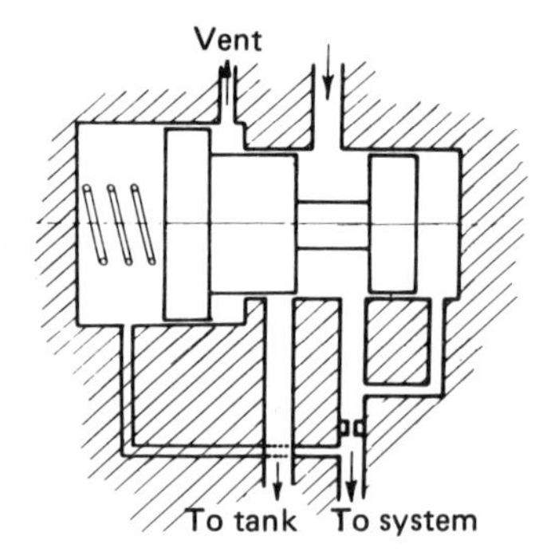


*Figure 2*

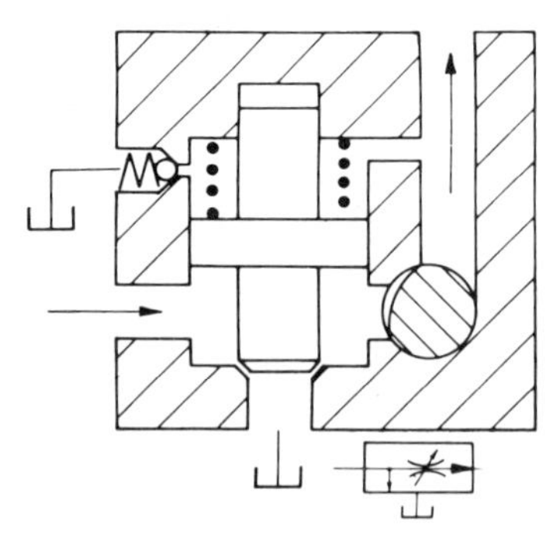

*Figure 3*
*Spill-off flow control valve.*

If there is no load on the actuator there will be no back pressure in the upper balance chamber. The relief valve will therefore open at the spring setting about 1.4 bar (20 lbf/in$^2$).

If a load is applied to the actuator, a back pressure will be created in the line and this will act on the valve plunger to increase the blow off pressure in direct proportion.

The pressure across the speed control orifice is, however, always maintained at the spring setting and the small ball valve set to blow off at the maximum safe pressure enables the valve to act as a safety valve also.

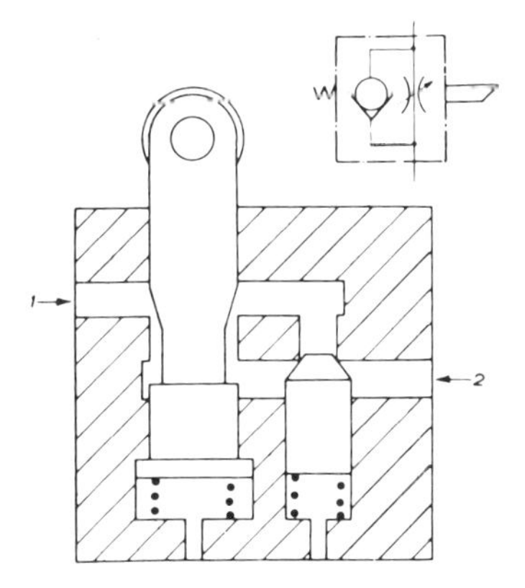


*Figure 4*
*Deceleration valve with integral check valve to give free flow on return stroke.*

This circuit has no control over negative loads and the movement under varying positive loads may be irregular, due to the intitial compressibility of the oil. This can be overcome, for many practical purposes, by placing a spring-loaded check valve in the exhaust line, which ensures a back pressure of 3.5 bar (50 lbf/in$^2$) or so.

**Deceleration Valve**

When working a cylinder at high speeds, it may be necessary to provide some means of slowing it down before the end of the stroke, so as to prevent shock. If the load is not too great, this can be done effectively by a simple *deceleration* valve as shown in Figure 4. The tapered plunger is depressed by cam on the moving part so that the amount of throttling can be controlled.

With another type, the valve is mounted parallel to the cylinder and comes into operation during the last half in or so of travel. the same effect can also be obtained with a *cushioned* cylinder.

If an attempt is made to decelerate too great a load with this type of valve, the pressure may build up dangerously and it is then necessary to employ a braking valve. The simplest form of braking valve is a relief valve which comes into operation at the appropriate point of the stroke. The energy is then dissipated by blowing through the relief valve.

**Cylinder Speed Control**

The speed of a cylinder can be controlled by means of a flow control valve fitted either in the inlet line *(metering-in)* or the outlet line *(Metering-out),* as shown in Figure 5. A flow control valve can, of course, be placed in both lines when it is

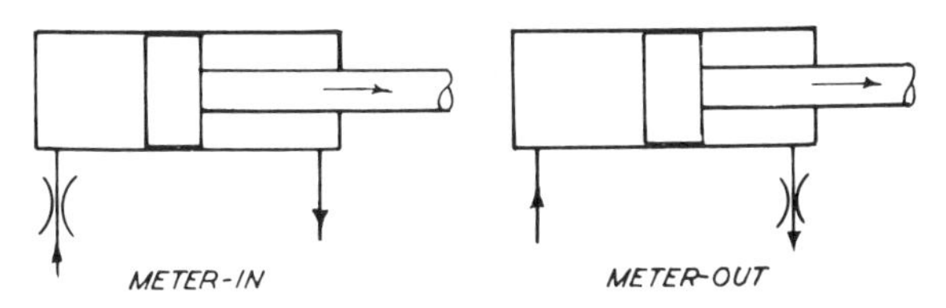


*Figure 5*
*Meter-in. Meter-out.*

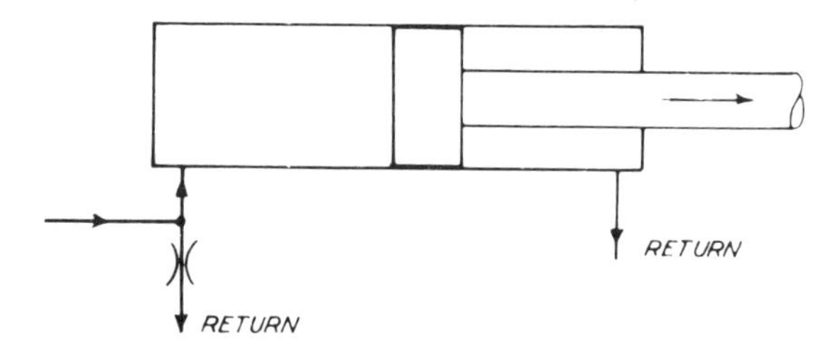


*Figure 6*
*Method of bleed-off.*

necessary to control the speed in both directions of the operation, but these would need to have reverse free flow check valves incorporated.

A further method is 'bleed-off' shown in Figure 6. General recommendations for the application of these alternative methods can be considered.

Meter-in accurately controls the volume of fluid entering the work cylinder. The pump must deliver more fluid than is required to give the desired cylinder speed, and the system relief valve must be set higher than the load pressure, to create a pressure drop across the flow control valve. This results in an inefficient system, if a fixed displacement pump is used, and could lead to overheating of the system fluid.

This kind of control can only be used in applications where a load is opposing motion, and should slight negative loads occur some back pressure must be provided. In fact, it is common practice to incorporate a back pressure of about 3.5 bar (50 lbf/$^2$) in the return from the cylinder to overcome cylinder 'stickslip' and to give the flow control valve compensator more stability. This back pressure, however, is also added to the relief valve setting.

Meter-out flow control meters the cylinder discharge flow and is used with over-running or negative loads. The relief valve setting does not have to be as high as with meter-in control due to the differential areas of the cylinder. However, with a 2:1 area differential, a pressure of twice the relief valve setting could be generated in the rod end of the cylinder if the load on the cylinder is removed.

This kind of control is not as accurate as 'meter-in' control as all the flow out of the cylinder is being metered and leakage through valves and across seals becomes a significant factor.

*Bleed-off* flow control is more efficient than meter-in and meter-out but is not as accurate in applications where the load varies. A controlled flow is bled off to tank, while the remainder of the pumped volume flows to the cylinder.

No flow takes place across the relief valve while the cylinder is extending or retracting, only when the cylinder movement ceases does the relief valve come into operation. The fluid in excess of the system requirements passes to tank at a pressure generated by the load, therefore generating less heat.

This type of control can only be used where the load is opposing motion and under lightly-loaded conditions a back pressure may have to be introduced in the exhaust line to achieve this. The accuracy of the system is dependent upon the volumetric efficiency of the pump as the controlled flow is passed to tank and control is mainly confined to systems where the load remains relatively constant because the pump volumetric efficiency varies with change of pressure.

# Directional Control

DIRECTIONAL CONTROL, as the name implies, is involved with controlling or changing the direction of flow of the fluid in a hydraulic circuit. This can be achieved by changing the direction of flow from a pump, by using a variable delivery overcentre pump in a closed circuit. These are often used where a large cylinder has to be continuously reciprocated without the associated pressure loss of intermediate control valves. But more often in small to medium sized systems directional control valves are used.

**Directional Control Valves**

Directional control valves provide the means of changing the direction of flow in a circuit, *eg* from one end of a cylinder to the other. They are usually spool valves although poppet and rotary valves may be used for specific applications.

Poppet or ball valves are normally used for non-return flow check valves which control the flow in one direction only. Pilot-operated check valves have an additional pilot piston to either pilot open the valve, to give full flow in both directions, or to hold the valve closed in both directions.

The pilot pressure required to open the valve is usually a fixed percentage of the pressure holding the poppet closed, the value of which depends on the relative proportions of the pilot piston and poppet areas.

Directional control valves are normally classified by the number of positions and the number of ports or 'ways' in the valve body. Standard configurations and symbols are shown in Figure 1. Valve symbols, as previously shown, comprise a number of squares, side by side, each square representing a possible *position* for that valve. Thus a two-position valve is represented by two squares, and a three-position valve by three squares. Ports are indicated in each square, but separately annotated to show the inter-connection in each position. The valve symbol also includes an indication of the method of actuation of the valve and return to its normal position.

**Two Position Two-Way Valves (2/2)**

A two-way or two-port valve is basically an on-off switching element and can be simply described as a shut off cock. Obviously a variety of designs can be used to provide this: *function gate, globe, rotary, slide,* poppet and spool valves. The first two are excluded, because their mode of operation is not suited to small sizes. Rotary or slide valves are practicable, with a capability of sealing in either

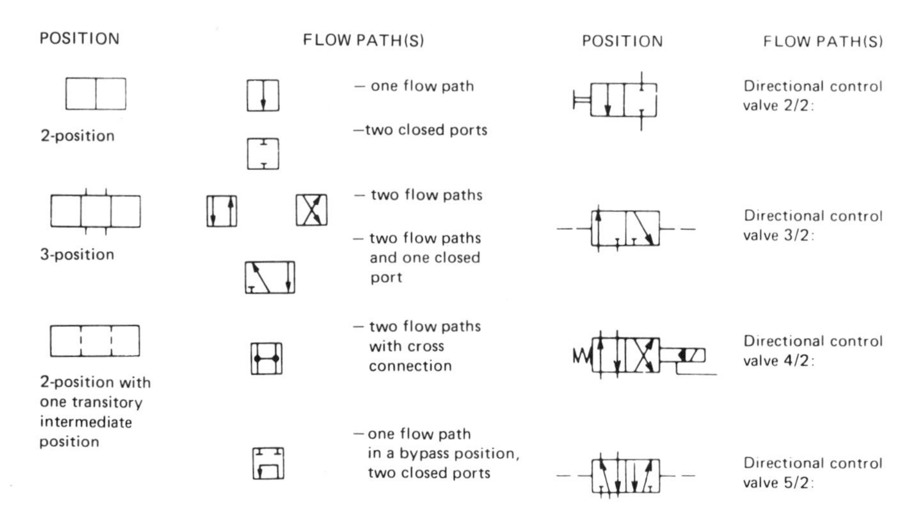


*Figure 1*
*Basic ISO symbols for directional valves.*

direction. Poppet valve cocks are generally capable of sealing in one direction only. Spool valves are the most logical choice for precision applications.

The simplest type of 2/2 valve is normally operated with a spring return. For a 'normally open' valve the spring holds the valve in the open position and for a 'normally closed' valve the spring holds the valve in the closed position.

**Three-Way Valves**

A three-way valve has three ports with five possible modes of inter-connection (see Figure 2). It is most commonly used as a two-position valve (3/2) for the control of a single-acting cylinder. It can, however, be used as a three-position three-way valve (3/3) with open centre characteristics (*ie* with a straight through pump to tank connection in mid-position) or as the operating mechanism for a two-way valve. Three-way selectors are also used for the control of double-acting differential cylinders.

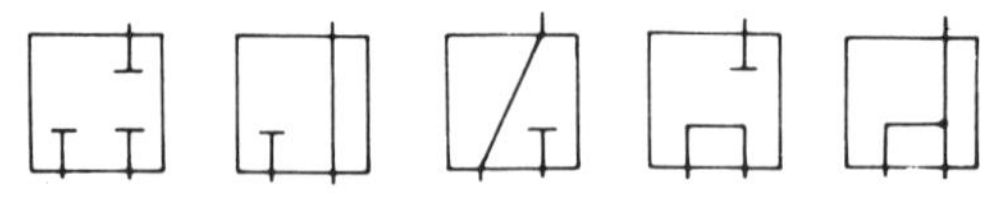

*Figure 2*
*Connections for three port selectors.*

Poppet valves are not generally used as three-way selectors, because an additional non-return valve would normally have to be included to prevent return flow passing through the inlet. An additional relief valve may also be necessary to accommodate thermal expansion of the fluid volume between the two valves.

**Four-Way Valves**

The four-way selector is the normal type of valve used for the control of reversing systems (cylinders, motors and pilot controls) and it is employed as a simple two-position reversing 'switch' to reverse the connections between two pairs of ports.

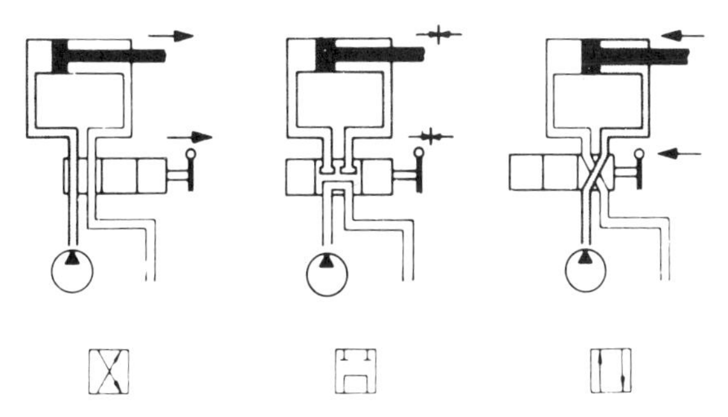

*Figure 3*
*The three 'working' positions for a three-position directional control valve.*

The changeover conditions may provide either open-centre, semi-open, or closed-centre characteristics (see Figure 3).

The four-way valve also offers numerous other useful configurations by alternative inter-connection of the four ports. This may be further extended by making the valve a three-position type (4/3) which again has its particular applications. Four-position four-way valves may also be used for a minority of special applications.

**Five-Way Valves**

A *five-way* selector is basically, a modified four-way directional control valve in which the two sets of return to tank ports in the valve bore are usually taken to separate external ports on the body. This provides independent speed control of exhaust flow, without the need for separate non-return valves, or can provide for dual pressure operation of an actuator. They are also used for interlocking circuits or circuit unloading.

**Six-Way Valves**

A *six-way* selector is, basically a four-way directional control valve with an extra two ports to provide pump unloading in the centre position. It is a type which has been developed specifically for mobile equipment applications, the flow path provided in the centre position saving unnecessary waste of power in the case of a continuously running pump by allowing simple unloading and reducing the heat generated in the system. Examples of various spool configurations are shown in Figure 4.

Six-way valves can be used as single units or more usually are banked together for mobile applications and are known as *mobile banked valves.* In this case the pump unloading port in the centre position is carried through a port cast into each section. There is a limitation to the number of sections which can be banked together due to an increase in pressure drop through each section and also due to the physical size of the banked valve and the strength of the through bolts on modular valves.

Where more than one banked valve is required, in different positions, from one pump system a 'carry-over' port is provided so that the valves can be piped in series.

Mobile directional valves are used to meter flow to the cylinder ports by shifting the valve spool less than fully off-centre. Shifting the valve only slightly 'cracks' the

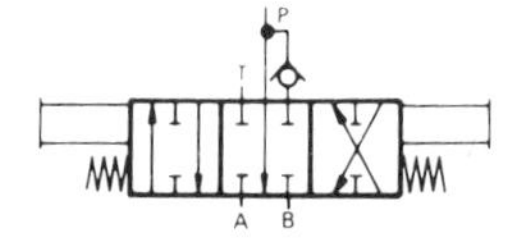


**"D" spool** – normally used to direct pump flow to either end of a double-acting cylinder. Both cylinder ports blocked in neutral position.

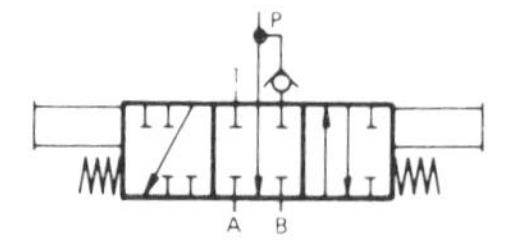


**"W" spool** – reverse operation to "T" spool.

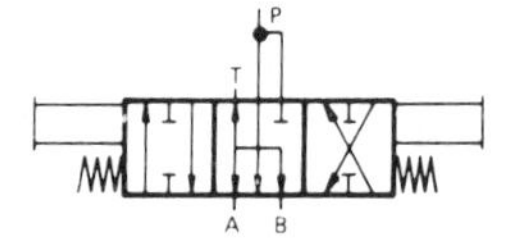


**"B" spool** – to direct flow to a bi-directional hydraulic motor. Ports partially open in neutral.

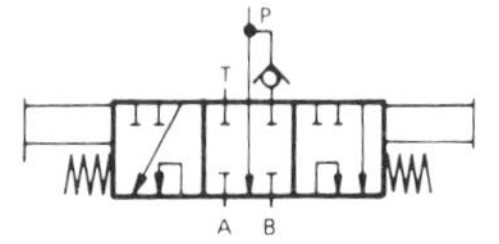


**"S" spool** – double-acting spool which checks oil back down bypass for additional functions.

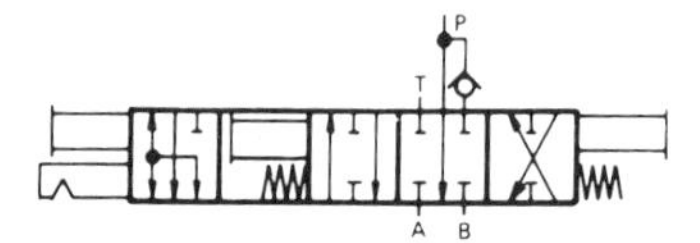


**"C" spool** – double-acting spool with fourth 'float' position which can provide free movement of a double-acting cylinder.

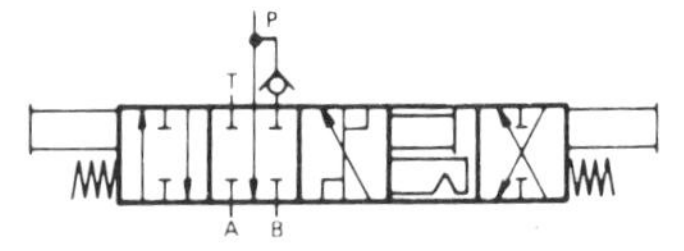


**"R" spool** – four-position regenerative spool allows cylinder exhaust at one end to combine with pump flow at the other for rapid rod extension.

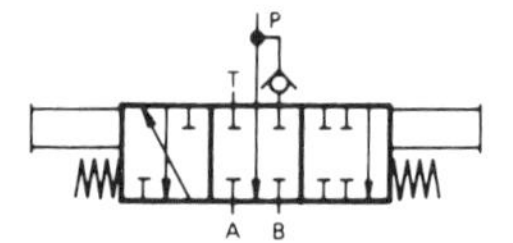


**"T" spool** – directs flow to one end only of a double-acting cylinder (*eg* for fork lift truck operation).

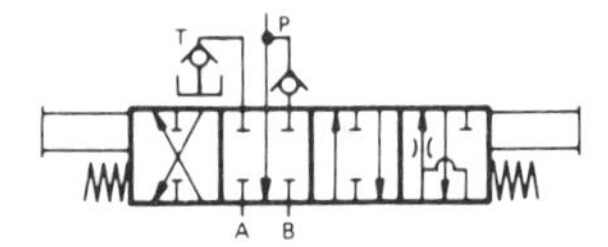


**"G" spool** – four-position regenerative float type action, normally used in conjunction with a back-pressure valve.

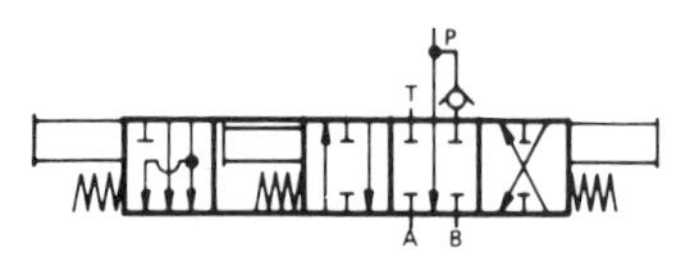


**"F" spool** – normally used to control dump cylinders.

*Figure 4*
*Examples of spool configurations and applications.*

pressure port to the cylinder port while only partially blocking the bypass. Thus the flow is divided sending partial pump delivery to the cylinder and bypassing the rest to tank.

The valves have infinite positioning between centre and wide open and they function both as directional valves and flow controls. The standard bypass is designed to handle the normal flow rates of the valve. Narrow bypass spools are used for better metering when the flow rate is low.

*Mobile-banked valve.*

**Proportional Control Valves**

Directional control valves can be produced similar to the mobile banked valves to meter flow in intermediate positions to perform a flow control function. Normally such valves are not pressure/load-compensated or temperature-compensated, thus the volume of fluid passing through the valve will be affected by changes in pressure drop and/or fluid viscosity and will not be proportional to spool movement.

For applications where consistent flow is required, independent of changes in supply pressure or load pressure, they can be associated with a pressure compensator to maintain a constant pressure drop across the valve, similar to a flow control valve.

Proportional valves are the modern choice for applications calling for fast control of velocity, acceleration and deceleration of actuators. As well as *manually-operated* valves, *solenoid-operated* valves are available with electrical position feedback control (proportional solenoid) as shown in Figure 5. Spool deflection is measured by an inductive position sensor and fed back to a proportional amplifier. The pilot stage is thus an electric position control system unaffected by load disturbances, while responding rapidly and precisely to input signals.

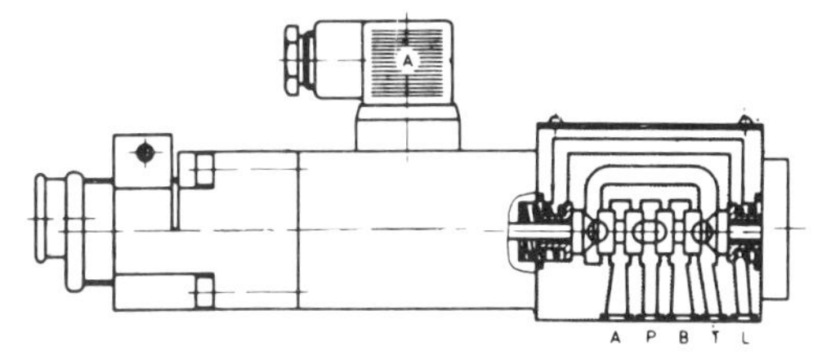


*Figure 5*
*Proportional throttle valve.*

Such valves also normally have built-in safety features so that in the event of supply failure, signal failure or loss of spool position feedback, the valve will automatically go to the centre or safe position.

Proportional valves are most commonly applied to *open-loop* applications, where in effect they give 'closed-loop' performance normally associated with a servo-system. They are not, however, suitable for *positional control* aplications *ie* to replace a servo valve with actuator feedback. Basic functional types are *proportional throttles* and *proportional selectors* (directional valves).

A basic requirement with such valves is that they should be used only with the electric controllers specified by the manufacturers for each valve.

The developments and growth within the electronics industry during recent years have affected the range of control systems available for static and mobile machines.

The electronic equipment now available from hydraulic equipment suppliers allows the user to obtain off-the-shelf packages. These can readily be incorporated into the user's own control system, to produce features economically that have previously been difficult or expensive to implement.

The electronic controllers used with proportional valves are basically solid state amplifiers which provide direct current to the solenoid, proportional to the command and/or feedback signals. Controllers use *Pulse Width Modulation* control circuitry in which signals are amplified and applied to the solenoid coil, giving an average coil current depending upon the percentage on time versus off time.

A ramp generator can be used to give smooth acceleration and deceleration of the mass by controlled ramping of the command signal. Standard accessories for use with proportional valves include an amplifier card, which may be produced in the Eurocard format and sized in accordance with DIN 41494 with card edge connections to DIN 41612, allowing it to be used in various rack systems.

Figure 6 shows a typical installation with power supply, control amplifier, command potentiometer and *closed-loop* valve with a 6-core shielded cable to a linear variable differential transformer on the valve.

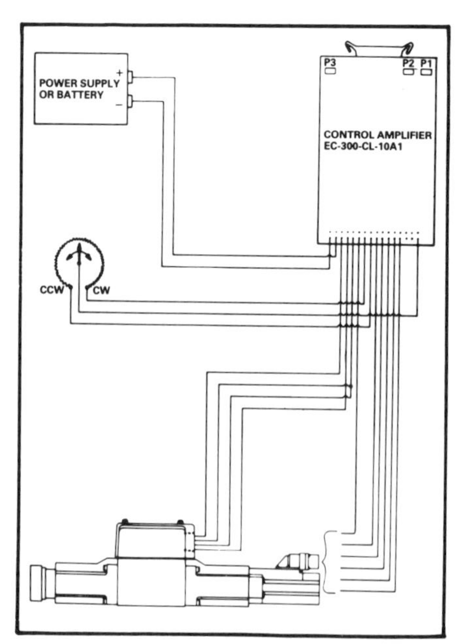


*Figure 6*
*Typical proportional control valve installation.*

# Servo Control

THE WORD *servo* is derived from the latin *servus* meaning a slave, so that a servo control system is a slave system using additional power to back up a control signal.

The correct definition of a *servo-system*, as applied to hydraulics, is a system which provides both power amplification and automatic correction for deviation, *ie* a closed-loop system with feedback (see Figure 1). The feedback is essentially an error signal, re-positioning the control valve to adjust the output to remove the error signal.

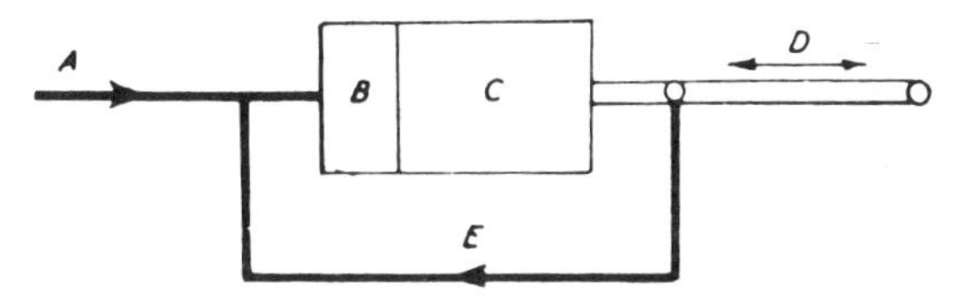


*A – input. B – servo-valve. C – actuator.*
*D – output. E – feedback.*

*Figure 1*
*Elementry closed-loop with feedback.*

The controlling valve is then specifically referred to as a *servo-valve* and is capable of accepting both an input signal and a feedback signal (error signal). The two are unbalanced as long as there is a positive or negative error signal. Thus the servo valve continues to respond to the combined signal, until the error signal falls to zero, *ie* the output quantity corresponds to that of the input signal quantity.

This is a true servo system, the main difference between it and an ordinary actuator circuit being the control valve employed, *ie* it is a servo valve capable of providing continuously variable flow with changing input signal. The latter is normally an electric signal, but feedback signals may also be derived hydraulically or mechanically.

As with simpler circuits, additional amplification may be introduced to accommodate a relatively weak input signal (see Figure 2). This amplifier is incorporated within the closed-loop, to accept both the command and error signals.

The system now provides both two-stage amplification and computation of

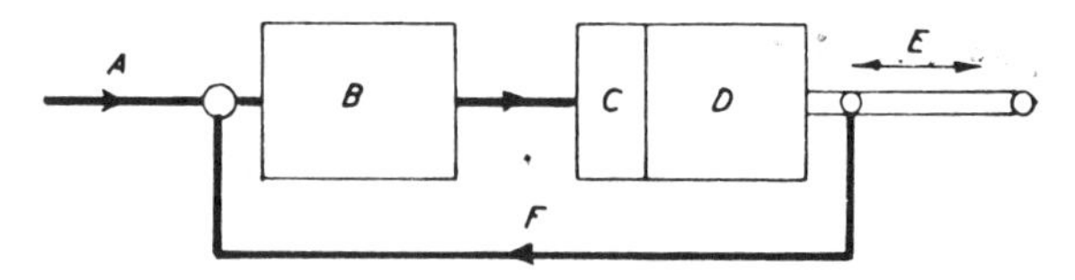


*A – input. B – amplifier. C – servo-valve.*
*D – actuator. E – output. F – feedback.*

*Figure 2*
*Basic circuit of elementry servo-mechanisms.*

position or linear response, with self-correction *via* feedback. Thus it will give true proportional response to the input signal, regardless of the effects exerted by external load. It also offers the advantage over the simpler system of being able to control high power levels at low input (signal) levels, both for the command and feedback signals. The combination of a servo-amplifier, servo-valve and actuator with a closed-loop feedback is generally referred to as a *servo mechanism.*

**Electrical Feedback**

Where the control valves are of the electro-hydraulic type the feedback signal can also be electric. The simplest way to do this is to connect the output movement to a potentiometer to generate a feedback signal directly proportional to the movement, which basically provides an analogue-type control. More sophisticated feedback signalling can be devised if required *via* a suitable transducer to eliminate the basic limitation of a simple analogue response (*ie* a tendency to hunt), or to provide an alternative function, *eg* compensation based on rate of change of either output to prevent over-shooting (derivative control), or of some other characteristic in the system (integral control).

This more sophisticated treatment is needed in modern hydraulic machine controls where forces or pressure must be infinitely adjustable, with regulated flows and speeds, controlled speed changes and exact cyclic repeatability. These features cannot be achieved in traditional hydraulics because manual and semi-automatic valves are severely limited in their flexibility by the relatively small number of control levels and by the presence of harmful transients normally associated with level changes. This is concerned with electro-hydraulic controls as a whole working on a closed-loop basis (Figure 3), rather than servo-operation as such.

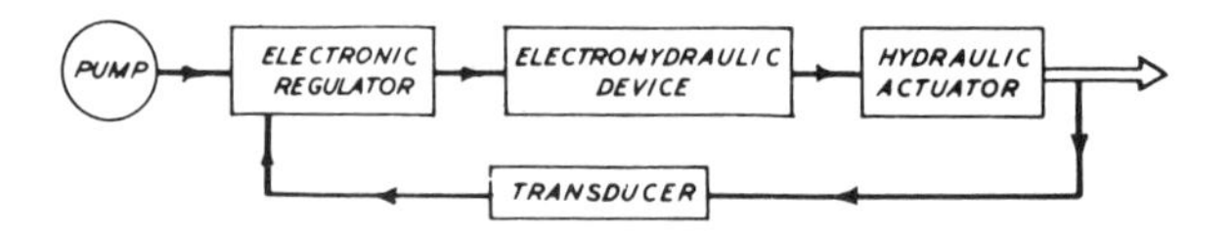


*Figure 3*

**Electronic Feedback**

The great advantage of electronic controls is that signalling rates are very fast. Also such controls are appreciably simpler to design and far more compact than mechanical or hydraulic feedback systems, especially when an increasing degree of sophistication is required. Proportional signals (voltage error signals) are readily derived from potentiometers, and simple resistor-capacitor networks can provide any necessary time delays or integral-derived signals.

Alternative control methods available for more sophisticated systems include:

(1) Pulse-length modulation – digital modulation of *bistable switching* valves:

(2) Differential pulse-length modulation – which avoids the high power loss inherent in pulse-length modulation systems and provides a closed-centre system adaptable to fluid valving techniques.

(3) Numerical control systems – involving the use of logic elements.

Whichever method is used, signals are usually amplified to yield a final output of the order of 0 to 15 milliamps dc for application direct to a solenoid-operated control valve or pilot valve.

**Mechanical Feedback**

Purely mechanical feedback systems are normally based on a differential lever mechanism, with manual input. Both the input and the actuator output movements are connected to the lever which is free to float under the action of feedback movement, automatically adjusting the servo-valve positon to compensate.

Actual synchronization of movement may be achieved more or less instantaneously, depending on the characteristics of the external load. Correction will also be initiated after movement has stopped, should the output member again be displaced by external load, with the input member held in its original (signal) position.

Mechanical feedback systems are relatively easy to design, but have distinct performance limitations where precise or complex response is required and also tend to be cumbersome as regards the length and complexity of the mechanical linkages involved. It is possible to simplify the linkage, *eg* by mounting the valve on the actuator with the differential lever moving with the actuator or mounting the valve on the moving part of the actuator and so dispensing with the differential lever entirely. In the latter case the input member is connected directly to the valve unit which is designed so that axial displacement of the input member produces an axial displacement of the output member in the same direction, thus automatically correcting for feedback movement. An example of this compact form of mechanical system is shown in Figure 4.

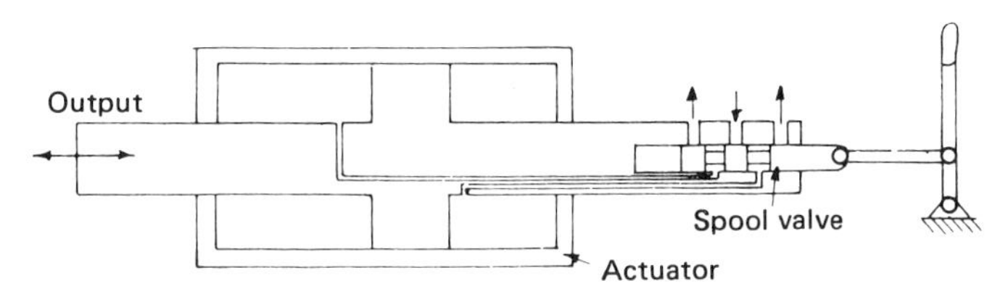


*Figure 4*

**Hydraulic Feedback**

Hydraulic feedback systems tend to be even bulkier because two cylinders of equal size are required, one being the actuator and the other an auxiliary cylinder which provides feedback, Figure 5. The body of the control valve is connected rigidly to the piston of the auxiliary cylinder. Thus any movement of this piston displaces the valve body in the same direction as the initial displacement of the valve plunger which is controlled by the input member.

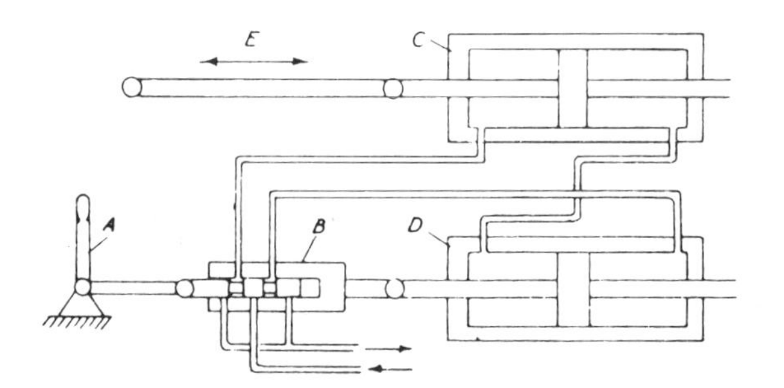


*A –input. B –servo-valve. C –actuator. D –auxiliary cylinder. E –output.*

*Figure 5*
*A simple servo mechanism with hydraulic feedback.*

Initial movement of the input member displaces the valve plunger to direct fluid into one end of the main actuator and an equal amount of fluid is displaced from this actuator into the auxiliary cylinder. This results in movements of the auxiliary cylinder piston, maintaining flow to the main actuator until a 'synchronized' position is reached, when the servo-valve is closed.

The basic advantage offered by hydraulic feedback is that it can be adapted to any size of actuator, is completely free from mechanical backlash and wear, and sensitivity and response can be very high. Lag can be reduced to as low as 1%, although a somewhat higher figure may have to be accepted in the case of fast movements, valve design being the critical factor.

**Electro-Hydraulic Servo-Control Systems**

These are two main types of electro-hydraulic servo-control systems in use, based on the employment of servo-valves. One is where positioning control is required. The other is based on *load monitoring*.

A positioning system works on the closed-loop system, described previously, with feedback. The performance of such a system can be improved by the addition of a velocity-control network to give positioning accuracy of the order of 0.025 mm (0.001 in) or better, depending on the resolution of the feedback components. It is also a relatively simple matter to extend such a system to multi-axis working with each axis being driven by its own *servo-actuator* controlled from separate channels.

A load-monitoring system is one based on monitoring the pressure in the actuator to maintain a pre-determined load (rather than position), regardless of actuator position. The sensor in this case is a transducer, sited on the load or, preferably, in the actuator. The transducer then feeds the *servo-amplifier* with a voltage signal, the value of which corresponds to the load. This is compared with a smaller reference signal. Any difference represents an error signal, with adjustment of movement until this is reduced to zero.

It is also possible to accommodate two variable functions in such a system – load variable (as described) and frequency variable. The latter is achieved by superimposing frequency oscillation to give varying frequencies which may vary from one per min to thousands of cycles per s. Under such conditions, although the load is held steady, the cyclic amplitude of the actuator decreases with the increase in cyclic frequency for a given power potential.

Load-monitoring control systems have a particular application for fatigue testing materials and components.

**Servo-Valves**

A basic difference between an ordinary solenoid-operated valve and a servo-valve is that, whereas an ordinary valve has only two or three positions (or possibly more in some cases), a servo-valve allows a continuous variation of flow with changing electrical signal. An electro-hydraulic servo-valve is thus defined as an *electrical input servo-control* valve capable of continuous control and it is commonly used to drive hydraulic actuators or motors in closed-loop servo-systems.

Typically, a flow control servo-valve consists of two stages: an electro-hydraulic pre-amplifier and a spool valve. The second stage is basically the same for the several types of servo-valve offered by various manufacturers; each manufacturer uses the same materials for both the sleeve and spool, the same heat treatment, the same diametral clearances and similar manufacturing and processing techniques. So, the only real basis for comparison and user selection of one type of valve over another is the first-stage configuration and the manner in which the inter-stage loop closure is accomplished.

**Servo-Actuators**

The actuator in practical servo-systems is an hydraulic cylinder, so a combination of a servo-valve and a conventional cylinder may be described as a servo-actuator. More specifically, however, the description is best reserved for cylinders which incorporate a suitable servo-valve or are designed for direct mounting of matching servo-valves. The advantage of such an approach (apart fom making the system more integrated) is that it can provide unrestricted flow between servo-valve and actuator. Provision is also commonly made for mounting of the master reference, *eg* a linear potentiometer, on the actuator itself.

**Types of Servo-Valve**

Basically there are only three first-stage configurations which are commonly used in the design of servo-valves, each of which incorporates a 'functionless' open-centre arrangement. The variations are:

(1) Half-bridge one-arm variable.
(2) Full-bridge two-arms variable.
(3) Full-bridge four-arms variable.

In all three types an 'elastic' feedback element can close a loop between the spool valve and the input torque motor.

A single nozzle flapper first stage is typical of the half-bridge one-arm variable configuration. The full-bridge two-arms variable is represented by the symmetrical double-nozzle flapper. Full-bridge four-arms variable is represented by the jet-pipe/receiver combination.

The single- and double-nozzle flapper arrangements normally have an operating clearance of about 0.025 mm (0.001 in) between the flapper and nozzle(s). Because, in the second case, the maximum flapper movement away from either nozzle is normally limited mechanically to about 0.05 mm (0.002 in) the contamination particle size which is double nozzle version can handle without malfunction is strictly limited. The movement of the flapper away from the single-nozzle is not tightly constrained; this type is, therefore, generally more tolerant of contamination.

A single-nozzle configuration, however, is inherently susceptible to null shift, *ie* to any change in the input current required to bring the valve to the condition where it supplies zero control flow at zero load pressure drop. Null shift may occur with changes in supply pressure, temperature or other operating conditions. The double-nozzle configuration is not inherently subject to null shift with temperature and supply pressure variations. A jet-pipe/receiver combination is most tolerant to contamination and is substantially insensitive to null shift arising in temperature and supply pressure.

**Flapper Valves**

Both *single-* and *double-*flapper valves operate on the same principle. The system consists of a fixed and a variable orifice in series, the variable orifice being composed of a nozzle with a flapper plate very close to it. Flow is metered between the flapper and the circumference of the nozzle. The area between the fixed and variable restrictors is connected to one end of a spool valve. In the case of a single-flapper, the other end of the spool is connected, through a half area piston, to supply pressure (or to a spring). The spool forces are balanced when the control pressure is equal to half the supply pressure.

A current in the torque motor in one sense applies a force to the flapper to move it towards the nozzle. This increases the restriction and decreases the flow through the nozzle, causing a decrease in flow and pressure drop across the fixed orifice. The control pressure increases and causes the spool to move. The feedback spring between the spool and flapper applies a force to the flapper to return it towards the null position. When the feedback force and torque motor forces are equal, the flapper is restored to its null position, the spool forces are balanced and spool stops.

A double-flapper valve works on the same principle, except that the flapper operates between two nozzle/restrictor systems with the control pressures applied to either end of a equal area spool.

**Jet-Pipe Valve**

The first stage of a *jet-pipe* servo-valve consists of a torque motor, a jet-pipe and a reciever. An electric current in the torque motor coils develops a torque at the armature. The jet-pipe is rigidly attached to the armature and rotates with it. A very small flow of high-pressure oil is fed by a flexible tube to the jet-pipe. As the high-velocity oil flows out of the end of the jet-pipe, it impinges upon the face of the receiver. Two small-diameter holes located side by side in the receiver are connected to either end of the valve spool. With the jet-pipe centred over the two holes, equal pressures are developed on either end of the spool, causing it to maintain its position.

When a signal is received by the torque motor, the resultant torque causes the jet-pipe to rotate off-centre and pressure unbalance occurs across the spool, causing it to move. As the spool displaces from null, it deflects the feedback spring, developing a force counter to the the input torque. This force returns the jet-pipe to null and the spool comes to rest in its new position. Spool displacement is thus proportional to the torque input. For practical purposes, a linear relationship exists between torque and input current. This flow-control characteristic is fully reversible and a servo-valve can be used as a three-way or four-way type of valve. Continuous control of spool position can be applied from one extreme to the other.

**Contamination**

In the earlier designs of servo-valves, which were mostly for missile applications, the object was to produce a valve of light-weight design with a good frequency response (above 100 Hz if possible) and a high sensitivity. This was partly achieved by the use of light-weight first stages and small torque motors.

Considerable trouble arose from the effects of working fluid contamination which resulted in the modification of some basic valve designs, to reduce the first stage susceptibility to contamination, and in the development of new designs with a higher level of contamination acceptance.

Contamination has two main effects: one is to reduce the reliability of the valve by clogging or silting and the other is to reduce its life by increasing the wear rate of the spool.

Although an improvement in the valve's ability to operate in relatively dirty oil can often be obtained by suitable re-design, resulting in an increase in reliability over the longer periods necessary for aircraft and industrial use, adequate filtration is still essential to reduce wear. Wear occurs mainly in the second-stage metering orifices where high fluid velocities and accelerations occur. This wear results in increased leakage flows, increased threshold and a general deterioration in performance.

**Filtration**

Although in *nozzle-flapper* valves the diameter of the nozzle is about 500 $\mu$m (0.02 in) and the diameter of the fixed orifice is about 140 $\mu$m (0.006 in), the clearance between the nozzle and the flapper is only 25 to 50 $\mu$m (0.001 to 0.002 in) and particles larger than this will cause jamming of the flapper, silting and consequent clogging of the nozzle, or a transient movement of the load until the particle is cleared. To prevent this, internal filters are usually fitted immediately upstream of each orifice and nozzle.

In double nozzle-flapper valves unbalance can occur due to filter blockage at one of the nozzles and to prevent this one valve type uses a common filter for both nozzles. This filter, however, cannot be placed immediately upstream of the nozzles and thus does not remove contamination which might be left in the passage during assembly.

Other filtration techniques have been used such as flushing the outside of the filter with the second stage flow and the use of filter inlets arranged at 90° to the main flow direction so that the momentum of the larger particles helps to prevent them from entering the filter.

For detailed information on contamination and filtration, consult the Handbook of FILTERS & FILTRATION published by The Trade & Technical Press Ltd.

# Control Methods

ALTHOUGH HYDRAULIC control valves have been divided into separate designations, such as pressure control, flow control and directional control, *etc*, these dividing lines are often crossed in practice. For example a spring-loaded check valve is called a directional valve, although it can be used as a simple relief valve. A pressure-compensated flow control valve is, in fact, a pressure control valve maintaining a constant pressure drop across an adjustable orifice, to maintain a constant flow rate.

Most hydraulic control valves are involved in controlling pressure drop and hence flow rate. In a constant pressure system, two or more pressure control valves can be used as directional controls by changing the pressure settings, allowing the flow to be directed along the flow-path with the greatest pressure drop.

In modern hydraulic systems the trend is towards using multi-purpose hydraulic control valves and it is the control method and the position of the valve in the circuit which determines its designation.

**Manual Operation**

Manually-operated valves are the simplest and most numerous valves to be found in hydraulic systems. The manual operation can be a direct movement of a valve spool or rotary spindle by means of a handle, lever or foot pedal.

Rotation of a screw-threaded spindle is used for the control of *shut-off* valves and *needle* valves and also for the setting of spring-loaded pressure controls. The spindle can be rotated by means of a knob or hand-wheel, which can sometimes be removed, with the spindle held in position by locknut, to prevent unauthorized operation of the valve or valve setting.

The force required to operate a valve spindle increases with the size of the valve due to the hydraulic forces within the valve. When these forces become too large to make manual operation practical, the valves are usually operated by pilot pressure from smaller manually operated valves. These can be either directly mounted onto the main valve or be remotely positioned with pilot pressure lines directing the pilot flow to the main valve.

**Mechanical Operation**

Valves are *mechanically-operated* by fitting a cam following roller, at the end of the spool, which follows a cam profile to depress the spool. Cylinder deceleration valves are usually mechanically-operated, with the cam profile designed to give the correct deceleration rate for a particular load. Directional control valves and even pressure control valves can be mechanically-operated in a system which relies on a positive mechanical positioning of the valve spool.

Flow control panels can be cam-operated in circuits controlling rapid traverse and feed cycles of cylinders. As well as being pressure- and temperature-compensated during the feed stages of the cycle, the unit can be be pressure-compensated during deceleration, which occurs during the first part of the spool movement. Each panel can provide up to four functions: rapid traverse, deceleration, course feed and/or fine feed from a single cam actuator (see Figure 1).

Functional symbols

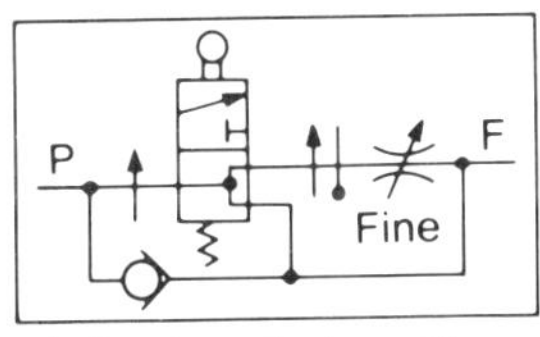


Single feed control panel

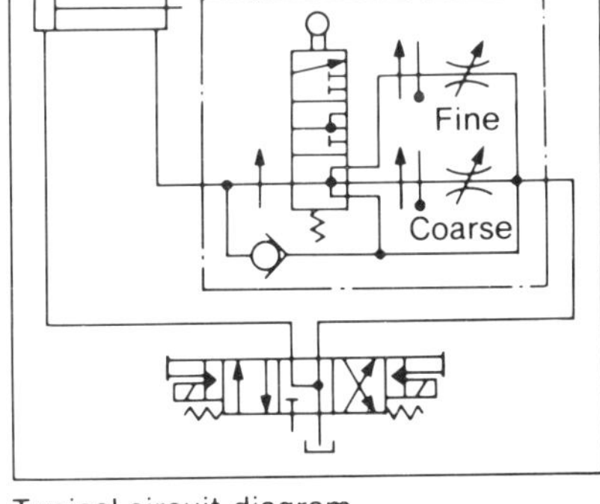


Typical circuit diagram

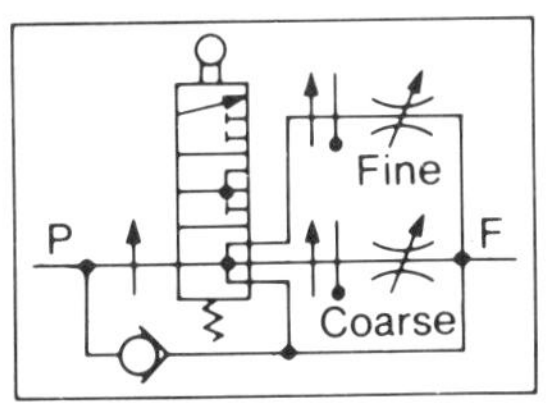


Dual feed control panel

*Figure 1*
*Mechanically-operated single and dual feed control panels.*

**Pilot Pressure Control**

Most *hydraulically-operated* valves are controlled by the system pressure or by a pilot or control pressure which is directed from the main system pressure.

In systems where the main pressure is allowed to decay while a control pressure is still required, at a pilot-operated valve, a spring-loaded back pressure check valve is usually positioned in the tank return line of the main valve to maintain a pilot pressure (see Figure 2).

Alternatively, in large volume high pressure systems, energy can be saved by employing a separate pilot system with external connections to the valves being controlled.

Hydraulic pilot-operated mechanisms are particularly adaptable for spool valves because the spool can be driven readily, like the piston in a cylinder, by an hydraulic force or pilot pressure applied to one end. Numerous variations on this theme are possible, with spring-centring or spring-biasing and with the incorporation of detents if required.

Operating speeds can be at least as high as (and can be made higher than) any mechanical operating system and speed control can be provided by incorporating a restrictor or throttling valve in an appropriate part of the circuit. This has the

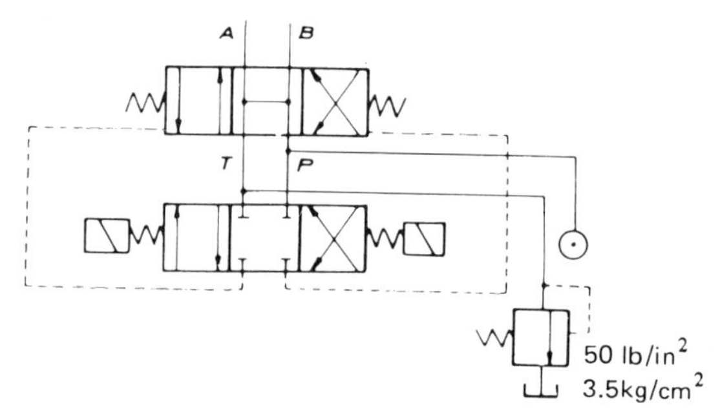


*A, B – to/from actuators.*
*P – pressure line.*
*T – tank (exhaust) line.*

*Figure 2*
*Pilot-operated valve circuit showing use of spring-loaded check valve maintain pilot pressure.*

advantage that high operating forces are available to overcome friction, although such a method of biasing will not have 'fail-safe' characteristics.

Pilot-operated valves are used both as units and integral with a solenoid-operated pilot valve. Figure 3 shows a four-way two position (springless)valve on which is mounted a block containing restriction and check valves in the pilot lines. By using these the rate of shift can be controlled and shocks in the system due to sudden pressure changes minimized.

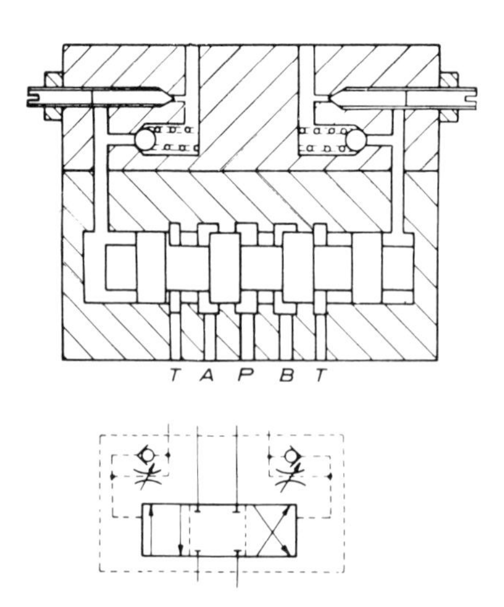


*Figure 3*
*Pilot-operated four-way valve with restrictors to limit speed of change-over.*

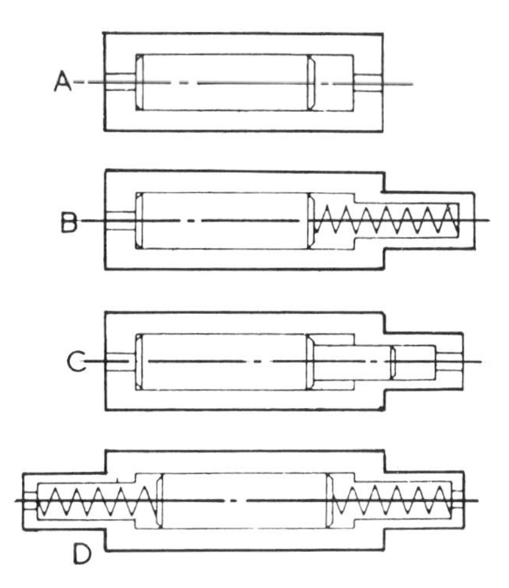


*Figure 4*
*Some arrangements of pilot-operated valves*
*A, B, C & D.*

The more common spool and spring arrangements are shown in Figure 4. The springless valves must be used with their axes horizontal and are suitable for pulse operation so that pilot pressure need not be maintained for longer than is needed to shift the spool.

**Electric Solenoid Operation**
Solenoid operation is used on either multi-position directional control valves, where the spool is moved into a positive position by the solenoid, or on *proportional control* valves where a spool or control element is moved proportionally to the applied current, as previously mentioned.

In a solenoid-operated directional spool valve an electric solenoid and spring are used to move the spool. The solenoid is generally used for a 'push' operation, utilizing spring-action for return motions.

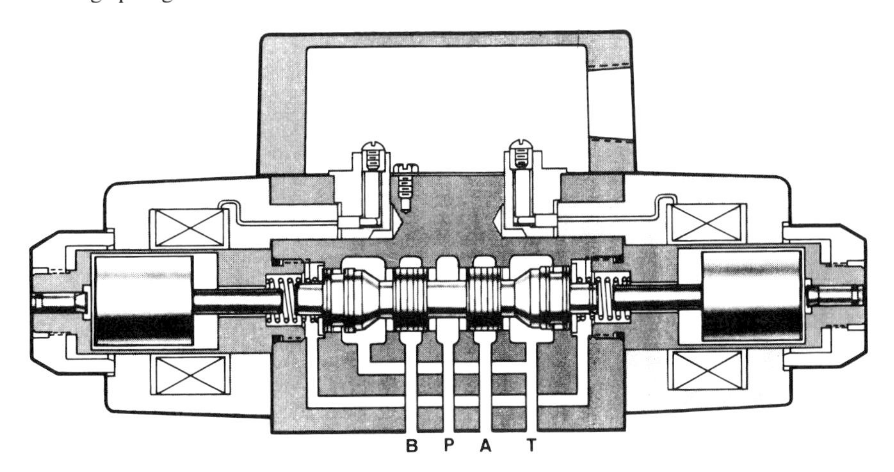


*Direct-acting double solenoid valve.*

The solenoid must be powerful enough to over-ride inertia and friction and also the spring and hydraulic forces. The latter may be extremely variable and not completely predictable, calling for generous margin in the power of the solenoid and springs.

The size of directly operated solenoid valves is generally restricted to flow rates up to about 45 lit/min (10 gal/min), *ie* 3mm (⅛ in) and 6mm (¼ in) nominal size valves. Many of these valves can be switched directly from static systems, the outputs usually being 24 V dc and 20 to 65 W, depending on the system.

For higher flow rates demanding larger port sizes and larger spool diameters, directly operated solenoid valves become excessively bulky and costly. This can be overcome by pilot operation where a small solenoid valve is used as a piloting first stage in a large body valve whose spool is moved by differential pressure applied to its terminal surfaces as previously described.

A typical layout of a miniature valve directly mounted on a pilot-operated valve is shown in Figure 5.

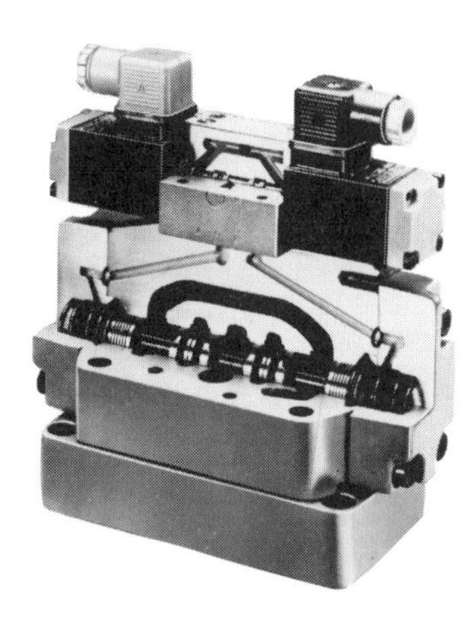

*Figure 5*
*Solenoid-controlled pilot-operated valve.*

*DC solenoids* are generally preferred to *ac* because dc operation is not subject to peak initial currents which can cause overheating and coil damage with frequency cycling or accidental spool siezure. *AC solenoids* are preferred, however, where fast response is required, or where relay-type electric controls are used. Response time with ac solenoid operated valves is of the order of 8 to 15 milliseconds, compared with 30 to 40 milliseconds typical for dc solenoid operation.

**Electro-Modulated Control**

Electro-Modulated Hydraulics embraces both electro-hydraulic controls and complete servo-systems operating as a closed-loop with feedback. Three basic 'blocks' are common to all such systems:(Figure 6).

(1) An electronic regulator.
(2) An electro-hydraulic device.
(3) An hydraulic actuator (*eg* cylinder).

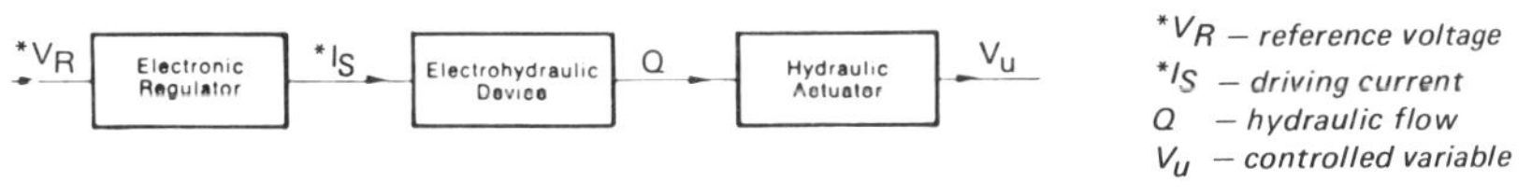


*Figure 6*
*Basic 'blocks' for electro-modulated hydraulics.*

In an open-loop *electro-modulated* system input is provided by a reference voltage fed to an electrical regulator which passes on this information in the form of a driving current applied to an electro-hydraulic device (*eg* an electro-servo valve). This in turn is responsible for the hydraulic actuator. The output thus corresponds with a value of the input signal.

For good response and correlation it is important that pressure, load on actuator and other system parameters are kept constant because the system cannot sense or correct any discrepancy between input signal and resulting output.

In a closed-loop electro-modulated system a transducer is added and limited to the output to provide a feedback signal directed to the input side. This signal may be indicative of position, velocity, force, or, if necessary, more than one output parameter (see Figure 7). The electronic regulator then compares the input signal with this feedback signal and varies its own output (driving current) to minimize or eliminate any discrepancy between input and output.

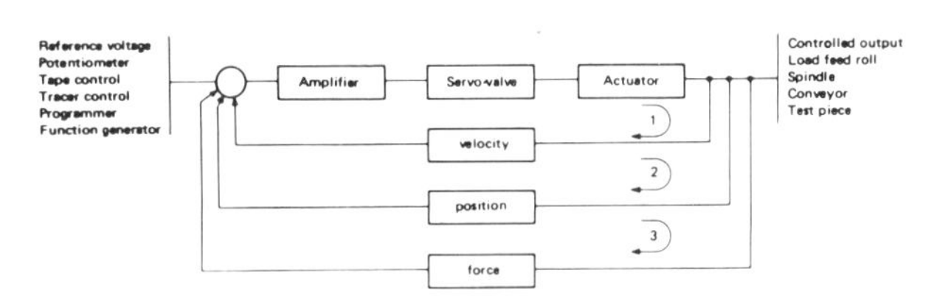


*Figure 7*
*Three main forms of control using electro-hydraulic servo-actuators.*

A further example of electro-modulation is shown in Figure 8. This is a conventional circuit, except for the fact that a main electro-modulating relief valve is included to control the pressure. The incorporation of such a valve allows the system to operate with pressure cycles of every complexity, with gradually increasing or decreasing levels rather than sharp pressure peaks.

**Valve Characterstics**

The characteristics of a servo-valve can be described by a number of parameters. The rated flow, customarily defined at 70 bar pressure, can vary from 0.5 lit/min, for a single-stage valve, up to 250 lit/min for a two-stage valve. The rated input at the rated flow can range from 4 to 500 mA, depending on valve and coil type. Linearity, a measure of proportionality between input and output, is usually better than 10%; a flow plot shows this (Figure 9). Separation of the two lines indicates the valve repeatability or hysteresis. The frequency response diagram (Figure 10) shows the ability of a valve to react to changes in input. This is an important characteristic and figures of 50 to 100 Hz are usual, while valves with a capability up to 400 Hz are available.

**Valve Types**

Control valves used in electro-modulated hydraulics are usually based on *torque*

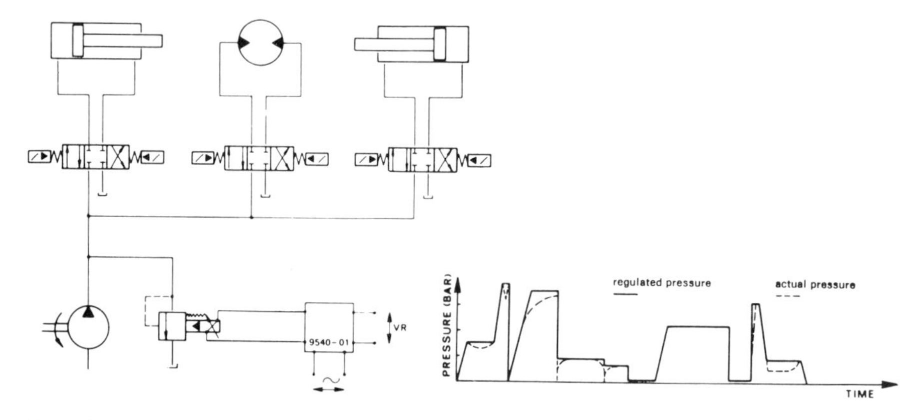


*Figure 8*
*Further example of electro-modulation.*

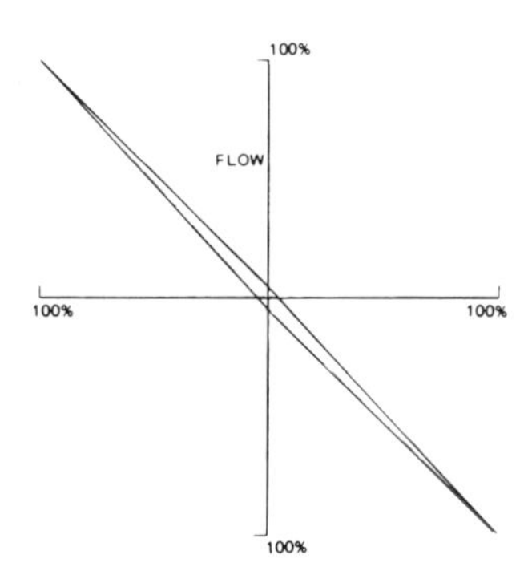


*Figure 9*
*Flow plot for servo-valve.*

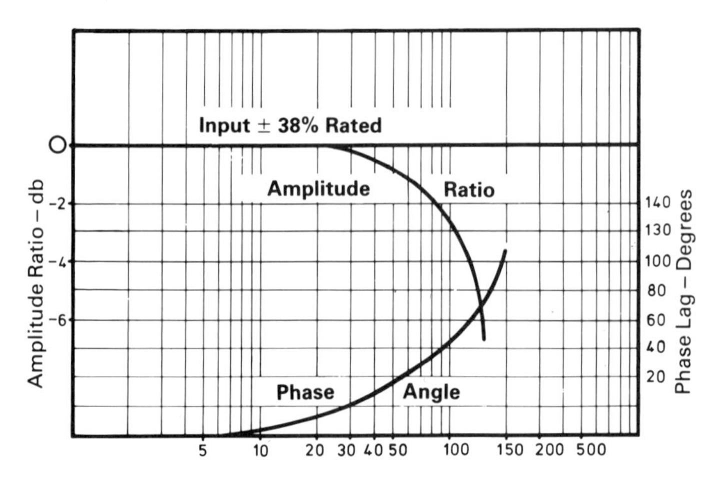


*Figure 10*
*Frequency response plot for servo-valve.*

*motor-operated* servo-valves, or specially designed proportional pilot valves. Choice is largely governed by the degree of control required over flow and pressure. The main types of valves are:

(1) Torque motor-operated electro-hydraulic servo-valves – particularly recommended in applications where it is necessary to keep a bi-directional control on flow and pressure (controlled feeding of one of the two branches of a hydraulic actuator, motor, cylinder, *etc*). The features of the torque motor servo-valves allow them to give excellent performance and to meet the hardest requirements, rapidity of response, accuracy, good-null point passing, *etc*. In order to avoid malfunction and to give long life without wear, efficient filtering system (5 to 10 $\mu$m) of the hydraulic fluid must be provided. The electric power required for operating the torque motor is low (0.1 to 0.2 W).

(2) Linear force motor-operated electro-hydraulic servo-valves – particularly recommended in applications requiring unidirectional control of pressure and flow. They are used as the first hydraulic operating stage of valves in which the output quantity (pressure and/or flow) can be directed by a pressure signal.

Solenoid-operated servo-valves have no critical filtering requirements (*ie* not higher than those of normal industrial components) and have the advantage that they can be used with normal type hydraulic components. Electrical power required to operate these valves is of the order of 10 to 25 W.

**Transducers**

The acccuracy achieved in a closed-loop electro- modulated servo-system depends primarily on the accuracy of the transducer. The main types employed are *linear-displacement transducers, pressure* transducers and *angular-displacement* transducers (see Table 1). There are also separate types of transducers used for measuring velocity, *viz:*

(1) *Linear-speed* transducers – obtained by transferring the linear motion into a rotary motion and measuring this motion, or by using the electric signal of a displacement transducer and determining the speed as derivative of the displacement function.

(2) *Pulse-speed* transducers – using a pulse transducer when the frequency of the pulses is proportional to speed. This can be used to measure either linear of angular speeds.

*Angular-speed* transducers – such as tacho-generators or alternators provide and electrical signal and a voltage proportional to the speed of rotation.

**Electronic Regulators**

Actual input signals may be derived from a reference voltage, potentiometer, tape control, tracer control, programmer or function generator. The transducer signal is rendered as a feedback voltage which may be a proportional signal *(analogue)* or a pulsed signal *(digital)*.

In either case the regulator is basically the same, except that with a digital error signal a digital-to-analogue converter has to be employed. In practice the electronics is considerably more complicated but digital control can be expected to

TABLE I – SUMMARY OF TRANSDUCER TYPES

| Main Type | Sub-Types | Remarks |
|---|---|---|
| Linear-displacement transducer | Capacitance | For small sizes only. |
| | Differential transformer | |
| | Potentiometer | |
| | Pulse | |
| Pressure transducer | Flat diaphragm | Insensitive to noise. |
| | Strain gauge | Good accuracy, wide range of response. |
| | Bellows | |
| | Bourdon tube | |
| Angular-measurement transducer | Potentiometer | |
| | Resolver | |
| | Pulse generator | Optical, magnetic, electronic, *etc.* |
| Velocity transducer | Linear | |
| | Pulse | Pulse frequency proportional to speed. |
| | Angular | Tacho-generators measuring rotational speed only. |

provide for greater resolution; also it can utilize high-threshold logic elements for virtually complete immunity from noise.

A block diagram of a modern digital command and closed-loop control circuits for a precision feed drive is shown in Figure 11. The operating principle of the servo-system relies upon integration of command pulses generated from an internal crystal clock and feedback pulses from machine-driven digitizers. This is accomplished by using a multiple array of reversible counters.

Errors derived from the feedback signals and the position and velocity commands are converted by a precision digital to analogue which is converted into an analogue voltage and is fed to a power amplifier controlling the input current to the electro-hydraulic servo-valve. Both proportional and derivative stabilizing terms are included for the optimum system depending upon machines and a variable frequency filter oscillator is also provided to minimize stiction, hysteresis and dead band effects.

The manner in which control functions are performed is summarized under the following headings.

**Position Control**

The pre-selection of start, transition and stop positions on the panel edge switch causes a pulse count to be stored in the respective parity detectors. Subsequently when the command is activated, pulses from the clock generator are counted into the *servo-logic unit* and command store until parity exists with the pre-selected count from the position switches.

Feeding the pulses to a reversible counter in the servo-logic unit results in error signals which drive the servo-valve to feed oil to the actuator and move the machine slide. Feedback pulses from table/spindle motion are produced by a rotary incremental digitizer driven through a rack and pinion mechanism and by feeding

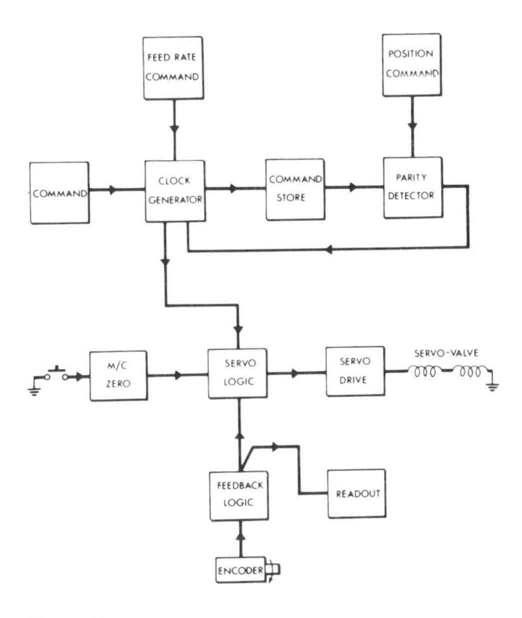


*Figure 11*
*Block diagram of digital command and closed-loop control circuits for precision feed drive.*

this also into the servo-logic unit counter, a proportional error signal is produced to move the table until feedback counts equal command pulses.

Typically a 5000 pulse/rev digitizer is mechanically and electronically coupled to achieve a single pulse count equivalent to 0.001 mm.

**Velocity Control**

The feed operation is also governed by the pulsed signals from the crystal clock; the output frequency of the clock is determined by the feed-rate setting registered on the feed-rate edge switches. When the clock pulses are received by the servo-logic unit counter and an error is produced, the digital-to-analogue converter generates a bias voltage which in turn causes the servo-valve to monitor oil to the actuator and causes the digitizer to move. Without feedback into the counter a linear ramp wave form would be produced by the converter and the servo-valve opening and hence actuator velocity would progressively increase. However, in practice, because pulses from the digitizer are negatively summed with command pulses, the voltage error is exponentially modified to settle at some steady value when feedback frequency exactly equals input frequency.

Actuator velocity is therefore controlled in direct proportion to the input frequency. Increasing frequency produces a higher analogue voltage, the actual value of which is dependent upon valve amplifier gain and the number of bits in the counter.

By introducing clock or drive frequency into the detection system, command parity is registered prior to actual axis position parity from the feedback digitizer and proportional corrections to servo-valve feed are introduced to minimize final approach errors. A schematic diagram of this circuit is shown in Figure 12.

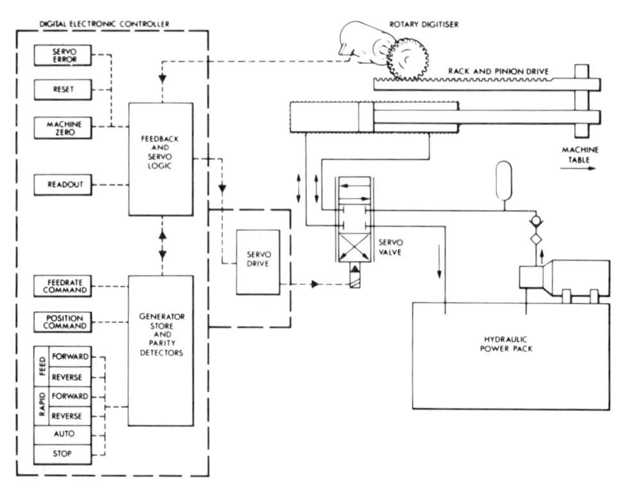


*Figure 12*
*Schematic diagram of digital precision drive.*

Another example of a digital-controlled system is shown in Figure 13. This is a flying cut-off drive. All electronic controls used are of the solid-state type, designed specifically for reliable operation in industrial environments. The digital controls are of a high-density integrated circuit type which has proved to be extremely reliable and maintenance free. Cut lengths and total number of pieces can be set on the operator's panel or can be fed into the system using punched tape or a computer output.

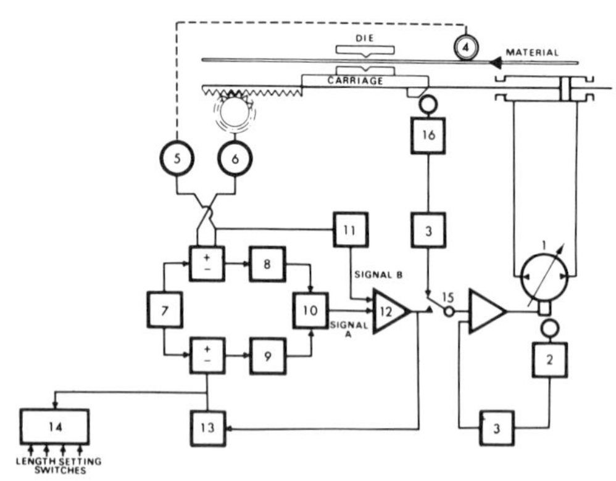


1–Servo controlled pump.
2–Linear variable differential transformer – pump.
3–Demodulator.
4–Measuring wheel.
5– Material digitizer.
6–Carriage digitizer.
7–Clock.
8–Feedback counter.
9–Command counter.
10–Comparator.
11–Pulse rate tachometer.
12–Summing amplifier.
13–Voltage controlled oscillator.
14–Length counter.
15–Reed relay.
16–LVDT – carriage.

*Figure 13*
*Digital position control.*

# SECTION 3

## Hydraulic Cylinders

TYPES OF CYLINDERS
CONSTRUCTION
PERFORMANCE AND SELECTION
APPLICATIONS

# Types of Cylinders

AN HYDRAULIC cylinder can be described as a *linear fluid actuator,* where the operating fluid is a liquid, usually an hydraulic oil. The descriptions *cylinder, jack* or *ram* are synonomous and largely peculiar to individual industries or traditions. All are essentially linear actuators and are similar in design, construction and function.

Basic types of hydraulic cylinders are shown by their appropriate symbols in Figure 1 in the chapter on Circuit Symbols, Section 1.

**Single-Acting Cylinders**
Single-acting cylinders are extended or retracted by applying fluid pressure to one side of the piston only to produce the power stroke. The return stroke, on release of fluid pressure is accomplished by means of some external force. Compression or tension springs are often used to achieve rod reversal and these may be mounted either inside or outside of the cylinder barrel.

Figure 1 shows compression springs mounted inside single-acting cylinders, one at the head end and one at the rod end, to extend and retract the piston rods, respectively. This type of cylinder is often used on fail-safe braking systems, the brake being applied by the spring force when the fluid pressure is removed.

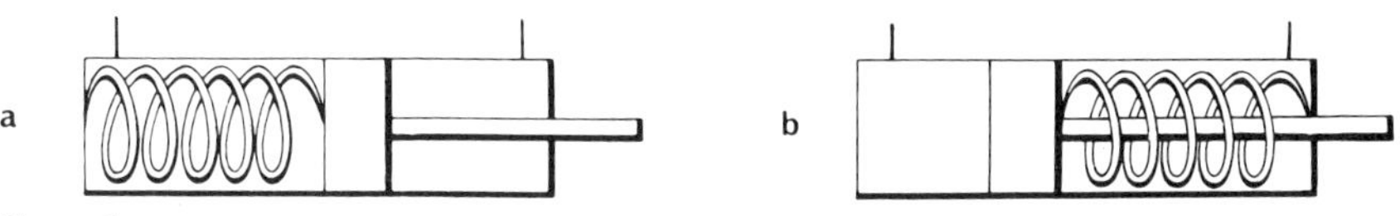


*Figure 1*
*Spring-loaded single-acting cylinders.*

A range of brake cylinders giving different braking forces can be designed, by changing the spring rates. The hydraulic force compressing the springs can be modified by opening up a flow path through the piston so that the pressure is acting on only the rod cross-sectional area instead of the full piston area.

Other single-acting cylinders which have rod hydraulically driven in one direction can take advantage of the force of gravity or a separate source of strain energy to reverse the direction of the cylinder rod. However, in order to prevent the effects of cavitation the non pressurised port must be allowed to breath freely, whether the fluid be air or oil.

In marine applications, where salt ladened air would be drawn into the non-pressurised end of the cylinder, it may be better to recirculate hydraulic oil from a small header tank, or low pressure accumulator, than risk the possibilty of corroding the cylinder bore.

**Displacement Cylinders**

Displacement cylinders are a form of single-acting cylinder with the conventional piston and rod assembly replaced by a plunger. They are also known as plunger type cylinders and are often referred to as *hydraulic jacks* or rams. The rod diameter is almost that of the cylinder bore diameter, apart from the space allowed for the gland bearing and seals. Stroke limiting stops should be provided in order to prevent the cylinder rod from being driven out of the barrel as shown in Figure 2.

Displacement cylinders are often mounted vertically and used for passenger and freight lifts and also for mobile equipment applications where the plunger is retracted by gravity.

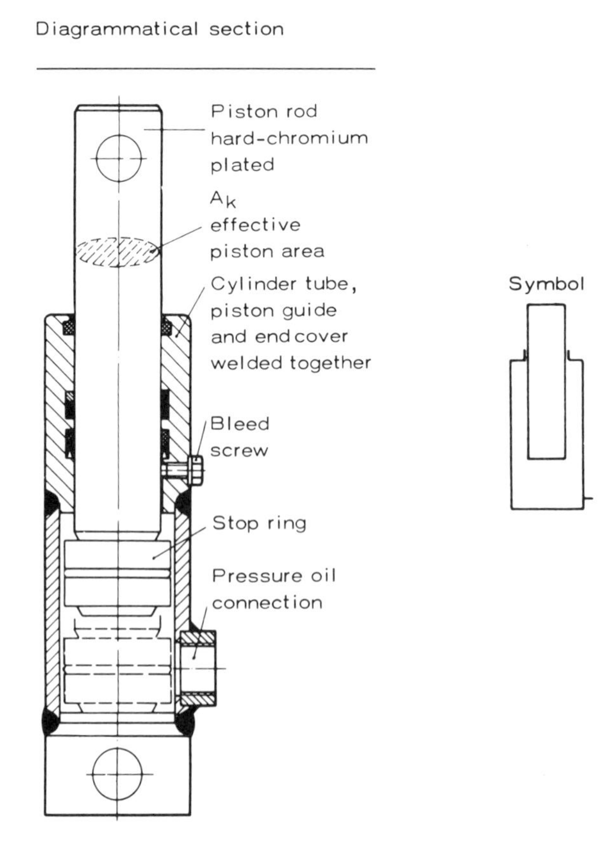


*Figure 2*
*Displacement cylinder.*

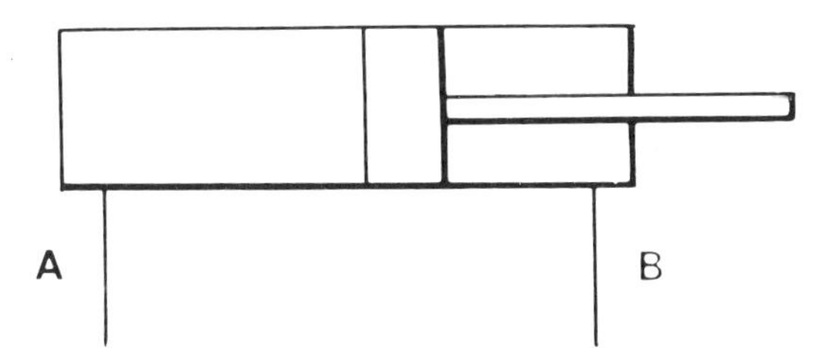


*Double-acting single rod cylinder.*

**Double-Acting Cylinders**

With a double-acting cylinder, ports are provided in each end so that the piston can be acted on by fluid pressure on both sides alternately. The individual ports thus function as inlet and exhaust ports alternatively, to provide the necessary fluid transfer in a reciprocating circuit.

*Single-rod* configuration is the more usual, although the through-rod or *double-rod* form may be adopted for greater rigidity or where exactly equal forces are required on both the outward and inward stroke. This feature of 'balanced' outputs is more significant with hydraulic cylinders than pneumatic cylinders because the rod section is usually more substantial and thus the force ratio with a single-rod double-acting cylinder is greater and fluid pressures are much higher.

The maximum output available is slightly less than that obtainable from a single-acting cylinder because when the fluid pressure is applied to the full piston area (outward stroke or extending) some back-pressure will be generated on the outlet side. Also, a *rod* seal will be required to prevent leakage when the piston is pressurized in the reverse direction, with consequently increased frictional resistance to motion.

**Through-Rod Cylinders**

Through-rod cylinders are also known as double-rod or *double-ended rod* cylinders and provide equal piston wear and equal forces in each direction, and 'two bearing' support for the rod. This type of cylinder is often used as a servo-actuator in servo control systems where it is advantageous to have equal forces and equal speeds in both directions for the applied pressures and flow rates.

Through-rod cylinders are usually double-acting, although single-acting operation can be obtained by pressurizing one side of the piston only and allowing the other side to 'breath' or circulate oil at low pressure.

**Telescopic Cylinders**

The telescopic cylinder is advantageous where a long stroke is required but the retracted length of the cylinder must be kept to a minimum. In this case the telescopic 'draws' are built up by using a hollow rod for the first section which acts as the cylinder for the second section and so on. Normally, however, the number of draws is limited to two, or three at the most.

Such a cylinder involves considerably more complicated and expensive construction and oversizing of the first stages, which add to its bulk. There is also the fact that the main bending loads are carried by the hollow rods acting as a cylinder, which must be proportioned acccordingly.

Double-acting cylinders of this type also tend to have a relatively low return force, because of the low effective piston area when working in this direction (see

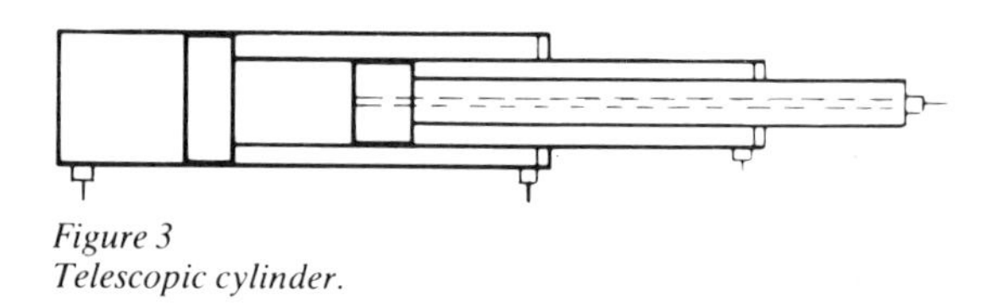

*Figure 3*
*Telescopic cylinder.*

Figure 3). For this reason the majority of telescopic cylinders are single-acting and are often used for tipper body systems on trucks.

**Differential Cylinders**

A double-acting telescopic cylinder is, in effect, a *differential* cylinder because it uses the volume of a hollow rod as a further source of power and is actually working as a multiple-volume device rather than a *single-* or *two-volume* device (characteristics of a conventional single- and double-acting cylinder, respectively).

The true differential cylinder, however, is used where differential outputs are required, when the ratio of piston area to the annular area on the hollow rod site determines the circuit function. Such units are often referred to as *multi-volume* cylinders.

Figure 4 shows a *three-volume* double-acting differential cylinder. There are three modes of working: inward and outward stroking of the piston and main hollow rod unit and independent extension *via* pressurization of the third volume. The only additional requirement in this latter case is that the appropriate port in the main cylinder must be opened or the cylinder will be hydraulically locked.

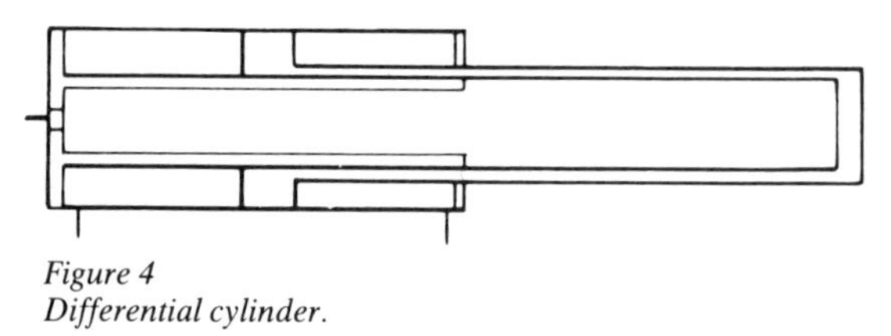

*Figure 4*
*Differential cylinder.*

It should be noted that the third mode of operation can be pressurized by the same fluid at the same or lower pressure if required or from a seperate source of pressurized fluid. Thus a cylinder of this type could provide for normal operation under hydraulic power, with emergency operation in the event of failure from another source (*eg* a compressed air bottle or accumulator) in one direction.

The same principles can be applied to give four-mode operation, with the second (or emergency) circuit being double-acting instead of single-acting.This calls for a *four-volume* cylinder, as shown in Figure 5. In fact, this really comprises two double-acting cylinders mounted concentrically in an integral unit.

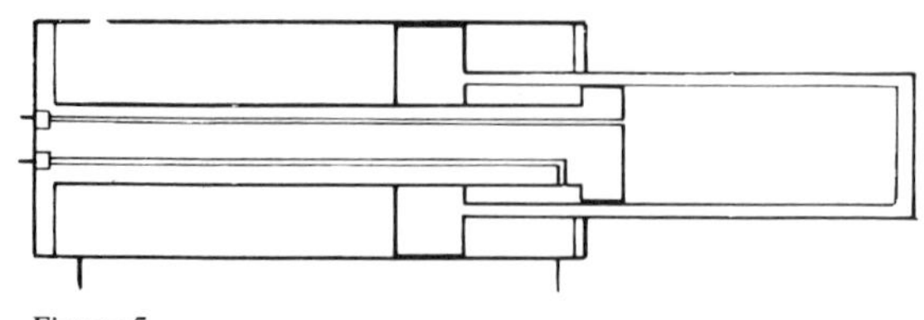

*Figure 5*
*Four volume cylinder.*

**Tandem and Duplex Cylinders**
*Tandem* cylinders comprise single- or double-acting cylinders mounted in line in a common construction. There are numerous possible arrangements, capable of providing separate movements and, because of this they are sometimes called *multi-position* cylinders. The simplest and most common type is a two-cylinder combination giving a three-position movement.

In the arrangement shown in Figure 6, two 'back-to-back' configurations are shown for providing a three-position movement.

In one case the tandem cylinder comprises literally two double-acting cylinders contained within a common cylinder casing but otherwise quite separate. In the other case the cylinders are physically separated, but their pistons are mounted on a common rod.

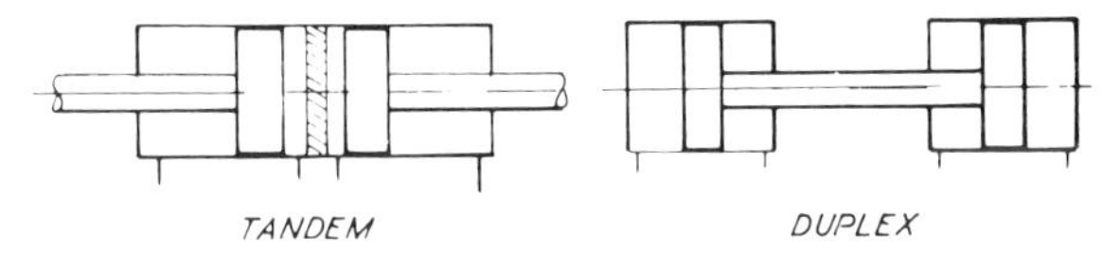


*Figure 6*
*Tandem and Duplex cylinders.*

Each cylinder, in effect, provides half of a complete movement, which can be used in combination to produce a full movement. Where the pistons of the two cylinders are mounted on a common rod the configuration is commonly called a *Duplex* cylinder.

A further type of three-position cylinder is shown in Figure 7. Here the two cylinders can be of different length, if necessary, for there is no rigid connection between the two rods. With P1 or P2 pressurized, the main piston is capable of performing a full stroke.

Pressurization of P1 and P3 would result in a half stroke or intermediate position being held. From this position the rod can be extended half a stroke by switching pressurization from P1 to P2; or retracted half a stroke by relieving P3 to act as an exhaust. The system is not necessarily rigid as P3 only is pressurized, however.

It will be appreciated that three-position movements can also be obtained from two-stage telescopic cylinders by working each stage independently, or multi-position movements by using a telescopic cylinder with more stages.

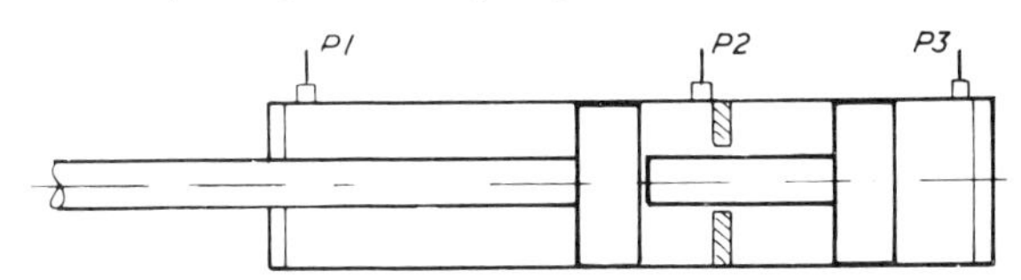


*Figure 7*
*Three-positioned cylinder.*

**Cushioned Cylinders**
The speed of hydraulic cylinder movement is usually governed directly by the pump delivery. Many cylinder applications demand fast movements, when the shock load can be very high as the piston reaches the end of its movement. To eliminate, or at least substantially reduce deceleration loads, the latter part of the piston movement can be 'cushioned'.

A typical cushion is based on providing dashpot dampening on the final movement by throttling the fluid escaping from the outlet port. The cushion end provides an extension of the cylinder with substantially reduced bore.

At the same time the leading face of the piston is fitted with a spigot of the same diameter which enters the cushion chamber as the piston approaches the end of the stroke, effectively sealing off fluid flow to the outlet port. The only escape for the fluid is then through a metered orifice, the throttling effect of which is usually adjustable by means of a needle valve (see Figure 8).

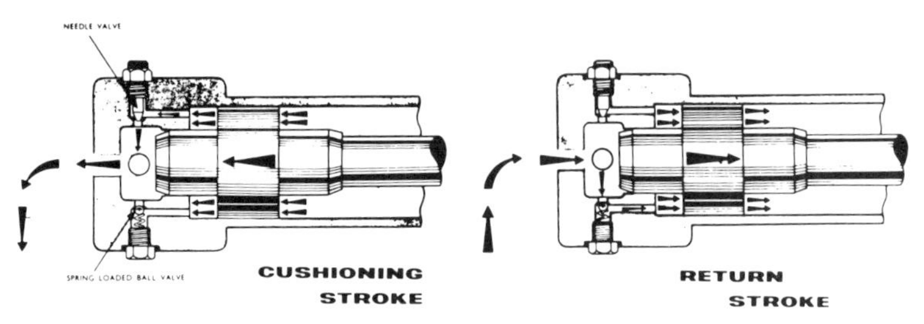


*Figure 8*
*Cylinder cushioning.*

This adjustment may be pre-set (for fixed orifice throttling); or left adjustable for the user to set the degree of throttling required to suit a particular application although this has certain dangers.

To initiate return movement, provision must be made to bypass the orifice, otherwise the fluid flow to the face of the piston would be so restricted that the necessary 'break-out' force might never be realized or at least initial movement would be very sluggish.

This usually takes the form of a spring-loaded non-return ball-valve, which bypasses the return throttling orifice for the initial return flow, until the spigot of the piston has left the cushion chamber. Not until this motion has been accomplished will the full output force be available from the cylinder.

The basic type of cushion may have diasadvantages for certain applications. Because the bleed orifice is fixed, either by design or by adjustment, a sudden change in piston velocity, which could affect machine operation, will occur at the beginning of the cushioning.

In such cases a relief valve may be employed instead of cushioning as the cushioning action will then be more moderate during the initial part of the cushioned stroke (see Figure 9). However, cushioning will become progressively 'harder' as energy is dissipated and the cylinder may well stop before the end of its stroke. This can be overcome by using a restrictor in parallel with the relief valve.

Cushioning may be applied to one, or both, ends of a double-acting cylinder. Cushioning is mostly used on fast-acting cylinders where speeds may range up to 0.3 m/s (1 ft/s). Standard cushions are not normally effective for speeds below 0.1 m/s (0.3 ft/s) and external deceleration valves or special cushions should be applied for cylinder speeds above 0.3 m/s (1 ft/s).

As a rough rule, the fitting of a cushion adds about one half of the cylinder diameter to the total length. However, there can be considerable differences in cushion lengths in individual manufacturers designs, and different cushions lengths may also be available for standard sizes of cylinders to suit specific requirements.

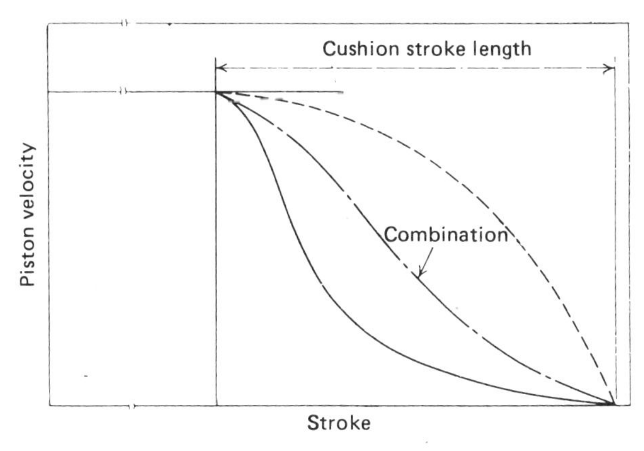


*Figure 9*

Production sizes of hydraulic cylinders are commonly available with or without cushioning,

Care must be taken when adjusting the cushion throttle. To be effective, the pressure developed in the cushion chamber must be higher than that of the fluid pressure on the piston. This pressure can be increased by increasing the throttling effect, but if this is taken to extremes, excessively high fluid pressures may be developed within the cushion chamber without access to a relief valve. This can be even more damaging than not having a cushion at all.

Essentially therefore, cushion throttle adjustment is something which should only be attempted by someone with expert knowledge of the subject and the particular cylinder construction.

**Locking Cylinders**

Because hydraulic oils are virtually incompressible, a linear actuator can be held in postion by cutting off the inlet flow while maintaining the 'working' fluid volume under the original pressure.

The 'hold' capabilities are then basically as good as the sealing of the cylinder, *ie* loss of 'hold' position is directly proportional to the internal leakage flow.

This system can be referred to as a *blocked fluid column*. Locking is effective as long as there is no internal leakage, but is not necessarily rigid. All real fluids are compressible and the locking capabilities of a blocked fluid column are thus dependent on load and particularly variations in load, as well as actual fluid pressure.

Given a high enough pressure a fluid will, in fact, act like a spring. The one great advantage of the blocked fluid column method, however is that it has infinite position locking capabilities.

Completely positive locking is best provided by mechanical latches. While these will only provide positive locking in a particular position, locking is rigid and there is the additional advantage that the system can be depressurized and the lock maintained for an indefinite period (see Figure 10).

Various types of mechanical latches are used as cylinder locks. The simplest is probably a straight forward latch applied to a suitable indent on the extended length of the rod. Such a latch can be automatically engaged, but requires manual or mechanical unlocking.

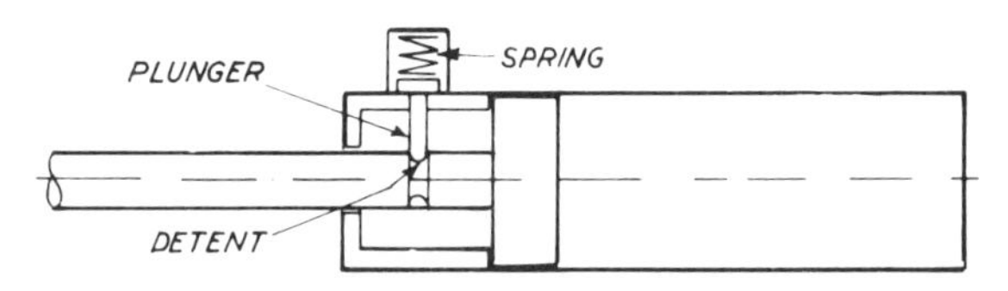


*Figure 10*

A *plunger-type lock* makes for a neater and well protected unit because this can be mounted inside the cylinder. The plunger can be spring-loaded to fall into a suitable indent in the rod when the locking position is reached. It can also be so shaped to be lifted by fluid pressure when the cylinder is pressurized for the return stroke, thus providing automatic disengagement of the lock.

Such a design needs careful attention because the plunger should lift and disengage before full pressure is applied to the piston otherwise the plunger may jam instead of disengaging. A plunger-type lock is also only suitable for locking relatively light loads.

A *collet-type* lock is normally used for locking against moderate to heavy loads. The collet, or radially expanding sleeve, may be moving or fixed, the locking action being provided by the radial expansion or contraction of the collet against a spring-loaded lock ring. Unlocking is normally automatic.

This form of lock is far more flexible in design because it permits the locking loads to be distributed over a relatively large surface area. Thus the collet size can be selected to lock the highest service loads likely to be experienced.

A further solution is offered by a design which is an hydraulically actuated mechanical device, based on the expansion of metal under pressure. It gives infinite positive locking, with zero backlash and high system stiffness.

The patented design is simple and introduces no moving parts into the system. A high tensile steel sleeve forms an interference fit on a hardened chrome and precision ground rod. The sleeve is enclosed in a barrel with sealed end caps, to form either an extension to the actuator, or fitted directly to the machine, as a hydraulically released mechanical lock.

The interference fit provides a zero clearance mechanical connection between the locking sleeve and rod in any position of stroke. When hydraulic pressure is introduced between the sleeve rod, the sleeve expands radially relieving the interference fit, allowing free movement with minimum inefficiency. The lock is re-engaged by removing the pressure, giving fail-safe operation in the event of pressure loss (see Figure 11). Capacity is proportional to sleeve length and rod diameter.

**Rotating Cylinders**

*Rotating* cylinders are primarily used for mounting on machine spindles or similar rotating movements. Such cylinders are invariably of compact length and short stroke, piston and cylinder being locked to rotate together by means of an internal pin on which the piston slides, with the piston rod connected to the machine spindle.

The blind end of the cylinder terminates in a distributor shaft, normally made of hardened steel, supported by a stationary distributor housing, which also acts as a bearing for the distributor shaft. The actual bearing surface may be plain, or roller bearings may be used.

The distributor shaft is drilled and ported to provide open passages to each end

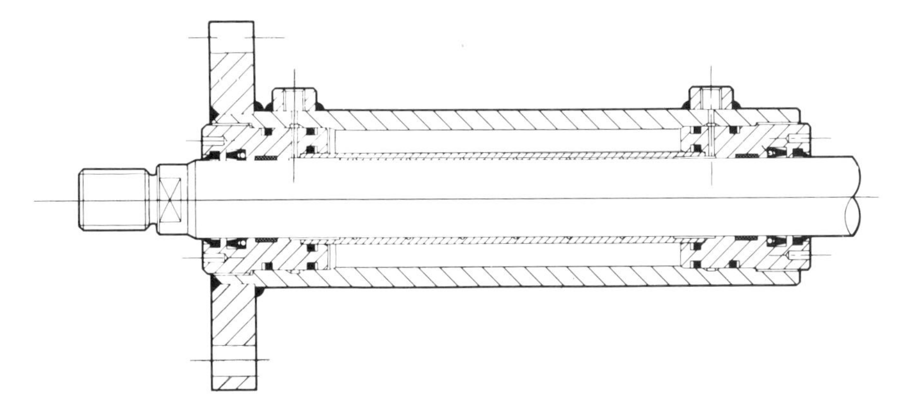

*Figure 11*
*Bear-Loc cylinder.*

of the cylinder. The shaft ports correspond in position to distribution rings formed in the distributor bearing surface; these rings being connected to external ports.

The design must also incorporate suitable seals, as necessary, although some leakage is desirable both to maintain adequate lubrication of the rubbing surfaces and conduct away frictional heat.

A slight venting action may also be necessary in order to avoid excessive back pressure being developed on pressure energised seals. Provision is often made in the design to collect leakage oil in a sump at the bottom of the distributor, from whence it can be returned to the reservoir (see Figure 12).

In general, rotating cylinders are usually designed for lower working pressure than most other types of modern hydraulic cylinders. Typical pressure ratings are

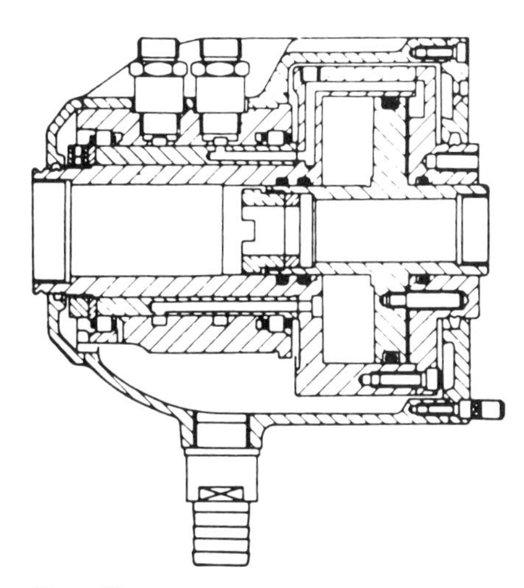

*Figure 12*
*Rotary cylinder with distributor gland.*

17.5 to 35 bar (250 to 500 lbf/in$^2$), although some designs may be capable of working at much higher pressures without excessive leakage developing. Rotational speeds may range up to 1000 rev/min or more. Problems associated with the design of a suitable distributor gland or sealing system increase considerably with increasing rotation speed and cylinder diameter.

Seal design is always a primary problem, for the most effective types of rotating shaft seals are not usually those which are capable of withstanding high pressure. Packed glands may be preferred. Also the cooling effect of the lubricant film may inadequate, in which case jacketing and water cooling may have to be employed.

**Miniature Cylinders**

*Miniature* cylinders are usually associated with pneumatics rather than hydraulics, but miniaturization of hydraulic components is attractive from the point of space and weight saving.

Specifically, to be described as miniature, a hydraulic cylinder needs to be smaller than the smallest standard hydraulic cylinder, *ie* 25mm or 1 in bore. Logically other components, especially valves, should likewise be reduced to a matching size. Cylinder forces available should then be of the order shown in Table 1.

**TABLE 1 – THEORETICAL MAXIMUM THRUST OF MINIATURE HYDRAULIC CYLINDERS**

| Cylinder Bore mm | SYSTEM PRESSURE – bar | | | | | | |
|---|---|---|---|---|---|---|---|
| | 35 | 50 | 75 | 100 | 150 | 200 | 210 |
| 5 | 7 | 10 | 15 | 20 | 30 | 40 | 41 |
| 7.5 | 15.5 | 22 | 33 | 44 | 66 | 88 | 92 |
| 10 | 27.5 | 39 | 59 | 78.5 | 118 | 157 | 165 |
| 15 | 62 | 88 | 132 | 176 | 263 | 352 | 370 |
| 20 | 110 | 157 | 235 | 314 | 470 | 628 | 660 |

Thrust in kgf

| Cylinder Bore in | SYSTEM PRESSURE – lb/in$^2$ | | | | | | | | |
|---|---|---|---|---|---|---|---|---|---|
| | 100 | 200 | 300 | 400 | 500 | 1000 | 1500 | 2000 | 3000 |
| 0.3125 | 7 | 1 | | | | | | | |
| (8 mm) | 7 | 14 | 21 | 28 | 34 | 65 | 105 | 130 | 210 |
| 3/8 | 7.5 | 15 | 22 | 30 | 37.5 | 70 | 112.5 | 140 | 220 |
| 1/2 | 17.5 | 35 | 52.5 | 70 | 87.5 | 170 | 262.5 | 340 | 525 |
| 3/4 | 40 | 80 | 120 | 160 | 200 | 380 | 600 | 750 | 1200 |
| 1 | 70 | 140 | 210 | 280 | 350 | 675 | 1050 | 1300 | 2100 |

Thrust in lbf

This shows that cylinder sizes commonly used in such systems range from 8 mm (0.3125 in) to 20 mm (0.75 in) bore, although larger sizes could also be accommodated. There is little advantage in making much smaller cylinder sizes as lower output requirements can be supplied more efficiently by working with reduced pressure.

Cylinders may be single- or double-acting. In the case of double-acting cylinders, miniaturization imposes certain limits on the output force available for the outward stroke because of the size of the rod required to accommodate the inward stroke force.

In other words, the rod diameter may need to be larger in proportion to the cylinder bore than in conventional hydraulic cylinders, *eg* the rod diameter used may be as much as 50% of the bore.

Some miniature cylinders are used as *clamping* cylinders for machinery fixtures and are shaped to either screw-in or screw-onto the fixture as shown in Figure 13.

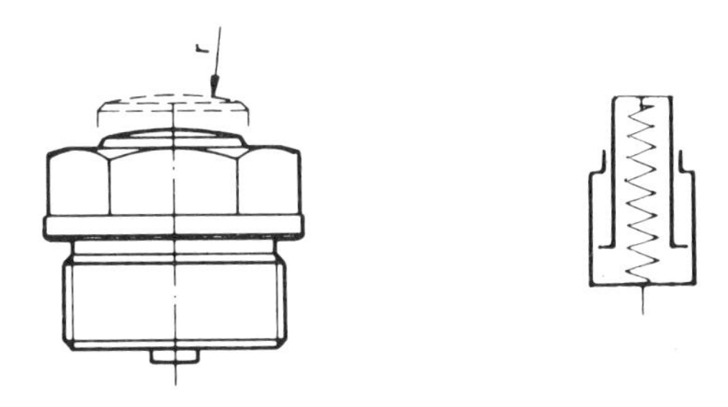

*Figure 13*
*Screw-in clamp cylinder.*

These cylinders are single-acting, equipped with return springs, and are used in high pressure hydraulic fixtures and many other hydraulic installations where only a very restricted space is available, for the generation of high forces with limited plunger movement. The cylinder bores range from 12 mm (0.47 in) to 24mm (0.94 in) with strokes from 2mm (0.08 in) to 20 mm (0.79 in).

# Construction

VARIOUS FORMS of cylinder construction are employed, *eg:* screwed-on ends, tie-rod, flanged and bolted covers, and welded construction.

All of them have some advantages and limitations. All-welded cylinders are generally cheaper to produce, but are basically non-serviceable. They are quite widely used on mobile equipment, however.

Cylinders with screwed ends have a slight advantage in diameter size over other types (except welded), but usually require a thicker cylinder to accommodate cutting the screw threads without weakening the ends of the cylinder.

*Welded cylinder with eye mountings and spherical bearings.*

Tie-rods construction is now widely used and length for length compares favourably with other types. It is the standard form of construction accepted by the Joint Industry Conference (JIC) with the *Society of Automotive Engineers* and specified in the NFPA Standard ASA – B93.3 1965.

A particular advantage of tie-rod construction is that it allows a very wide range of different mounting styles and optional features from a standard range of components. It is now generally accepted as a standard for the machine tool industry (see Figure 1).

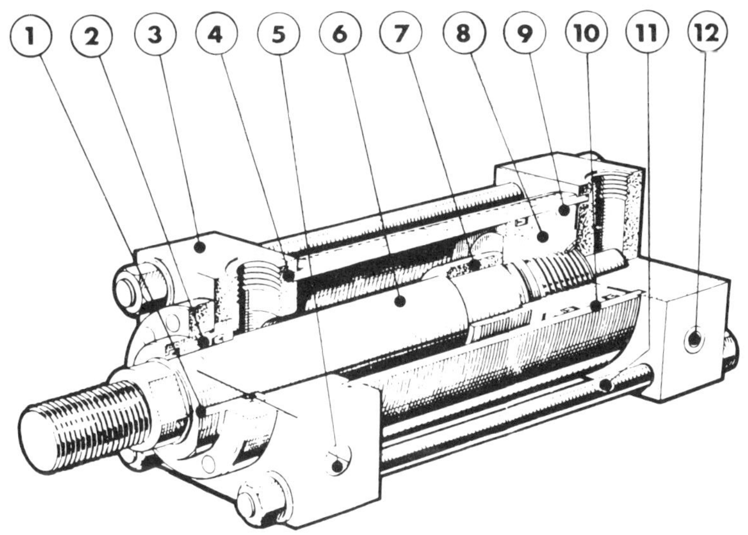


*Figure 1*
*Cylinder specification. 1. Gland seal fitted with combination of heavy duty wiper seal and gland seal. 2. Cartridge type gland can be removed externally. 3. End covers produced from square steel. 4. Fully trapped tube seals 5. Cushion speed adjuster needle pin. 6. Piston rod from precision ground steel, hardened and chrome-plated/alternately stainless steel. 7. Floating head cushion. 8. Piston head manufactured from high tensile iron. 9. Piston sealing can be either cast iron piston rings or positive lip seals. 10. Cylinder body produced from seamless steel tube, honed in the bore. 11. Tie rods from high tensile steel. 12. All cylinders are fitted with non-return valves for high initial acceleration.*

The use of flanges welded to the cylinder ends with bolted-on covers is one of the strongest forms of construction but not necessarily superior to tie-rod construction. Its disadvantage is that it is more costly to produce and extreme care has to be taken during welding to avoid distortion of the cylinder barrel.

*Cast* cylinders, however, can be produced with cast-iron flanges. This type of construction is still used for some heavy-duty cylinders, particularly in larger sizes.

**Materials**
The favoured material for hydraulic cylinder tubes is *cold drawn tubing*, which is available in a wide range of standard sizes, with high bore finishes and excellent geometric and mechanical properties, enabling them to be used without further finishing. The majority of such tubes are drawn from low carbon mild steel.

The strength of these tubes, as drawn, can be of the order of 6300 bar (40 tons/in$^2$) ultimate, with a 0.1% proof stress in excess of 4700 bar (30 tons/in$^2$). This may be further enhanced by further cold working, such as bore polishing. Also the more the material is hardened by cold working the better its machinability.

However, machining tubes in a hardened condition is also likely to produce distortion and ovalness. Thus as a general rule cold drawn cylinder tubes should not be further worked by machining unless absolutely necessary; or, if machined, should be stress relieved. This will noticeably reduce the ultimate and 0.1% proof stress of the material, but also produce a more clearly defined yield point or limit

## TABLE 1 – DRAWN CYLINDER TUBE MATERIALS

| Material | Condition or type | Ultimate Tensile Strength (min) | | 0.1% Proof Stress | | 0.2% Proof Stress | | Design Maximum Permissible Stress | |
|---|---|---|---|---|---|---|---|---|---|
| | | bar | lb/in² | bar | lb/in² | bar | lb/in² | bar | lb/in² |
| Low carbon steel | As cold-drawn | 3550 | 55000 | 3200 | 45000 | | | 1250 | 18000 |
| | After deep polishing | 6000 | 85000 | 5000 | 70000 | | | 1750–2100 | 25000–30000 |
| Stainless steel (304) | Annealed | 7000 | 1000000 | | | 2400 | 34000 | 2300 | 33000 |
| | Half-hard | 8800 | 125000 | | | 6000 | 85000 | 3000 | 42000 |
| | Hard | 10000 | 150000 | | | 8300 | 118000 | 35000 | 50000 |
| Tungum alloy | Annealed | 4650 | 66000 | 2600 | 37000 | | | 1550 | 22000 |
| | Precipitation-hardened | 4700 | 67000 | 32000 | 45000 | | | 1750 | 25000 |
| Aluminium alloy | 61S–T6 | 3200 | 45000 | | | | | 1000 | 15000 |
| Titanium | DTD 5013A | 4700 | 67000† | 2350 | 33500* | | | 1550 | 22000 |
| | DTD 5063 | 6300–7700 | 90000–110000 | 4700 | 67000* | | | 2100 | 30000 |

† Maximum *Minimum

## TABLE 2 – FABRICATION METHODS FOR CYLINDER TUBES

| METHOD | MATERIALS | SERVICE | REMARKS |
|---|---|---|---|
| Casting | Cast Iron<br>Centrifugally Cast Iron<br>Steel<br>Aluminium Alloy<br>Bronze<br>Titanium | Conventional<br>Conventional<br>Conventional<br>Lightweight<br>Water Fluids<br>High Strength/Weight and/or High Service Temperature | Bores require machine finishing (straight through tube castings preferred for precision cylinders) |
| Cold Drawn | Low Carbon Steel<br>Stainless Steel<br>Aluminium Alloy<br>Brass<br>Titanium<br>Tungum | Conventional<br>Corrosive or High Pressure<br>Lightweight<br>Water Fluids<br>High Strength/Weight Ratio and/or High Temp Service<br>High Strength/Weight Ratio and corrosive or hazardous ambients | May be suitable as drawn for general applications. Honed bores preferred for general duties Honed and polished bores required for precision cylinders or higher pressures |
| Welded | Steel | Large Cylinders | Bores require machine finishing |
| Forging | Steel | Large Heavy-duty Cylinders | Bores require machine finishing |
| Cold Forming | Steel<br>Aluminium<br>Titanium<br>Stainless Steel | Conventional<br>Lightweight<br>High Strength/Weight and/or High Temperature<br>Corrosive or High Pressures | Good surface finish as formed but not necessarily uniform. Usually requires honing and polishing to finish bores |
| Precision Hollow Extrusion | Steel<br>Stainless Steel<br>Titanium | Conventional<br>Corrosive or High Pressure<br>High Pressure and/or High Temp & Lightweight | Length: Bore limited to about 8 : 1 Formed with integral end cover. Good surface finish as formed. Cost less than flow turning |
| Flow Turning | Aluminium<br>Steel<br>Stainless Steel | Lightweight<br>Conventional<br>High Pressure or Corrosive | Good surface finish as formed |

for the maximum permissible design stress. Some typical data are summarized in Table 1.

Drawn cylinder tubes are also produced in other materials, such as aluminium, brass, stainless steel and titanium. Hot drawn or hot rolled tube is seldom used because this method of production normally results in lowered mechanical properties, low dimensional accuracy and a very poor bore finish. To obtain a suitable bore finish such tubes would have to be bored out and honed and polished. Welded construction may, however, be used for large steel tubes, suitably bore finished.

Typical variations in materials and techniques are summarized in Table 2. Here it will be noted that a machined finish (*eg* honing) is generally necessary where the fabrication method produces an irregular or inconsistent bore finish, although direct measurement may appear to indicate a satisfactory fine finish.

Only flow turning or extrusion is likely to give a consistent 800 $\pi$m (20 $\pi$in) finish or better and even this will be more irregular than the same degree of finish produced by honing.

*Hollow extrusions* are essentially cylindrical components open at both ends or with one closed integral end. Open ends can be bulged, flared and flanged. Closed ends can be made in a variety of shapes and the wall thickness can be varied along the length. They are manufactured in plain carbon or low alloy steels and aluminium alloys or, where design permits, in higher alloy, *Nimonic* or stainless steels. The ability to move metal gives strength where it is needed.

In particular, with the one-piece closed end configuration, the integral end is formed as part of the warm extrusion process which produces a grainflow following the contours of the cylinder, giving greater mechanical strength than a fabricated component and improving fatigue life.

Spades for pivot points, hexagons and squares to facilitate assembly and many other external and internal shapes can be incorporated in the design of the extrusion (see Figure 2). The open end can be thickened, bulged and flanged to allow, *eg* screw threads to be machined without weakening the component or to provide metal for porting into.

*Casting* was the original method used for production of high pressure cylinders and is still used for larger high pressure cylinders (forging may be used for even higher properties). Cast or forged cylinders have the advantage that one end cover can be formed integral with the cylinder tube, thus saving one joint and a source of leakage.

In the case of the cast cylinder the blind end is normally dome-shaped, with generous radii and fillets for stress relief at abrupt changes in contour. Forged cylinder blind ends are usually flatter, but with similar attention to the elimination of stress raisers.

The disadvantage of this form of construction with an integral end cover is that it is more difficult to bore finish. It is easy to leave machining marks in the blind end of the tube and honing is always more difficult applied to a blind tube. For this reason, and in view of the importance of surface finish for satisfactory seal performance, cast cylinders may be cast as open tubes so that they can be finished by through machining and polishing.

**Surface Finish**

The bore surface finish in the as-drawn condition can vary between 400 to 3200 $\pi$m (10 to 80 $\pi$in) CLA, depending on the drawing technique employed. There are also

*Figure 2*
*Hollow extrusions for cylinder manufacture.*

likely to be extensive local variations, but a nominal 800 to 1200 πm (20 to 30 πin) maximum roughness can be held by suitable production control.

A honed finish is capable of producing a surface finish of 400 to 600 πm (10 to 15 πin) CLA with a minimum of local variations. For an improvement on this honing must be followed by polishing, 'deep polishing' being capable of providing a uniform finish as fine as 200 πm (5 πin) CLA.

The finer the finish the more costly the production and so some compromise may be reached depending on the duty for which the cylinder is intended and particularly the type of *piston* seals employed.

A surface finish of 800 πm (20 πin) CLA may be considered adequate for a general purpose cylinder using *rubber-impregnated fabric* seals or an even coarser finish could be used with *leather* seals. For *elastomeric* seals a minimum surface finish of 600 πm (15 πin) CLA is desirable in order to achieve satisfactory seal life.

The higher the pressure and the surface speed and the more important it is to minimize internal leakage, the finer the finish required. A 400 πm (10 πin) finish would generally be considered satisfactory for precision cylinders operating at pressures up to 140 bar (2000 lbf/in$^2$), with a 280 to 320 πm (7 to 8 πin) finish for cylinders operating up to 350 bar (5000 lbf/in$^2$). For virtually perfect sealing at pressures of 140 to 350 bar (2000 to 5000 lbf/in$^2$) a surface finish of 160 to 200 πm (4 to 5 πin) is required.

**Bore Protection**

Bore surfaces may be protected by nickel or chromium plating (subsequently polished) or by other protective treatments. This is seldom necessary in the case of cylinders intended for use with oil fluids, although plating may be used to give a harder and more scratch-resistant bore surface. In that case hard chrome plating is the usual choice. With water based fluids some form of protection treatment may be considered essential to combat corrosion, even though the fluid may contain corrosion inhibitors.

For working in corrosive conditions the outside of the cylinder tube may be protected, either by surface coating or treatment or by jacketing in a corrosion-resistant material.

The use of plastics for tube construction is a possibility in such cases, providing the cylinder is reasonably small and the working pressure not too high. Plastic cylinders have been used successfully for marine applications as an alternative to stainless steel or a resistant aluminium alloy, where the cylinder is exposed to salt spray.

**End Covers**

End covers are usually made from the same material as the cylinder tube. They do not present any particular problem in design and material construction, because they can easily be made thick enough for the job they have to do. The main problem is the method of fastening them in place to provide a tight, high pressure seal.

One popular method of fitting, particularly in the case of smaller cylinders, is to make the end cover in the form of a cap which can be screwed on, the assembly also including a gasket-type seal. The main objection to this design, however, is that threading will weaken the walls of the cylinders. This has to be allowed for when selecting the cylinder wall thickness.

Machining work is also involved on a finished cylinder tube, which can produce ovality at the ends. Lesser objections are that this form of cover attachment produces a rather bulky cover and also one where it is difficult to ensure positive alignment of the heads.

A screw-fastened end cover may used where there is sufficient thickness of the tube wall to accommodate a tapped hole together with a simple O-ring seal (see Figure 3). While making a neat assembly this is less satisfactory from a maintenance point of view, with the possibility of stripping an internal thread or breaking a corroded screw.

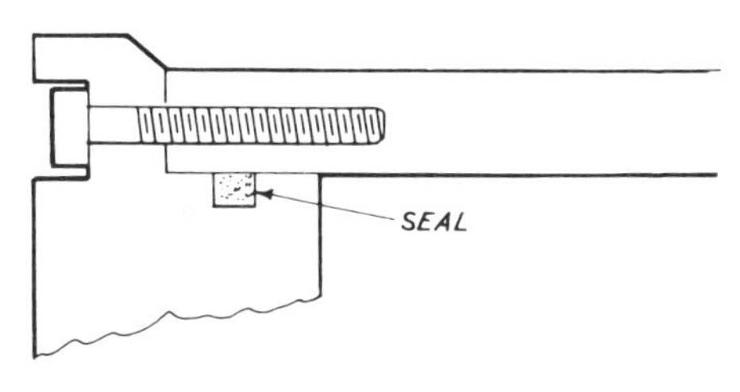


*Figure 3*

Welded assembly is even neater (see Figure 4) and obviates the need for any separate seal. This form of fitting is, however, largely limited to mild steel constructions which can be easily welded and is mainly confined to small light cylinders.

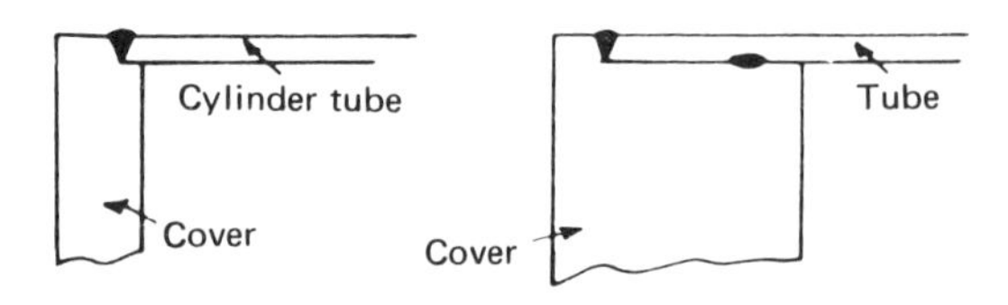


*Figure 4*

Fitting the cylinder tube with a welded-on ring at each end is a popular method for medium size cylinders. The tube ring, in effect, forms a flange to which the end cover can be bolted. The tube ring may be inset to accommodate a relatively shallow end cover with a gasket type seal, or flush with the end with a plug-fitting end cover which can be sealed by an O-ring (see Figure 5).

The latter method is generally to be preferred because it brings the weld area into a blanked off length of the cylinder tube so any distortion produced by welding is not significant.

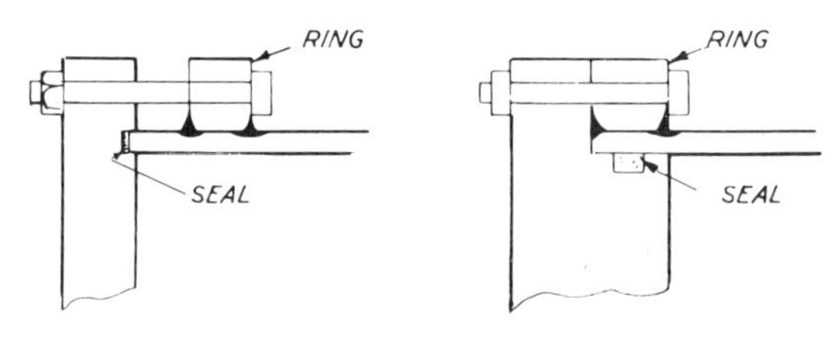


*Figure 5*

An alternative method is to use a freely fitted tube ring as a flange, locating against a wire ring welded to the cylinder tube (see Figure 6a). This is a simple form of fitting suited to light-duty cylinders with adequate tube wall thickness.

The preferred method (see Figure 6b) uses a wire locking ring fitted into semi-circular grooves machined in both the cylinder wall and end cover, the wire being introduced through a hole drilled in the cover tangential to the groove, and fed into position by rotating the end cover. This again has limitations from a servicing point of view and is suitable only for relatively light-duty cylinders.

Cast cylinder tubes are commonly formed with an integral flange, when the end cover is simply fitted by bolting up together with a *gasket-type* seal.

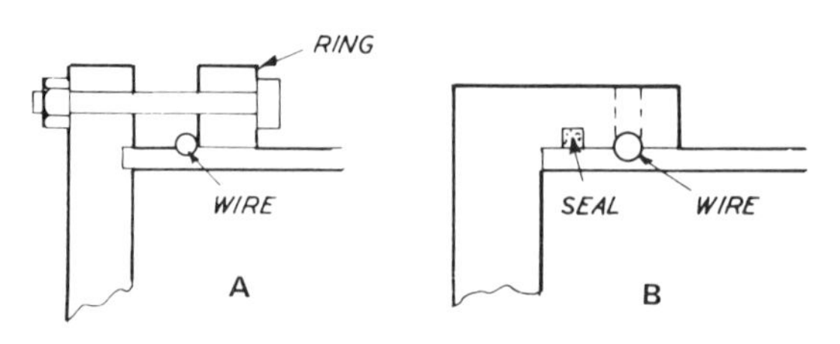


*Figure 6 a & b*

The clamped flange is an attractive alternative (see Figure 7) because the flange need only be of small diameter. This is backed by a freely fitting ring, to which the end cover is bolted. The end cover can seal on a gasket or, more usually, can be

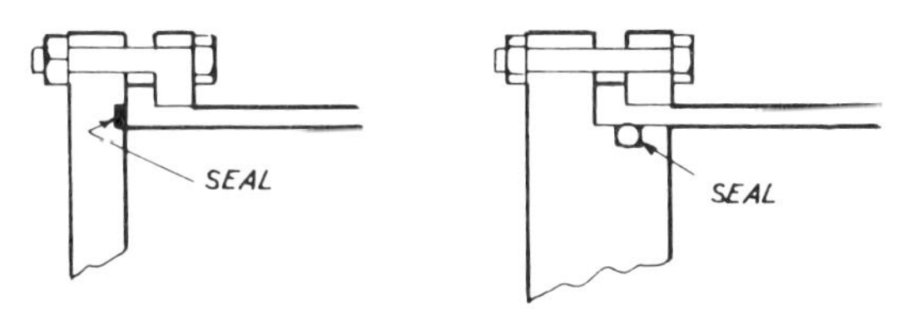


*Figure 7*

plug-fitting to seal on an O-ring. This type of construction is suitable for quite heavy-duty cylinders and high pressure ratings.

Probably the most positive fitting of all for high pressure working is tie-rod construction. The end covers are usually square in shape, with holes drilled in each corner through which high tensile steel tie rods are fitted and bolted up. Again the cover can seal on a gasket or be plug-fitting to accommodate an O-ring.

The particular advantage of tie-rod construction is that it is relatively inexpensive and no working or weakening of the end walls of the cylinder tube is involved. It does, however, increase the bulk and weight of the cylinder and there can be problems with regard to the stability of the tie-rods on long stroke cylinders.

It should also be noted that the plug-fitting end cover is less susceptible to leakage at high pressure due to tie-rod elongation than a simply fitted cover which depends on a gasket-type seal.

Further variations on end cover fittings may be found on various proprietary cylinders, these having been developed by individual manufacturers, are preferred solutions for the duties for which such cylinders are designed.

**Rod Bearings**

The end cover must provide adequate bearing surface for the piston rod on the working or 'open' end of the cylinder or at both ends in the case of the through-rod cylinders. The bearing length must also incorporate a seal or gland, which is usually located on the inner side, and commonly a *wiper* seal on the outer side. The length of the bearing must also provide adequate support for the rod, although rod diameters on hydraulic cylinders are usually sufficiently large to give good rod rigidity.

The choice of rod seal varies according to different manufacturers and different duties. Simple O-ring seals may be regarded as satisfactory for light-duty cylinders, although for pressures above 103 bar (1500 lbf/in$^2$) this would normally be associated with back-up rings to prevent extrusion of the seal into the clearance space.

Proprietary composite seals, based on the O-ring form, are generally to be preferred for higher pressure working, with the advantage of having no more friction than conventional O-rings.

For larger cylinders, or heavier duty types, *flexible lip* seals would normally be preferred, either as single seals in a matching groove or as seal sets assembled in a gland.

**Pistons**

Choice of material for the piston is generally limited to cast iron or steel, but more recently sintered iron or steel is also used. Aluminium alloy pistons may be used for lightweight cylinders, and pistons for use with water-based fluids may be of aluminium, brass or bronze; or again iron or steel with a plated or protective coating.

Pistons may be of one-piece, two-piece or three-piece construction, depending on the type of seals used. The piston may be attached to the rod by a nut or nuts, or sometimes simply threaded in place on the end of the rod, or welded to the rod. For light-duty cylinders it may be simply located by circlips.

Simple ring seals permit the use of one-piece pistons although, in the case of double-acting pistons, venting may have to be provided between the seal grooves. Modern designs of composite rings are available which provide excellent sealing up to the highest pressures likely to be required, with very low friction and wear.

Metallic piston rings may be used for higher temperature applications, although they are generally inferior in performance to flexible seals. Elastomeric seal materials are available with a suitable temperature rating for use with most conventional hydraulic fluids up to the maximum service temperature rating of these fluids.

*Chevron, U-cup* and proprietary flexible seal sections are used on larger and heavy-duty cylinders, either singly or in sets, mounted back-to -back in the case of double-acting pistons. Pistons construction must allow for easy dismantling and accurate reassembly of such seals.

Some pistons also incorporate an inset bearing or strip of bronze or PTFE. In the latter case there is no metal-to-metal contact between the piston and cylinder bore.

**Piston Rods**

Piston rods are normally of hardened steel, ground and polished, or chrome plated and polished. Stainless steel rods may also be used in corrosive atmospheres, but are less scratch-resistant than hard chrome plating.

A very smooth rod finish is desirable as this reduces wear on the rod seals to a minimum.

Polishing is virtually essential to produce a satisfactory surface finish and texture pattern, which can have a marked effect on oil leakage as well as seal wear. The surface pattern produced by grinding alone is generally unsatisfactory, and in the case of centreless grinding, may actually produce a spiral pattern promoting oil leakage.

**TABLE 3 – TYPICAL STANDARD AND OVERSIZE PISTON ROD**

| Cylinder bore (inches) | Piston Rod Diameters (Inches) | | | Pressure Ratings (lbf/in$^2$) | | Test pressure |
|---|---|---|---|---|---|---|
| | Standard | Oversize | 2:1 | Heavy-Duty Service | 4:1 Safety Factor(yield) | |
| 1½ | ⅝ | | 1 | 3000 | 2200 | 4500 |
| 2 | 1 | | 1⅜ | 3000 | 2150 | 4500 |
| 2½ | 1 | 1⅜ | 1¾ | 3000 | 1900 | 4500 |
| 3¼ | 1⅜ | 1¾ | 2 | 3000 | 2200 | 4500 |
| 4 | 1¾ | 2 | 2½ | 2700 | 1700 | 4500 |
| 5 | 2 | 2½,3 | 3½ | 3000 | 1900 | 4500 |
| 6 | 2½ | 3,3½ | 4 | 2700 | 1750 | 4500 |
| 7 | 3 | 3½,4 | 5 | 3000 | 1950 | 4500 |
| 8 | 3½ | 4,5 | 5½ | 3000 | 1900 | 4500 |

Because the diameter of the rod can be relatively large, a hollow rod may be preferred for weight saving, particularly in the case of a horizontal cylinder. Normally, however, solid rods are used since they are easier to manufacture.

Rod sizes for standard production are normally standardized at about half the bore diameter, giving a ratio of areas of 4:3. With such geometry back-pressure effects are generally negligible for ordinary working. Standard cylinders are also usually offered with alternative sizes of rod: smaller in diameter for light-duty where the area ratio is usually of the order of 7:6, and larger in diameter for heavy-duty application where the ratio of areas may reach about 2:1 (see Table 3).

# Performance and Selection

THE HYDRAULIC performance of a cylinder has been defined in Section 1 relating to its output thrust and speed of operation. It must be remembered, however, that an hydraulic cylinder not only performs the function of a linear actuator but also becomes a structural member of the machine in which it is incorporated when extended under pressure. In this form the overall tensile, compressive and bending loads must be considered as well as the stresses developed due to the internal pressure.

**Strength of Cylinders**

Where the diameter:thickness ratio of the cylinder tube is greater than 16:1, the stress produced in the wall material due to internal pressure can be determined from the simple formula uniformly distributed hoop stress:

$$S = \frac{PD}{2t}$$

where S = hoop stress
P = internal pressure
D = internal diameter of tube
t = wall thickness.

This can be rendered in the form of working formula, expressed as a solution for wall thickness required:

$$t = \frac{P_W D}{2 S_m} \times F + c$$

Where $P_w$ = design working pressure
$S_m$ = maximum permissible material stress
F = design factor of safety
c = an additional thickness allowance for corrosion.

The additional corrosion allowance (c) is normally ignored, the necessary margin being accommodated in the safety factor employed.

For thick-walled homogeneous tubes the stress is no longer uniformly distributed through the tube walls, when the maximum hoop stress is given more accurately by:

$$S = \frac{D^2 - 2t + 2t^2}{2t(D - t)} \times P$$

A simpler formula, written as a solution for wall thickness is:

$$t = \frac{D}{2}\left(\sqrt{\frac{Sm + Pw}{Sm - Pw}} - 1\right) \times F + c$$

Again the added value (c) may be ignored and included in the safety factor (F).

The stress produced in cast cylinders may also be determined from the above formulae, typical values for maximum permissible material stress being:

| | | |
|---|---|---|
| cast iron | – | 280 bar (4000 lbf/in$^2$) |
| high duty cast iron | – | 420–550 bar (6000 to 8000 lbf/in$^2$) |
| cast steel | – | 840 bar (12000 lbf/in$^2$) |
| cast aluminium alloy | – | 550 bar (8000 lbf/in$^2$) |
| cast brass | – | 420 bar (6000 lbf/in$^2$) |
| cast bronze | – | 420 bar (6000 lbf/in$^2$) |

The safety factor allowed in such cases is normally generous.

The strength of low carbon mild steel is substantially increased by cold working, with a potential ultimate tensile strength in the order of 6300 bar (40 tons/in$^2$) after cold drawing and deep polishing.

Cold working also tends to promote better machining of the material but at the same time increases the tendency towards distortion when machined. Stress-relieved tubes, with minimum distortion on machining, are more difficult to machine.

On the other hand, stress relief does avoid ovalness, which might otherwise result from machining and is particularly advisable in the case of thin-walled tubes with a diameter: thickness ratio in excess of 16:1.

All cold-drawn tubing is subject to ovalness. However, very close tolerances can be held by controlled production, followed by honing the bore to finish, provided the diameter: thickness ratios of the tube is less than 20:1. For higher diameter:thickness ratio it is generally impossible to hold very close tolerances because of the thinness of the walls.

**Cylinder Mounting**

In chosing a suitable mounting the type of application is the major factor to be considered. The first consideration is whether or not the thrust line and the mounting are coincidental, and the second is if the equivalent strut length of a long stroke cylinder can be improved by selecting the correct mounting method. Figure 1 shows different types of mounting and the following factors affecting selection of suitable mountings should be considered.

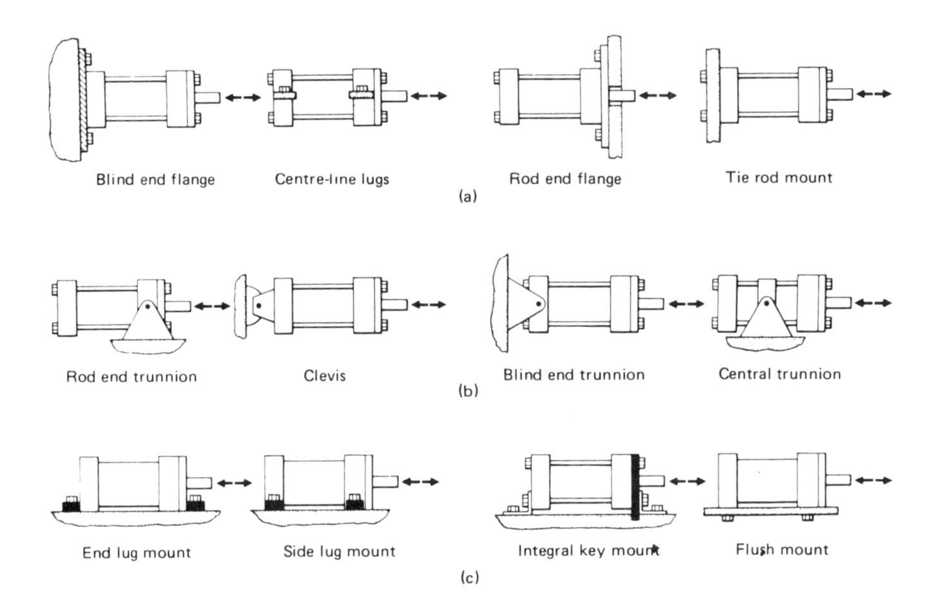


*Figure 1 (a,b,& c).*

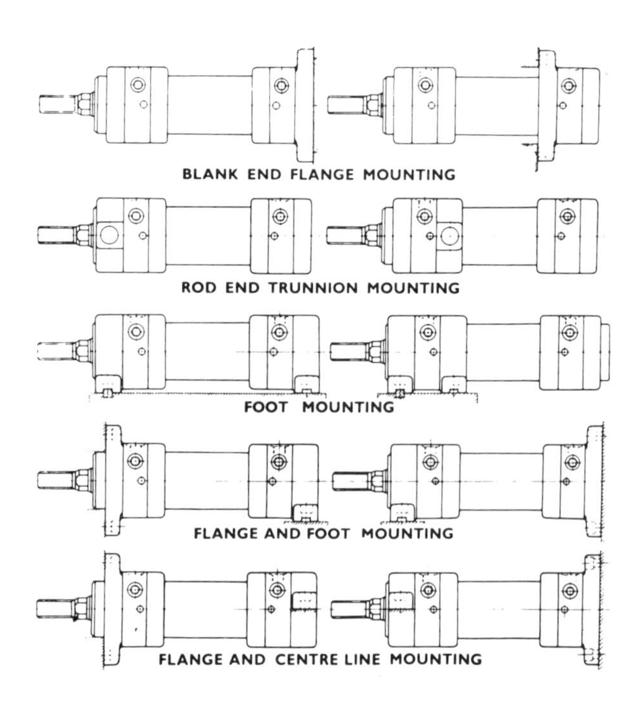


*Figure 2*
*Cylinder mounts.*

Cylinders with mountings not on the centre-line tend to sway, probably resulting in extra wear and shorter life. The frame holding the cylinder must be strong enough to resist bending moments which would be imposed on the cylinder.

A pivot mounting is the best solution when the object moved by the piston rod travels in a curvilinear path, but a fixed mounting-type cylinder should normally be used when essentially linear motion is required. Pivot mountings can be used at either end of the cylinder or along its barrel.

Cylinder strength must be considered against stroke length. Long stroke, pivot-mounted, centre trunnion-type cylinders can usually have smaller rods without danger of bending compared with cylinders with fixed mountings which may require extra support to avoid sag or buckling. The major force applied to the machine may result in tension or compression of the rod.

A suitable mounting for thrust loads is the blank (rear) end flange type, with the rod in compression, while a rod (front) end flange mounting is better when the rod is in tension.

If misalignment is possible between a cylinder and whatever it operates, it may be necessary to provide for compensation by selecting a suitable mounting.

If the misalignment is primarily in one plane, the simple pivoted centre-line mounting will give the necessary compensation.

The basic cylinder mounts are:

| | | | |
|---|---|---|---|
| (1) | Foot | – | usually fitted to the blank and rod ends of the cylinder. |
| (2) | Centre-line | – | less common and confined to specific designs. |
| (3) | Flange | – | either rod (front) end or blank (rear) end. |
| (4) | Nose | – | an alternative to front end flange mounting. |
| (5) | Pedestal | – | similar to a flange mount for fitting at either or both ends of a cylinder, but incorporating a foot to bolt down to a surface parallel to the cylinder. |

All the above provide rigid mounting and may also be used in combination, *eg* flange and foot, flange and centre-line, *etc,* (see Figure 2).

| | | | |
|---|---|---|---|
| (6) | Clevis | – | providing a single-forked pivot mounting at the blank end. |
| (7) | Trunnion | – | providing a pivot mounting point at either end, or at the centre of the cylinder. |
| (8) | Tongue or eye | – | providing a pivot mount at the blank end in conjuction with a matching forked bracket. |
| (9) | Pivot | – | virtually the same as a trunnion mount except that the cylinder is fitted with a pair of pivot pins only for pivotal mounting, not necessarily in a trunnion. |

All pivoting mounts are normally, but not invariably, mounted on the centre-line of the cylinder.

Plain cylinders, sometimes referred to as 'flush-mounted', are commonly designed to accommodate a variety of different mounts in standard productions, except in the case of fixed centre-line mounts which are normally an integral part of the cylinder construction.

Piston rods are normally finished in one of the following forms:

(1) Plain.
(2) Screwed.
(3) Clevis (female).
(4) Tongue or eye (male clevis).

A screw-threaded end is more usual because this enables a choice of end fittings to be mounted.

Self-aligning or swivelling joints may be required at one or both ends of the cylinder and the end fittings or bushes must be designed accordingly. Standard eye-type fittings can generally tolerate misalignment of up to about 3°, although this may need to be compensated for by flexible packings or washers. Spherical bearing eyes are generally used where the degree of misalignment is considerable.

The majority of cylinder manufacturers aim to provide for as many alternative mountings as possible on standard production cylinders so that most likely applications can be catered for without resort to expensive special designs. They should be consulted as to the suitability of particular mounts for a specific application in any case of doubt.

**Critical Rod Lengths**

Cylinder rods can be stressed as rigid rods, provided the length of the rod does not exceed ten times its diameter. The stress formula in this case is:

$$\text{material stress} = \frac{F}{A}$$

where F = compressive or tensile force or load
A = cross-sectional area of rod.

For general working the maximum permissible material stress may be based on the ultimate tensile stress of the material and a suitable factor of safety. This will then give adequate rod strength in tension or compression.

If the rod length exceeds ten times its diameter, then it may be subject to buckling under compressive loading. Adequate strength in tension can then no longer be taken as indication of adequate strength in compression.

The case of compressive loading must be analysed separately, when the rod is considered as a column. The material stress then depends on the method of end fixing which determines the equivalent strut length.

*Euler* buckling load $$P = \frac{\pi^2 EI}{L^2}$$

For circular section rods, moment of inertia $$I = \frac{d^2}{4} \times A$$

Therefore buckling stress $$\frac{P}{A} = \frac{\pi^2 Ed^2}{4L^2}$$

where P = critical buckling load
A = rod cross-sectional area
E = modulus of elasticity of rod material
d = rod diameter
L = rod length in consistent units.

If the above value for the buckling stress is taken for the rod end rigid and guided then equivalent strut lengths will modify the formulae as follows:

buckling stress $$= \frac{\pi^2 Ed^2}{8L^2}$$ for rod end pinned

$$= \frac{\pi^2 Ed^2}{64L^2}$$ for rod end free.

Most manufacturers adopt factors for equivalent strut lengths of cylinder rods, depending on the cylinder mounting and the rod end fixing as shown in Figure 3. Typical limits or critical rod lengths, can also be expressed graphically, as in Figure 4.

All such computations are based on the assumption that there is no eccentric loading. In the case of horizontal cylinders, these must be supported to counteract bending moments. Where bending loads are actuallly present on the rod in compression, the critical length is substantially reduced.

| CYLINDER MOUNTING | RIGID ROD END GUIDED | PIVOTED ROD END GUIDED | FREE ROD END |
|---|---|---|---|
| FRONT FLANGE 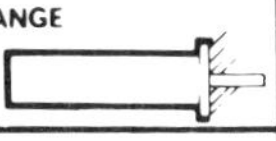 | EQUIV L = STROKE / 2 | EQUIV L = STROKE / 1 42 | EQUIV L = STROKE x 2 |
| REAR FLANGE | EQUIV L = STROKE / 1 25 | EQUIV L = STROKE x1 12 | EQUIV L = STROKE x 3 2 |
| FOOT  | EQUIV L = STROKE / 2 | EQUIV L = STROKE / 1 42 | EQUIV L = STROKE x 2 |
| REAR EYE OR PIN | EQUIV L = STROKE x1 12 | EQUIV L = STROKE x1 6 | |
| TRUNNION 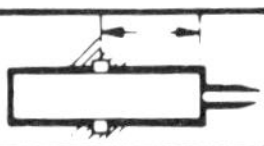 | EQUIV L = STROKE · 6x / 1 42 | EQUIV L STROKE · 6x | |

*Figure 3*
*Equivalent strut length (L) or cylinder rods .*

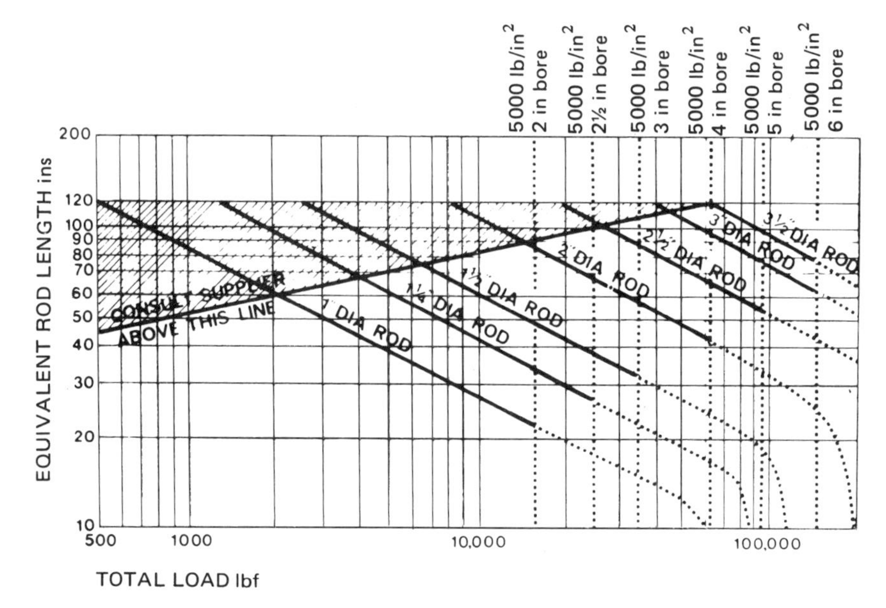


*Figure 4*
*Safe range for rod lengths.*

In certain applications the bearing length, with the rod of a long cylinder fully extended, may be increased by the fitting of a stop tube (see Figure 5). The way in which the cylinder is mounted and the manner in which the rod is supported at the extremity of its movement will determine whether or not a stop tube can be advantageous.

**Installation**

Correct installation can be controlled to a considerable extent while the layout of a machine or plant is being arranged. Some of the following points should be considered at this early stage.

The inherent elasticity of a cylinder can mean the difference between success and recurring trouble. If high shock loads are expected the cylinder should be mounted to take advantage of this elasticity.

Fixed mounting cylinders should be keyed or pinned, provision being made at the design stage of the machine. If the appropriate member of a machine is thick enough to take key-ways, cylinders with integral key-mounts can be provided.

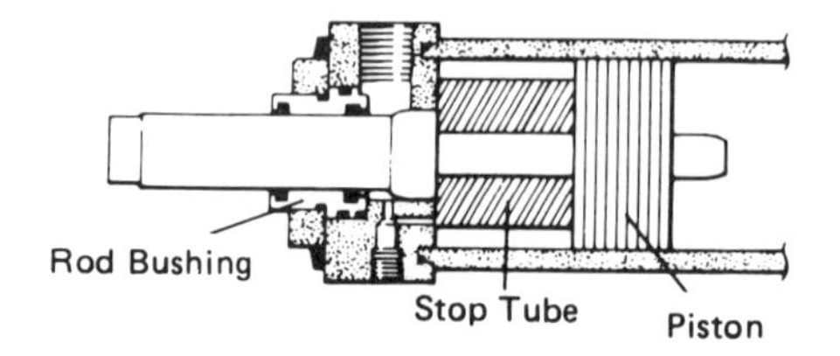


*Figure 5*
*Cylinder with stop tube.*

Separate keys to take shear loads are common. These should be at the correct end of the cylinder; at the rod end if major shock loads are in thrust and at the blank end if they are in tension. Only one end should be keyed to the machine to avoid losing the advantages of cylinder elasticity.

Temperature and pressure effects should also be considered. Locating pins can be used instead of keys to take shear loads and maintain alignment, again at one end or the other but not both. Whatever the problems of the machine designer there should not be pinning across corners.

In fixing a cylinder rod to a machine member, the rod should not be rotated more than necessary to avoid danger of scoring the cylinder body. The danger is greater if there is misalignment when the rod is rotated.

Rod end knuckles, cam surfaces, *etc,* should not be permanently fastened to piston rods during installation, as the joints would have to be broken during replacement of glands. Care must be taken to ensure that paint, dirt, *etc,* is kept from the piston rod and that no damage is caused to exposed rods.

**Standardization**

It has been mentioned that the form of construction of the tie-rod cylinder has been standardized in the USA by the JIC, the SAE, and has been incorporated in the NFPA standard ASA – B93.3 1965.

In Europe, hydraulic cylinder internal diameters and rod diameters, sizes of threaded rods and ports, nominal pressures and piston strokes have been standardized in the CETOP recommendation R10H. In this document the basic sizes and pressure steps are stated from which the values for standard cylinders should be chosen.

The diameters are graduated in standard figures such that the resulting output forces in turn lead to other standard figures. For this standard the usual piston areas have been summarized for given area ratios. The great number of possible area ratios has been reduced to eight after considering the most usual ratios and the requirements of different types of users. These recommendations have been incorporated in the International Standards ISO 3320 – 3322.

Mounting dimensions have also been standardized for three cylinder size ranges, giving dimensions for rear and front flange mountings, rear eye or clevis, trunnion mounting and front flange spigot for each cylinder bore size.

The size ranges and relevent standards are: 160 bar medium series (25mm to 500mm bore) – CETOP R58H, BS 6331 Part 1 and ISO 6020/1; 160 bar compact series (25mm to 200mm bore) – BS 6331 Part 2 and ISO 6020/2; 250 bar series (50mm to 500mm bore) – CETOP RP73H, BS 6331 Part 3 and ISO 6022.

# Applications

THE APPLICATIONS of hydraulic cylinders are as numerous as the application of hydrostatic power, so only the major applications will be considered to show the developments in this field.

**Static Applications**

Cylinders used in static applications are few of the truly hydrostatic machines where fluid flow does not need to be considered. They are used to support static loads on a column of fluid under pressure and, although the term *static* is used, there is always some movement due to compressibility of the fluid and minute leakages across seals.

The application where static cylinders have been most successful is for hydraulic pit props in mines. Hand-set individual props comprised the first major incursion of hydraulics into the mines, as a direct replacement for timber or rigid steel props with great improvement in safety and ease of installation. They lend themselves readily to a variety of methods of mining and many millions have been produced at remarkably low cost. They are still extensively used, although now being superseded by powered supports.

A typical prop is shown in section in Figure 1. Basically, it embodies two steel cylinders fitted one inside the other, which can be extended by hydraulic pressure derived from a pump incorporated in the assembly.

Provision is also made for quick release of pressure, while a relief valve in the hydraulic circuit also ensures that when the hydraulic pressure in the prop exceeds a specified figure the prop yields.

As the diagram shows, the main or pressure cylinder is enclosed by a guard tube. The inner cylinder, which slides in the pressure cylinder, forms a fluid reservoir as well as containing the pump and valve mechanisms.

The pressure cylinder and inner cylinder operate together as a hydraulic ram, with a main bearing located at the open end of the pressure cylinder. This bearing also forms part of a out-stop should the prop reach the limit of its travel. Included in the bearing assembly are a metal scraper ring and wiper ring to prevent dirt and water from entering the pressure cylinder.

A piston head is welded to the bottom of the inner tube and carries: the pump cylinder, the main release valves and the relief valve capsule plus a gland ring which forms a high pressure seal, an anti-extrusion ring and a piston ring. Valves and rings

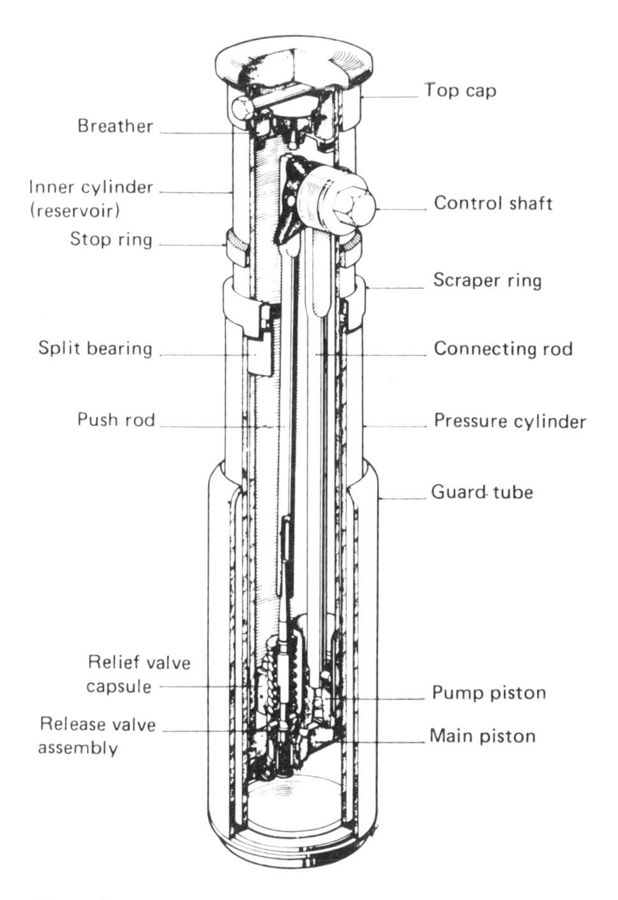


*Figure 1*
*Hydraulic pit prop.*

are retained by a detachable plate.

A control shaft is mounted transversely in a housing at the top of the inner tube, linked *via* a crank to a connecting rod to the pump and push rod to the valve assembly.

An oscillatory motion imparted to the control shaft *via* a suitable handle operates the pump for setting the prop in position. Lifting the handle right up past the normal suction stroke position, first compresses the spring in the valve assembly and then operates the relief valve mechanism for withdrawing the prop.

Being fully enclosed units (except for a breather), hydraulic props may use mineral-oil fluid, although many are designed specifically for operation on oil-in-water emulsions for coal mining applications.

Static cylinders are also used for civil engineering applications to support or even lift buildings during construction. Single-acting displacement cylinders are normally used due to their rigidity and the inclusion of a permanent pressure guage at the base of the cylinder gives a good indication of any change of pressure and hence load.

*High-tonnage single- and double-acting lift cylinders.*

**Mechanical Handling**

*Mechanical handling* covers a whole range of subjects even when mobile and vehicular applications are excluded. Static mechanical handling systems are finding an increasing number of applications in industry, civil engineering and process plants, *etc.* Examples include:

*Scissor lifts* – These come in a variety of types and sizes. The advantage is that the load may be lifted from a very low minimum height to several times that height. Sometimes automatic control is used so that sheet material may be loaded or off-loaded a sheet at a time and the top of the stack will remain at a constant height. In the simplest form a feeler or finger is used to sense the top of the stack. This is connected to a valve which controls the flow of oil into or out of the supporting hydraulic cylinder.

*Passenger* or *goods lifts* – Hydraulic power has many advantages, such as low initial cost when used only as far as the first or second floors. In such applications the single-acting displacement ram is accommodated in a bore-hole directly under the centre of the floor of the lift cage. For greater heights a hydraulic ram is mounted in the side of the lift shaft and operates a jigger mechanism which, with a cable multiplies the movement. This hoisting system has the added advantage that there is no mechanism at the top of the lift shaft. The motor, pump and control equipment are usually housed in the basement, where any noise can be controlled.

*Machine tools* and *automation* – Hydraulic cylinders have been used so extensively in this industry that the JIC tie-rod construction cylinder has been now generally accepted as a standard for the industry. The particular advantages of hydraulic cylinders applied to machine tools are:

(1) The elimination of costly lead screws and the attendant anti-backlash equipment always associated with mechanical drives.

(2) Extremely smooth movement under infinitely variable speed control.
(3) The availability of proportional control response *via* proportional control valves and electric signalling.
(4) Ready adaptation to automation through manual control; or sequential or combination control *via* logic control circuits.
(5) Simple and easily replaceable power cylinders and accurate positioning using servo-control systems.

The cylinders are used mainly for feed slides and clamping operations, and for transfer lines and component handling.

*Industrial robots* – Hydraulic cylinders are used on industrial robots to compete with pneumatics and D.C.Electrics.

Pneumatics gives a cheap and simple power system but does not allow easy control of either speed or position, D.C.Electrics are convenient and precise but there is a power limitation. Hydraulic power is highly reliable and there are a number of positive advantages of this system over the others, the most important being the greater overall power and response capability. The availability of particular hydraulic components, such as the cylinder for linear actuation is also very useful. These cylinders are usually specially designed for the machine to enable the gripping, clamping and swivelling motions to be performed.

The functional control system can be as basic as a programmable plug board, consisting of a matrix where the rows correspond to each axis and a stepping switch energizing each row in turn to operate the robot through its cycle. Electronic computers are clearly the most powerful and attractive means of control.

Here the remarkable reduction in cost brought about by the more recent development of micro-processors opens up many possibilities. Robots can be programmed to carry out specific tasks, take alternative actions dependent on results from the sensory feedback and, through communication links, interact with other robots.

**Hydraulic Presses**

It has been stated that the basic laws of hydrostatics were first applied to hydraulic presses, so it is not surprising that this is a major application for cylinders.

Hydraulic presses vary enormously in size, geometry and work capacity. They include general-purpose and specific-purpose types, the latter commonly designed to meet an individual user's requirements. The 'size' of a press continues to be specified in metric tonnes, in imperial tons or US tons.

Modern hydraulic presses may be designed to operate at working pressures up to 400 bar (5800 $lbf/in^2$) or more, using oil fluids and with modern high speed pumps directly coupled to the driving motor. An overload control is included so that when a given pressure is attained, delivery ceases and the ram is reversed. Pressure is dependent firstly on pump operation and secondly on the reaction encountered by the ram. Ram pressure builds up to overcome resistance until either a pre-determined resistance or a pre-determined ram position is realized. Pressure is thus under positive control, as is the ram speed.

Both *upstroking* and *downstroking* presses are made but the preferred layout, except where operational considerations conflict, is the downstroking press with all possible hydraulic gear mounted on top of it. This arrangement keeps floor space required to a minimum, although it does pre-suppose adequate roof height; the tank should be immediately above the cylinder, so simplifying the pre-fill valve.

The pump can then be mounted inside, above or alongside the tank.

Double-acting rams avoid the complication of having additional return cylinders. Metal piston rings may reduce maintenance costs, although piston head packings have been omitted altogether and the consequent slight leakage on the return stroke tolerated.

If a double-acting ram is used gravity has to be relied upon to speed up the approach; the weight of the suspended parts is normally sufficient to ensure this, and indeed the speed may even need regulating.

The superiority of hydraulic presses for deep drawing is well established. Presses equipped for single-, double-, and triple-action drawing are now made in a variety of sizes.

The conventional method of double-action drawing (see Figure 2) uses a pressure plate controlled by rams, usually four in number, clustered around the main ram. These are arranged to apply the pressure plate before the main punch makes contact. The force on the pressure plate can be varied as pressing proceeds and made proportional to the resistance to the punch ram.

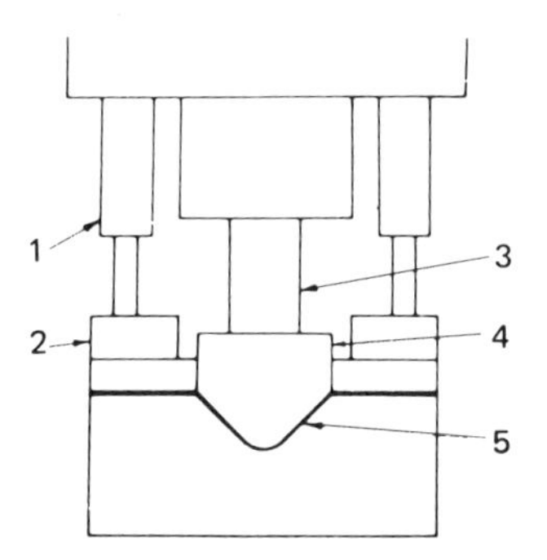


*1–pressure plate rams. 2–pressure plate. 3–main ram. 4–punch. 5–die.*

*Figure 2*
*Double-action drawing.*

It is often more convenient, however to invert the punch and die (see Figure 3) and the blank is laid on the pressure plate which is carried down against the resistance of the bottom ram. The bottom cylinder can be connected to an accumulator or to the main hydraulic system. With the conventional stationary bottom cylinder the main ram and press frame must be suitable for a load equal to the sum of the forces required for drawing and for the pressure plate. The size of press can be appreciably reduced by fitting the pressure plate cylinders to the crosshead carried by the main ram as shown in Figure 4. The pressure plate force is now completely self-contained and the main ram need only be large enough to give sufficient force for the drawing tool.

With this method it is necessary to provide a separate pressure supply to the pressure plate rams and they cannot be connected directly to an accumulator. An alternative construction, giving the same effect, has a single ram below the press tool supported from the crosshead by tie bars.

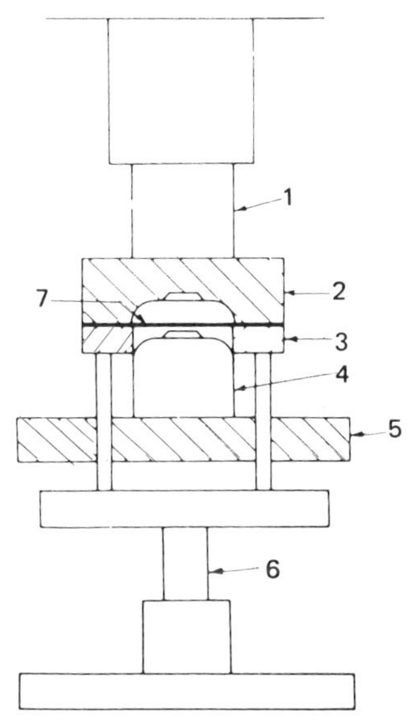


*1–main ram. 2–die. 3–pressure plate. 4–punch. 5–press bed. 6–bottom ram. 7–blank.*

*Figure 3*
*Double-acting press with die cushion acting from below.*

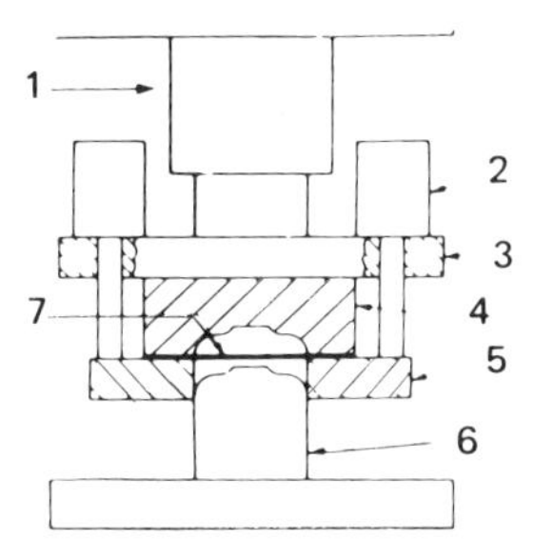


*1–main ram. 2–pressure plate cylinders. 3–moving crosshead. 4–die. 5–pressure plate. 6–punch. 7–blank.*

*Figure 4*
*Double action drawing with gripper cylinder carried on cross-head.*

*Special valves* are required in hydraulic presses for dealing with the high pressures and high flow rates encountered and also for dissipating the energy stored in press frames and cylinders and in the fluid itself due to compressibility effects. If such energy was released through a conventional spool valve there would be severe shock forces generated. Also, the heat generated and the high oil velocity would lead to damage to the valve seat.

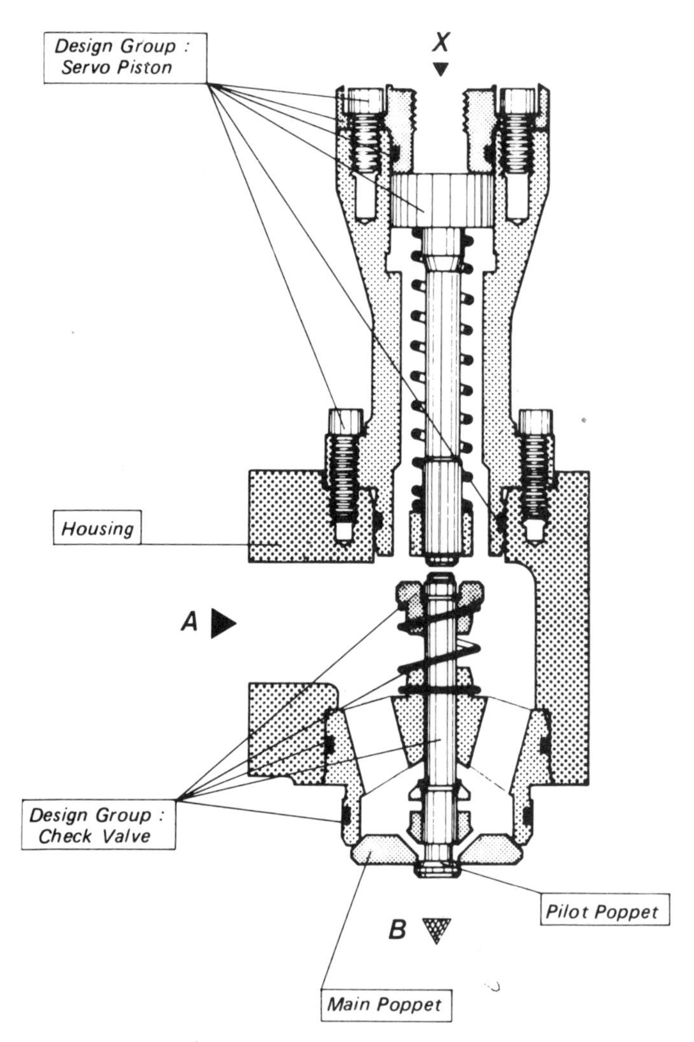


*Pre-fill valve model with decompression feature.*

A special decompression valve must therefore be used and this is often incorporated in the main directional-control valve. It is usually of the four-way type, with ports for high-pressure supply, exhaust to reservoir, main cylinder, and return annulus or subsidiary cylinders. The valve is designed to allow the high pressure compressed oil to decompress in a controlled manner before the main valve spool is moved.

*Injection moulding machines* are modern forms of presses where high speed and high temperatures are involved. Hydraulic cylinders are used for the opening and closing of the moulds and for the injection action of the plastic material.

Mould actuation needs to be quick but at the extremes there must, of course, be dampening to prevent mechanical shock. The ideal flow diagram for the closure cylinder is dependent on the particular type of mould closure mechanism. In all cases, however, the following characteristic phases are present:

(1) *Initial movement* – high pressure to overcome inertia forces and provide initial acceleration to the moving parts.

(2) *Middle phase* – low pressure because the movement is at constant velocity with, in effect, just frictional forces to overcome.

(3) *Closure phase* – high pressure required for closing movement of the toggle mechanism and for holding the mould halves closed at end of cylinder stroke.

The injection action is accomplished by means of cylinders and in this case, the significant hydraulic parameter which needs regulation is the pressure level. The injection cylinder moves at the same time that plastic material flows into the mould, which occurs at a pressure of between 1000 bar and 2000 bar at the injection nozzle.

At least two hydraulic pressures should be provided during the mould filling operation. An initial pressure level is needed to fill the mould up to 85 to 95% of its capacity, while a second pressure level is needed to accomplish the final mould filling operation and also to offset plastic volume reduction due to in-mould cooling. More complex moulds and plastic materials, having particular characteristics require that three or four different pressure levels be provided during mould filling.

These hydraulic pressures are applied to the cylinder in accordance with the mould filling characteristics desired and are, therefore, in proportion to the position of the injection cylinder. Because the load on the cylinder, besides being high, is also for all practical purposes, constant, there is a proportional relationship between the hydraulic working pressure and the injection velocity until the mould is filled.

**Mobile**

Miniature hydraulic cylinders have been used on vehicles since the introduction of hydraulic *brakes* and were followed by hydraulic *power steering* cylinders. A master cylinder of approxiamately 25mm (1 in) diameter is operated by a foot pedal through mechanical linkage so that a maximum pedal pressure in the order of 45 kgf (100 lbf) will produce a fluid pressure in the range of 40 to 60 bar (600 to 800 lbf/in$^2$). This master cylinder supplies fluid to four slave cylinders simultaneously which operate the brake shoes on the drum or disc brakes.

Where the brake-actuating forces required are greater than can readily be obtained through a manipulation of the various factors involved, it is desirable to apply boost to the manual input. This normally involves the employment of a booster cylinder which supplements the pedal pressure above a pre-determined level of input effort.

Higher input pressures are then accompanied by 'boost' pressure, resulting in a substantial increase in the force applied to the master cylinder and consequent increase in system pressure without requiring excessively high pedal loads. Such a system also retains substantially the same 'feel' as a simple system, with braking effect directly proportional to pedal pressure.

A typical vacuum booster unit, which is connected to the inlet suction of the engine, used in conjunction with an otherwise conventional hydraulic circuit, is normally referred to as a *vacuum hydraulic* servo system.

Power-assisted steering, in its basic form, comprises a hydraulic booster cylinder

*Mobile vehicle cylinder application .*

with one end fixed and the other linked to the steering arm, fed by an individual hydraulic circuit with its own pump and control valves.

A typical system comprises a hydraulic pump driven by the engine, an oil reservoir, an actuating cylinder incorporating a directional control valve, flow regulating and pressure relief valves. These two valves may be incorporated in the same block which may be integral with the pump/reservoir unit. The actual arrangement of components can vary with the type of vehicle and space available.

Hydraulic cylinders are more readily associated with *mobile plant* for mechanical handling and earth moving duties where large cylinders are visable on the machines. Mechanical handling vehicles include: *fork-lift trucks, side-loading* trucks, *straddle carriers, mobile cranes, etc,* while *bulldozers, loaders, digger, excavators, graders* and *tractors* all come under the title of earth moving vehicles. The cylinders used are usually of the cheaper welded construction to keep the cost of the vehicles down but with attention paid to hard chromium plated cylinder rods for protection from harsh environments.

Lift trucks employ *hoist* cylinders and *tilt* cylinders and current systems use a two-speed lift approach which can be achieved by a double-acting lift cylinder in which the annulus and head sides can be inter-connected by a valve.

Low speed is achieved by supplying only the head side of the cylinder and thus exhausting the annulus side, while high speed is attained by inter-connecting the head and annulus sides of the cylinder and therefore pressurizing the annulus and adding its exhaust oil to the supply at the head side. The lifting capacity is, of course reduced in the proportion of the rod area to the piston area and the lifting speed increased inversely.

Side-loading lift trucks are designed to carry long loads and complete containers

*Lift truck container handling cylinders.*

*Lift truck hoist cylinders.*

on larger size vehicles. The mast is mounted on one side of the chassis and moves across to pick up or deposit the load outside the wheelbase of the vehicle. The load is moved in and lowered on to the deck for transportation. To give stability during lifting and lowering, hydraulically-operated stabilizer jacks are used.

Most mobile cranes in the 15000 kg (15 ton) class are hydraulically actuated and considerable inroads are currently being made into the 30000 and 60000 kg (30 and 60 ton) and even larger classes. Cranes demand greater control than any other materials-handling vehicles. Safety, as in other handling equipment, is of paramount importance.

Telescopic mobile cranes normally have *derrick* cylinders, a telescopic cylinder to extend and retract the jib and hydraulic motors for the hoist winch and slewing. Lattice jib or strut cranes will normally have two or three hydraulically-driven winches controlling the jib and hook through wire rope connections. Both types require stabilizer jacks or outriggers to confer the stability necessary for working at extended radii round the machine.

The type of system, widely used on bulldozers, loaders and diggers, is the so called 'open' system. Here the particular pump feeds a set of selectors in series as shown in Figure 5.

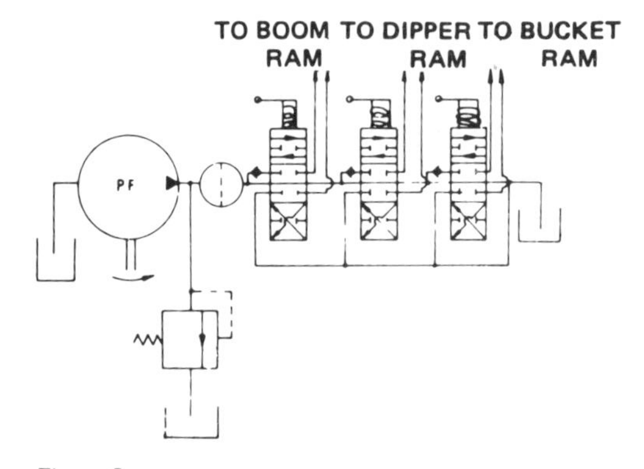


*Figure 5*
*Open series circuit.*

There are at least three positions and in the mid setting the output from the pump is fed directly through them back to the reservoir. Moving any one then cuts off the flow and allows the pressure to build up and operate the particular service.

If the selection is complete then the selectors further downstream are inoperative but if the selection is only partial then it is possible to operate services simultaneously. Again, in the neutral position the service connections to the cylinder can either be blanked, maintaining it in any desired position for booms and buckets on loaders, or it can connect the service connections together on some track motor controls. It is also common to incorporate check valves to prevent the inadvertent release of pressure on the operation of the subsequent service.

The basic method on which the draught and position-controls of most modern tractors are based still follows the original *Ferguson* system where implements were mounted directly on the tractor, lifted and lowered by hydraulic rams, with the further refinement that hydraulics also provided a means of controlling the manner in which the which the implement moved through the ground.

The built-in hydraulic system, with the increasing use of the tractor as a mobile power source, has enabled designers to provide power for cylinders used on tractor-trailed equipment such as *tipping trailers,* and has led to the introduction of many services for which hydraulic power has replaced human muscle power.

**Marine**

Hydraulic cylinders have traditionally been associated with marine applications since the 1850s. It was during this time that I.K. Brunel used 18 water hydraulic jacks for the sideways launch of the *Great Eastern* into the Thames from the Isle of Dogs, applying a combined thrust of 4500 tons.

About the same period W.G. Armstrong (later Lord Armstrong) was busy with the development of the hydraulic crane for dock and harbour work. An important device used by Armstrong was the hydraulic jigger, which was jack with a set of pulley sheaves mounted at each end. By means of a chain or cable the stroke of the ram was multiplied by the number of pulleys in the sheaves to reduce the length of the cylinder.

Today hydraulic cylinders are used on board ships for steering gear, hatch covers, loading ramps, deck machinery, cranes, lifts, and stabilizers.

The main problem encountered with marine hydraulic systems is deterioration due to working in a salt-ladened atmosphere. Many hydraulic systems and their associated equipment and components deteriorate in the marine environment simply out of neglect because the operators are very often ignorant of the requirements of the systems for regular servicing and maintenance. Corrosion is, of course, the chief cause of deterioration.

The corrosive effects of sea water are well known. Carbon steel is eaten away

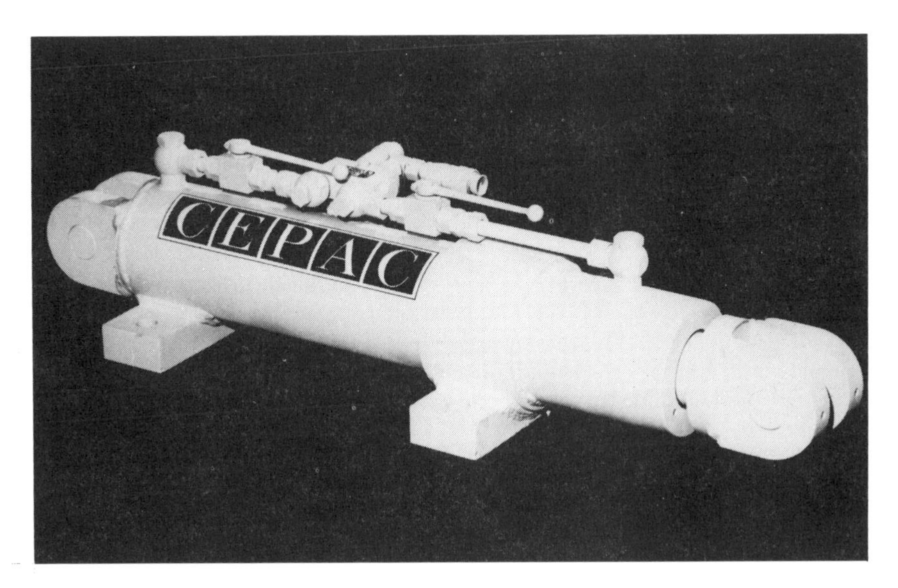


*Sub-sea cylinder with isolating valves and pilot-operated check valves.*

almost visibly. Chrome-plating, which gives a hard surface resistant to seal wear, is porous and permits corrosion of the base metal and resultant blistering and flaking of the chrome. This in turn causes damage to the seals as the blisters foul the seal lips and allows oil leakage from the cylinder.

The porosity of chrome-plating can be reduced by the use of a suitable intermediate plating treatment before the chrome is applied. Unfortunately, the best of these intermediate materials are rapidly attacked by hydrogen sulphide which is present in many crude oils.

An obvious material to resist corrosion by sea water, crude oil and petroleum products, is stainless steel. Several different types are available. None are completely corrosion-resistant but the most resistant is 18/10/3 (BS 316 S16 or AISI Type 316).

The cylinder rods can be protected by suitable covers. Only bellows-type rod covers can provide complete protection against dirt or solid contaminants being drawn back into the rod seals when the rod is retracting.

The problem is more severe with hydraulic cylinders because the emergent rod will be covered with film of tacky oil, to which solid particles will readily cling. Wiper seals alone cannot provide complete protection, although the protection they offer is usually adequate for most purposes.

The fitting of protective covers is thus largely confined to cylinders operating in particularly contaminated atmospheres containing a high proportion of solid and/or abrasive particles.

The considerations applied to shipboard hydraulics must also be applied to cylinders used for static installations for dock and harbour gates and for flood prevention barriers.

The largest flood prevention barrier recently constructed in the UK is the Thames barrier at Woolwich which is just downstream from the place where Brunel's Great Eastern was launched.

The barrier spans the river Thames with rising sector gates and falling radial gates all operated through linkages by hydraulic cylinders. Each gate is operated by cylinders at each end, the largest being 1.1m bore by 3.13m stroke with a 0.4m rod diameter (3.64 ft bore by 10.27 ft stroke by 1.31 ft stroke rod diameter). The ability of the hydraulic equipment to operate the gates, some weighing 1500 tonnes, from one end only provides an adequate degree of back up to deal with unexpected failures.

**Special Cylinders**

Special cylinders are designed to fit applications requiring special features or to meet the requirements of systems with positioning or feedback control.

A *cable* cylinder (see Figure 6) is capable of providing a long stroke within the nominal overall length of the cylinder itself. This is accomplished by connecting the piston to an endless cable passing around external pulleys. Force output is taken by

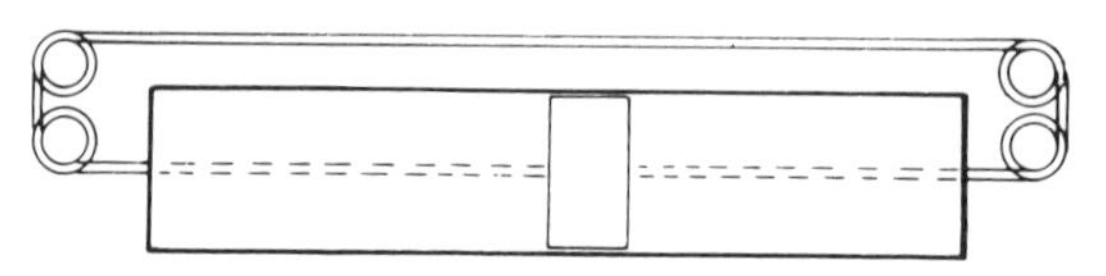

*Figure 6*
*Cable cylinder.*

a suitable connection to the external cable. Cable cylinders present sealing problems with high fluid pressure and are more usually rated as pneumatic rather than hydraulic cylinders.

*Aircraft actuators* are virtually the standard choice for powered control systems and utility services on modern aircraft, largely because of the highly favourable power/weight ratios attainable with system pressures of 210 to 280 bar (3000 to 4000 lbf/in$^2$).

System pressures have become more or less standardized at 210 bar (3000 lbf/in$^2$) in the USA, although 280 bar (4000 lbf/in$^2$) is also now used – higher pressures being favourable to the power/weight ratio, although at the expense of increasing component stresses. Thus, light alloys have been largely excluded and jacks are fabricated from high tensile steels or titanium.

With servo-actuators the servo-valve is usually mounted directly onto the cylinder housing. A system pressure of 280 bar (4000 lbf/in$^2$) is considered by some authorities as the optimum with regard to equipment rate and response rate with servo-controls.

*Feedback devices* are fitted to servo-cylinders to give a signal of the piston's position, speed, or acceleration. A linear displacement transducer has been developed, which is inherently digital in nature, operating at low currents compliant with intrinsically safe requirements. The transducer in non-contacting and has been designed to operate in a variety of fluids.

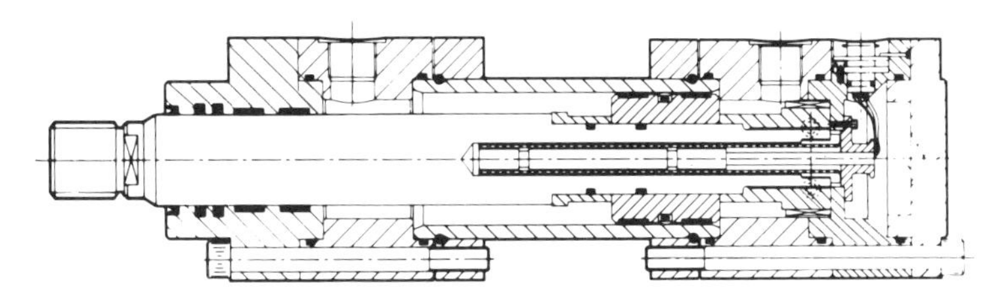

*Linear actuator fitted with 100μ resolution transducer.*

Based on transformer technology, the device operates in mutual inductance between spatially wound coils each having relative linear co-axial displacement. The main components are cylindrical in nature and small in diameter, thereby in convenient form for integration within the moving piston rod.

In operation the female measurement coil, secured within the rod, displaces linearly relative to the fixed probe attached to the cylinder end cap. Power is provided by two drive coils spaced either side of the male detector coil. The drive coils commutate the measurement coil.

The spatially wound detector coil, during movement, scans the pitches of the measurement coil, sensing the spatial excursions of the piston rod. Measurement accuracy depends solely on the coil pitch accuracy but, as several coils are enmeshed and searching at any given position of the stroke, any pitch errors are averaged out and minimized and cumulative errors cannot arise.

Another system utilizes high-frequency sound impulses (ultrasound) generated in a solid-state transducer to find the range from the transducer to the piston face. In its function, the ultrasonic pulse generated by the transducer element travels through the hydraulic fluid to the piston face and reflects back to the element. The time taken for the pulse to propagate from the element and back again is used to compute the piston position, the principle being similar to present day sonar.

Where previously such a transmission of ultrasound would create a problem, ascertaining precise readings with changes of temperature and pressure in the hydraulic fluid, the system uses a 'test-target' method which overcomes the problem.

The 'test-target' is a small wire placed at a known distance form the transducer element inside the transducer housing. Using this comparator many times a second, the correct speed of sound in the fluid is known. Therefore, the correct position of the piston can always be found.

*Hydraulic amplifiers* combine the 'muscle' of hydraulics with precision of electronic control by converting the torque of a low power *stepping* or *servo-motor* to linear motion while amplifying the force hydraulically.

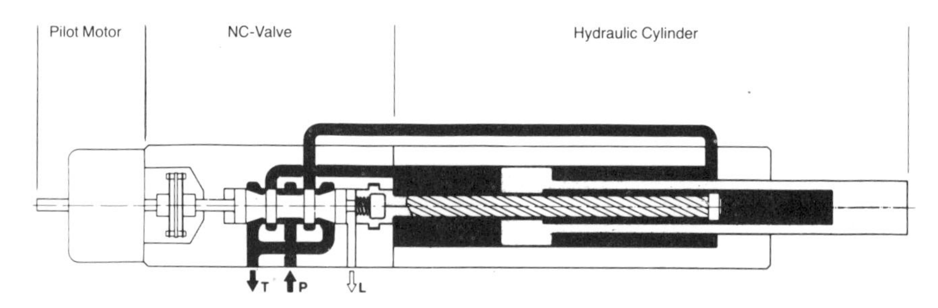


*Hydraulic linear amplifier.*

A *pilot* motor is connected *via* a torsionally-rigid, axially-flexible coupling to the control spindle. Rotation of the motor causes a spindle to screw into or out of a nut or threaded bush. This displaces a *servo-spool* allowing pressure and flow to one side of the actuator, displacing the piston. The resultant movement of the piston rotates the control spindle tending to return the valve spool to its original position.

Continued rotation of the motor will, therefore, establish a spool position which will allow the actuator to move at a speed in proportion to that of the motor. Stopping the pilot motor will cause the valve spool to be closed, preventing further movement. It is possible by controlling the frequency and number of pulses to the stepping motor to establish precise control over velocity, position, acceleration and deceleration. By installing additional external feedback devices it is possible to achieve the same degree of control using dc pilot motors.

*Electric* cylinders or actuators are made on the same principle as the hydraulic amplifiers with a dc motor driving a leadscrew through a roller nut to extend or retract a piston rod. The same positional accuracy and control can be applied but without the output forces supplied by the hydraulic amplification. The actuator is made to the same compact dimensions as a hydraulic or pneumatic cylinder to allow it to be interchanged for specific applications.

*Pressure intensifiers* comprise basically, two cylinders of different size with a common piston unit (see Figure 7). Their use is to provide a supply of high pressure

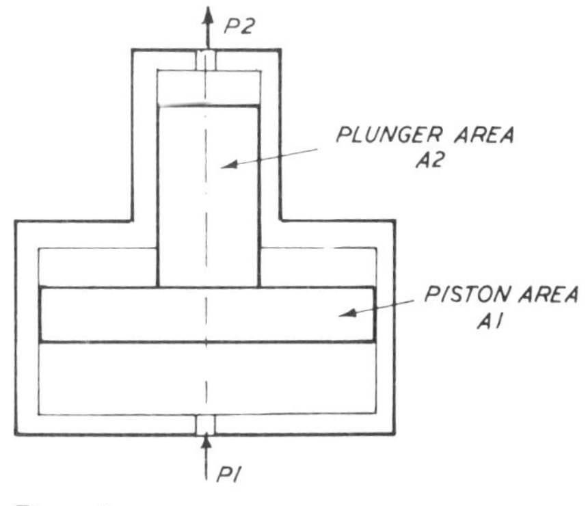


*Figure 7*

fluid from a lower pressure supply or, alternatively, as a simple source of second pressure in a dual pressure system working from a single pump. Neglecting frictional losses, the intensification ratio is equal to the ratio of the piston areas.

Theoretically, at least, pressure intensification ratios of several thousand are possible, although very much lower ratios are normally used in practice, except for highly specialized applications. The chief disadvantage of using too high a pressure intensification ratio is the reducing volume of high pressure fluid available (the volume flow being determined by the small cylinder bore and stroke) and the marked compressibility which is likely to be present with highly pressurized fluids.

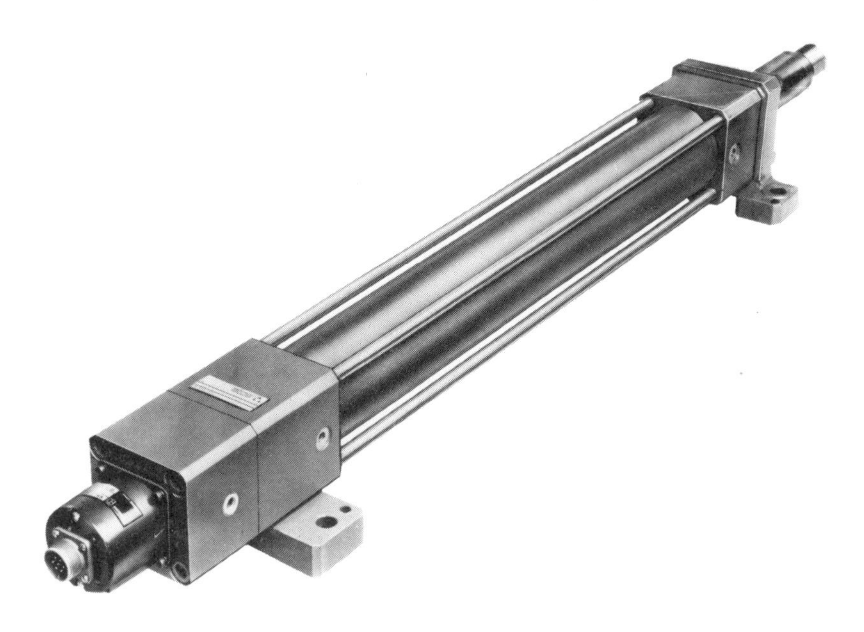

*Hydraulic positioning cylinder with rotary transducer.*

*Swing* cylinders are designed so that a clamping arm, fitted to the end of the rod, swings through an angle of 90° when the rod is extended. The arm then swings back into the clamping position when hydraulic pressure is applied and the rod retracts to perform a clamping action. This simplifies the clamping of awkward shaped work-pieces in automatic machining operations.

*Figure 8*
*Triple-acting swing cylinder.*

The box-form units, shown in Figure 8, have been developed to give production engineers extra tooling capabilities to meet increasingly complex demands of machining processes. The design permits the setting up of work-holding fixtures in which the clamping arm retracts fully below the level of the machine tools work-table. This then leaves the surface area completely free for work-piece loading and unloading movements. It also enables work-pieces to be clamped into position from the inside.

# SECTION 4

## Hydraulic Valves

TYPES OF VALVES
CONSTRUCTION
VALVE OPERATION
VALVE SELECTION
INSTALLATION AND COMMISSIONING

# Types of Valves

IT HAS been shown that valves can be classified by the function they perform, *eg* pressure control, flow control and directional control valves, *etc*. They may also be classified by mode of operation, *eg* manual, mechanical, pilot-operated, electro-hydraulic and servo *etc;* and also by their general geometry and method of mounting, *eg* pipe mounting (in-line), surface mounting (gasket), sub-plate or manifold mounting, and modular mounting. In the latter case the construction of the valve is designed to meet the mounting method required.

Basically hydraulic control valves can be divided into three classifications: (1) *seated* valves, (2) sliding spool valves, and (3) *variable orifice* valves.

**Seated valves**

Seated valves include spring-loaded ball and poppet valves, used as check, pressure control and directional control valves, as well as screw down shut-off valves.

Poppet valves have the advantage of high response and relative insensitivity to contamination. They are also well suited to high pressure duties and so may be preferred for specific applications. They have low leakage, can be made to seal properly for long periods of time and are relatively cheap to manufacture. They are less suitable for large valve sizes, however, because the opening load becomes excessive and calls for the use of a pilot valve system.

When the limiting size of orifice through the selector has been established, a poppet valve of the smallest possible diameter consistent with this throat area is chosen (with other passages slightly greater than the throat area). The lift required is dictated by the poppet angle. With a 45° valve, a lift of about half the valve throat diameter is required for full opening, with no valve stem in the throat. Push-rod, rather than stem, operation is generally to be preferred because the stem size, and thus the restriction in the throat, can then be reduced to a minimum. Pilot systems are widely used for reducing the valve opening force required, flow-operated valve mechanism generally offer the most benefits.

Figure 1 shows a section through a pilot-operated relief valve with two spring-loaded poppets. Flow entering port (A) is blocked by the main poppet at low pressures. The pressure at (A) passes through orifices to the pilot cone and also to the top of the main poppet. There is no flow through these sections until pressure exceeds the spring setting of the pilot cone.

The degree to which the spring is pre-loaded corresponds to the operating

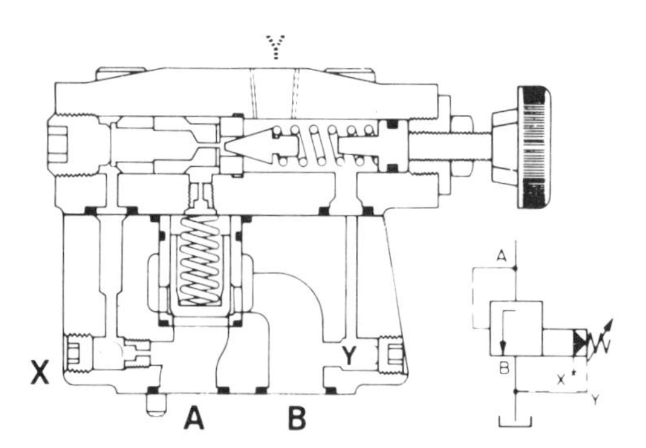


*Figure 1*
*Pilot-operated relief valve.*

pressure of the valve. If the pressure rises above the set pressure, the pilot cone is lifted from its seat, releasing a small pilot flow to tank *via* port (B). This results in a pressure drop across the main poppet. The main poppet opens, passing only enough flow from (A) to (B) to keep inlet pressure at the set value.

The main poppet closes instantly when the inlet pressure drops below the set value of the pilot section. The pilot spring cavity normally drains to port (B). External drain from port (Y) is optional. The valve may be vented or remotely controlled from port (X).

Most pilot-operated check valves are poppet-type seated valves and are used to hold a loaded cylinder on a column of pressurized oil due to the low leakage characteristics. Problems can occur due to the fast response of the valve, where large inertias are involved and high peak pressures are generated when the valve is closed. It is advisable in this case to incorporate some form of relief valve between the pilot-operated check valve and the cylinder to relieve the high peak pressures.

**Spool Valves**

Sliding spool valves are generally used for pressure and directional control purposes. There must be an annular clearance between the spool and the body so there is always some leakage flow across the spool which must be taken into consideration.

Spool valves operate on a sliding principle, so design normally follows the basic requirements of all slide valves *ie:*

(1) Pressure-balanced ports are required, so that there is no net pressure force acting axially on the spool.

(2) Valve diameter should be a minimum consistent with suitable stiffness.

(3) The valve body or sleeve must have adequate rigidity.

(4) Friction forces must be minimized and are largely controlled by material selection for rubbing/sliding parts.

(5) Annular flow should be symmetrical to avoid radial unbalanced forces which could increase friction.

(6) Bernoulli forces arising from changes in fluid momentum must be minimized.

Parameters (5) and (6) are largely controlled by the detail design of the spool.

*Greater reliablity and a smooth and precise control is achieved with the inclusion of on-board electronics and fully integrated electro-hydraulic proportional valves.*

Spool cushioning passages can be built into the valve as shown in Figure 2. These equalize hydraulic forces on the ends of the spool and cushion the spool shift. When the spool is shifted, the fluid displaced from one end of the spool is transferred to the other end through the passage which is designed to provide a cushioning effect and balance the spool.

Forces may also be set up due to the changes in fluid momentum through the valve, generally described as Bernoulli effects or Bernoulli forces. Thus, typically, there may be a reduction in pressure on the valve spool at the controlling edge, leading to a force being generated producing unbalance or tending to close the valve. At the same time, if backlash is also present in the system, Bernoulli forces may produce high frequency 'chatter' of the valve spool.

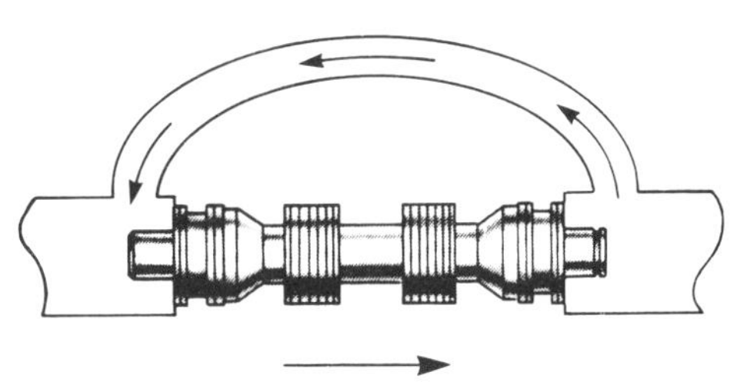

*Figure 2*
*Spool cushioning passage.*

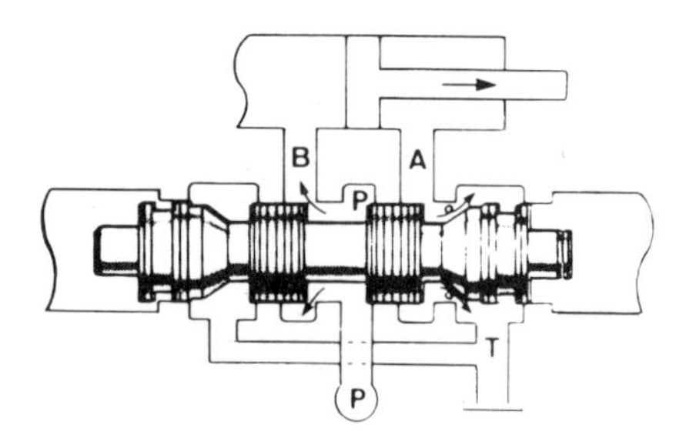


*Figure 3*
*Spool force-balancing contour.*

The hydraulic unbalancing effects of fluid momentum between the cylinder and tank ports of a valve can be minimized by contouring the spool shape as shown in Figure 3. Flow forces that are developed at the conventional square land orifice (P to B) are partially compensated for by the force balancing contour on the outer spool lands (A to T).

Accurate sequencing of land opening and closing also provides maximum axial stability, as shown in Figure 4. In this example, it is important that flow path (A to T) is opened before the path (P to B) to prevent pressure intensification which could upset axial balance and limit valve friction.

An advantage of spool valves over poppet valves is that the speed of the main spool can be controlled and the spool profiled so that high inertia loads can be decelerated gradually without high pressure peaks in the system.

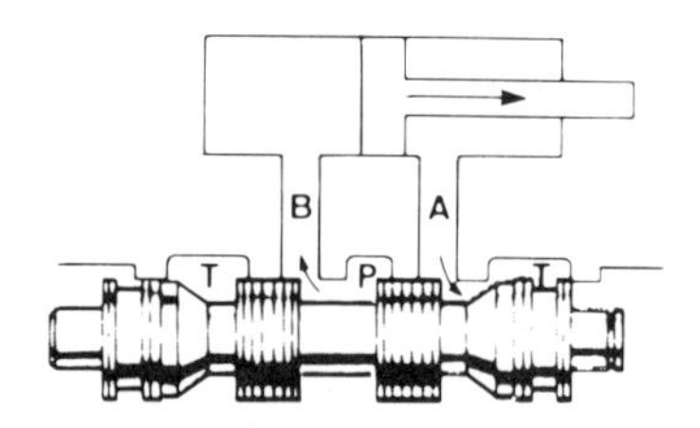


*Figure 4*
*Spool land sequencing.*

As previously described, it is often advantageous, with very large spool valves, to use a smaller pilot spool to direct flow to the ends of the main spool in a controlled manner, as shown in Figure 5. The pilot pressure can be tapped from the main pressure port (P) or supplied from an external source and piped to port (X). The latter course can be advantageous where fluctuating pressures are generated in the circuit and would not then affect the pilot control of the main spool.

**Variable Orifice Valves**

Although seated valves and sliding spool valves create a variable orifice while they are opening or closing, the term will be applied to valves where an orifice size can be selected to control the pressure drop and hence the flow through the valve.

*Needle* valves are used without any compensation for changes of load or viscosity

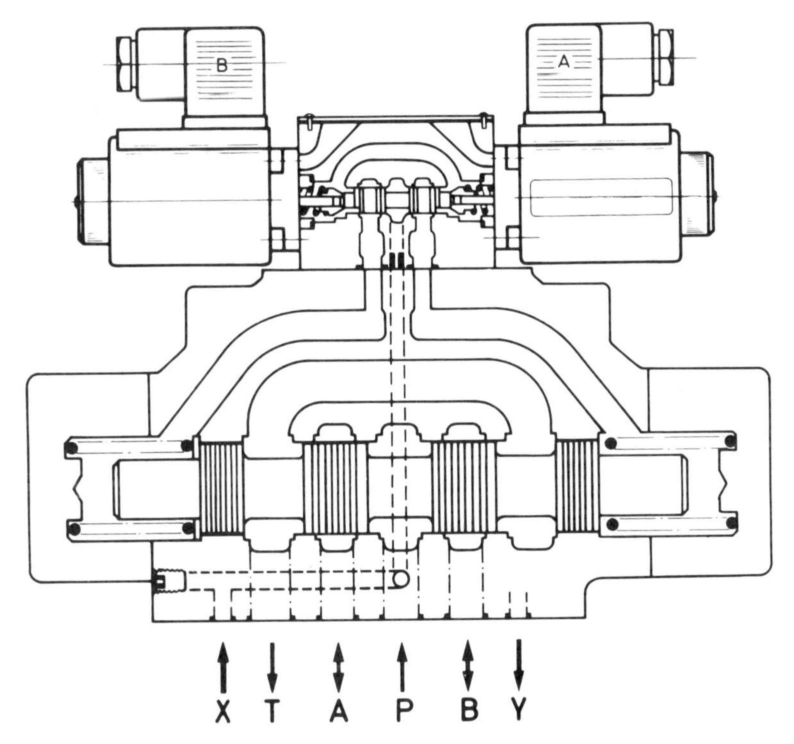


*Figure 5*
*Solenoid-controlled pilot-operated valve.*

and also for screw down shut-off valves. The fineness of control depends on the angle of taper of the needle and the axial movement relative to the axis of a concentric orifice, thus controlling the effective opening of the orifice.

As well as a simple screw-down valve, needle valves can be supplied at an oblique angle to the line of flow, offering a more direct flow path. Another form can be obtained where the controlled outlet flow is at right angles to the main flow.

Ball valves are used in hydraulic systems as shut-off valves, or to divert flow to or isolate flow from part of the system. Basic geometry involves a spherical ball located by two resilient sealing rings in a simple body form (see Figure 6). The ball has a hole through one axis, connecting inlet to outlet with full bore flow when aligned with the axis of the valve. Rotating the ball through 90° completely closes the flow passage with positive sealing *via* the sealing rings. Sealing is equally effective in both directions.

When the ball valve is open there is a very low pressure drop across the valve and the 90° movement of the hand lever can be used to indicate the position of the valve to show at a glance which part of the circuit has been isolated.

Ball valves are also produced in multi-port configurations, thus normally requiring a larger size of ball to accommodate multi-port drillings. These ports can be proportioned to give positive lap or negative lap as required.

Rotary valves rely on close contact being maintained between a rotating port plate and a back-up member and are rather more difficult to produce with adequate

*Figure 6*
*Ball valve with O-ring seals.*

sealing for high pressures. Also high operating forces may be required unless the elements are pressure balanced.

They are, therefore, more usually applicable to lower pressure systems, where the relative simplicity of the configuration and its flexibility as regards porting configuration can be used to advantage. Some designs are produced for higher pressures, although the provision of a suitable *rotary* seal on the spindle remains a problem.

*Rotary spool* valves are usually used as the restricting element in pressure-compensated flow control valves. The spool is either notched to produce a 'new moon' shaped orifice between two ports at right angles or an helical-shaped spool end connects two parallel ports. The aperture is gradually opened as the spool is rotated and the fact that a sharp edged orifice is formed gives the valve a low dependence on temperature change.

Details of the spring-loaded pressure compensator spool, required with these valves to provide a variable orifice, to give a constant pressure drop across the valve are given in the chapter on *Flow Control* in Section 2.

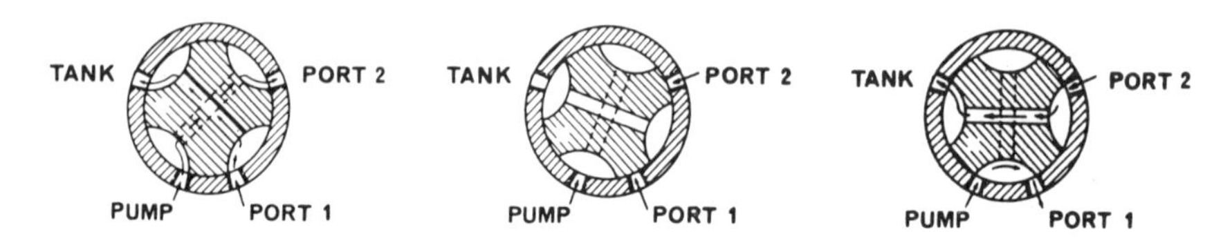


*Figure 7*
*Four-way rotary spool valve.*

Rotary spool valves are also used as four-way selector valves. These consist simply of a rotor closely fitted in a valve body. Passages in the rotor connect or block the ports in the valve body to provide the four flow paths as shown. A centre position can be incorporated if required.

The valves can be actuated manually or mechanically and are capable of reversing cylinders or motors. However, they are used principally as low pressure pilot valves to control other valves.

# Construction

**Pipe-Mounted Valves**

PIPE-MOUNTED valves originate from the early industrial hydraulic systems where the valves were mounted on solid pipework. They were usually made of cast iron with steel spools and poppets for oil hydraulic systems. The ports were usually tapped with female taper pipe threads for pipe sizes up to 38 mm (1½ in), nominal bore and flanged connections were used for 50 mm (2 in) nominal bore pipe sizes and upwards.

*Pipe-mounted* valves are still used today, for in-line check and relief valves where it is convenient to fit them to pipeline systems, although parallel pipe threads are usually preferred in conjuction with a seal washer.

The main disadvantage with pipe-mounted valves is that the pipe joint has to be broken and the pipeline disturbed each time a valve has to be replaced, which could introduce contamination into the system. The pipe coupling and sealing device may also need to be replaced.

Many pipe-mounted valves are therefore designed with detachable end caps so that spools, poppets, valve seats and springs can be replaced with the valve remaining in position on the pipeline. This still does not solve the problem of contamination entering the system, especially in dirty environments, nor enable a quick change of valve to be made with a minimum of downtime in an emergency situation.

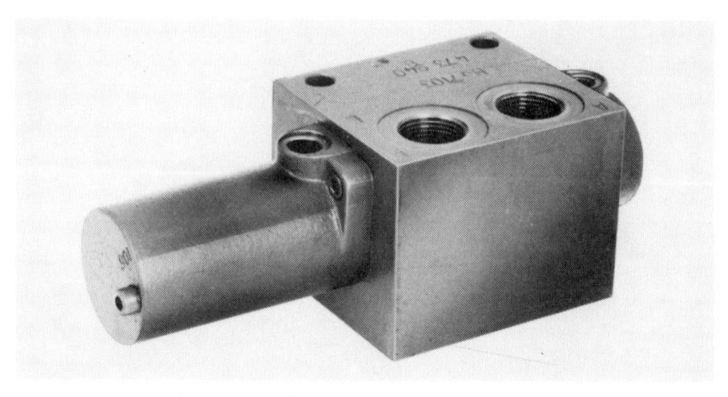

*Pipe-mounted control valve.*

*Mobile multiple banked valve.*

Pipe mounting is mainly used today on mobile multiple banked valves, although in no way must the valve be supported by the pipework but by the brackets which are either cast into or bolted onto the valve block. There are two basic forms of construction for this type of valve, one being made of sections each containing its own spool and service ports. When a multi-spool valve is required, these sections are bolted together. The second is known as a *monoblock construction* in which the valve body is cast with all passages, spool bores and service ports in one piece.

Sectional construction has the advantage that various standard sections can be stocked to provide the valve configuration required for a specific application. Generally valves are purchased already assembled with spare sections. The user can then add or subtract sections to make up a valve to suit his requirements. Equally, faulty sections can easily be replaced.

There are, however, disadvantages with sectional construction. The faces of individual sections, which must be absolutely parallel and flat, can easily be damaged when handled or placed in store, Any scratch makes it difficult to obtain perfect sealing between sections when they are bolted together.

*Mobile monoblock valve.*

Another disadvantage is the critical torque range required when tightening up the assembly bolts. If they are not torqued up tight enough, leakage can occur between the sections. If over-tightened, problems can arise with the spools.

Sticking spools can also cause trouble when the valve stack is mounted on a machine. This type of construction is generally less rigid than a monoblock design and if the mounting face is not flat and rigid, or if the mounting bolts are not tightened properly or torqued evenly, it is possible to create sufficient twisting in the stack to cause the spools to stick. In addition, because of the problems just mentioned, even greater difficulty would be experienced in servicing the valve under field conditions.

To minimize these problems, the sections are usually designed thicker, and thus have greater overall width compared with the monoblock design, but the small size of an individual section is much easier to handle through the manufacturing process and in the event of faulty manufacture or material there is less scrap cost.

**Surface-mounted valves**

To overcome the problems of replacing pipe-mounted valves, without disturbing the pipelines, *surface-mounted* valves were introduced, which could be bolted down onto the ground surface into which all port connections are drilled. These are also known as *gasket-mounted* valves, although these days, most valves are sealed by O-rings which are contained in recesses in each port on the valve's ground mating surface.

Manufacturers usually provide steel subplates with one ground surface to mate with the valve ports and the other surface tapped with pipe threads to enable the valve to be piped up to the hydraulic system.

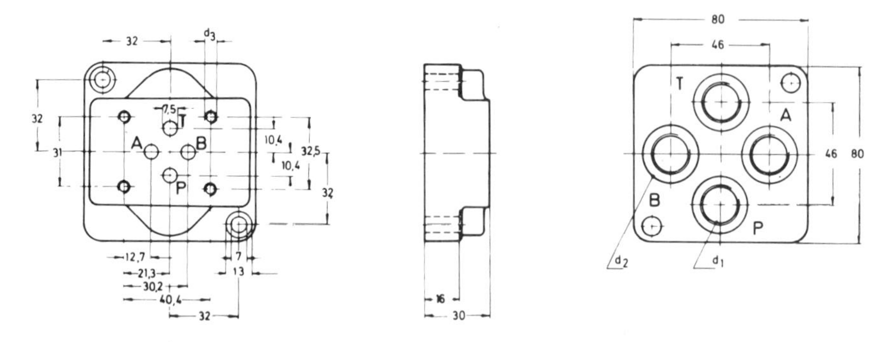


*Typical valve subplate.*

The subplates can be panel mounted vertically with the valves mounted on the front side of the panel and the piping connected on the rear side of the panel to make a neat arrangement. Pressure gauges and selector valves can also be mounted on the panel to form a control station.

Alternatively, subplates can be mounted horizontally either singly or in groups to position the valves at the required places on a hydraulically controlled machine.

Originally each manufacturer made the valve and subplate interfaces to his standards so that there was no interchangeability between different manufacturers valves.

*Sandwich design valve.*

However, standardization of valve mounting faces has now taken place through CETOP in Europe and the ISO so that interchangeability of different manufacturers valves can take place. The mounting surfaces of directional valves were standardized in Europe first under CETOP recommendation R 35H which laid down the port position, dimensions and mounting bolt threads for each size of valve.

The size range is designated from 00 to 12 which was originally evolved from the valves nominal port size in ⅛ in. increments. The current recommendation lists valve sizes of: 02, 03, 05, 07, 08 and 10, and manufacturers would refer to an 03 size valve as CETOP 03 valve in their literature.

A further provisional recommendation has been issued by CETOP in RP 69H to cover mounting surfaces for flow control, pressure control and check valves and these include the position of location pin holes so that the valve is mounted correctly in relation to its port positions.

Standardization has been carried out in the USA by the NFPA and collaboration between the national standards organizations is allowing international standards to be drawn up by the ISO.

**Modular Valves**

Although the subplate-mounted valves overcome the problem of quickly replacing valves, it was found that making the pipework to the rear of the subplate became labour intensive. This involved pipe bending and making pipe connections in a restricted space with the added problem of curing initial leaks at the pipe joints.

Hydraulic circuits usually have many separate 'legs' each with a directional valve to select the operation of a cylinder or actuator. Each 'leg' usually requires valves to control pressure, flow and direction as well as the selector valve and these were all originally pipe or subplate-mounted.

Modular valves are constructed as a *sandwich* with parallel ground surfaces through which all the service ports are drilled and are mounted between the subplate and the directional valve as shown in Figure 1. Many sandwich valves can be mounted together vertically on one subplate and are known as *stacked* circuits or assemblies. The number of valves in a stack was limited by the length of the holding down bolts but manufacturers now supply special *bolt extenders* which are built up with the stack and act as jig during assembly.

On one system each extender is a hexagon rod with male and female threads at

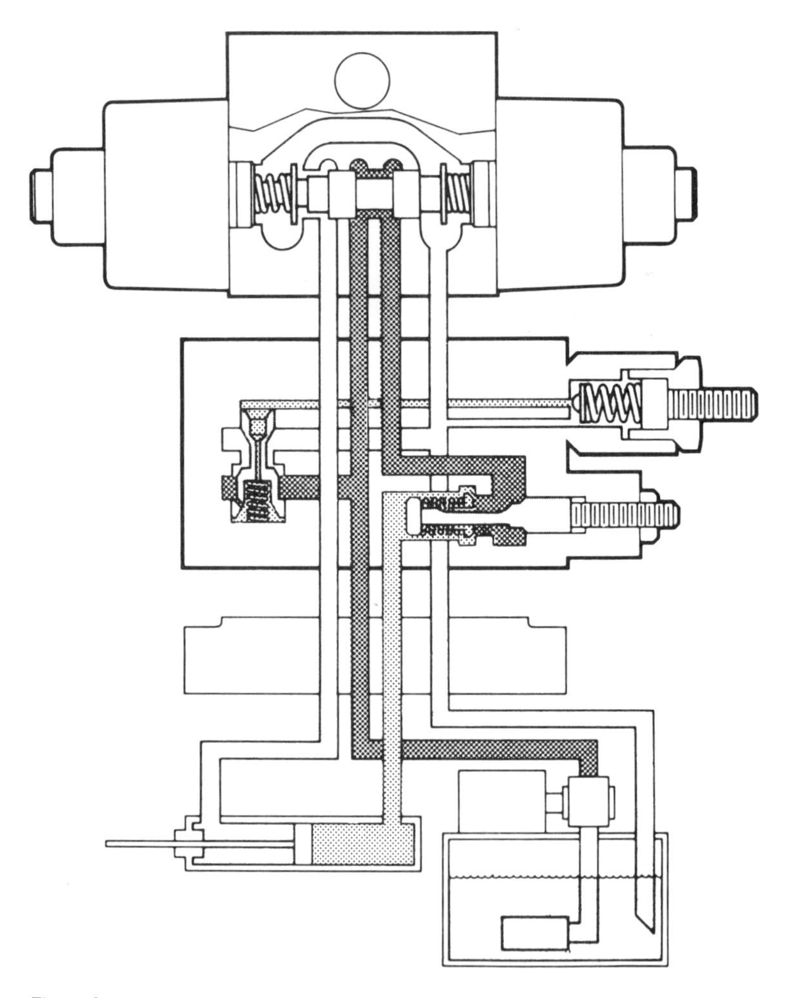

*Figure 1*
*Principle of circuit stacked valves.*

opposite ends. The plain length equals the standard depth of the sandwich module. To assemble the stack the first four extenders are screwed into the subplate and the first module is placed over them. A further four extenders are then screwed into the first four and the next module is placed in perfect alignment with the first. The stack is built up in this way until the final component, usually a directional valve, is secured to the extenders using standard socket head screws.

Modular valves can also be stacked horizontally by using modular subplates with parallel ground surfaces at right angles to the valve mounting surface. These subplates which have the pressure, tank return and service ports drilled through them can be bolted together in conjuction with inter-connecting plates and reducing plates as shown in Figure 2.

The ultimate result of using modular subplates is, of course, that valve stacks can be made of combined horizontal and vertical stacking assemblies.

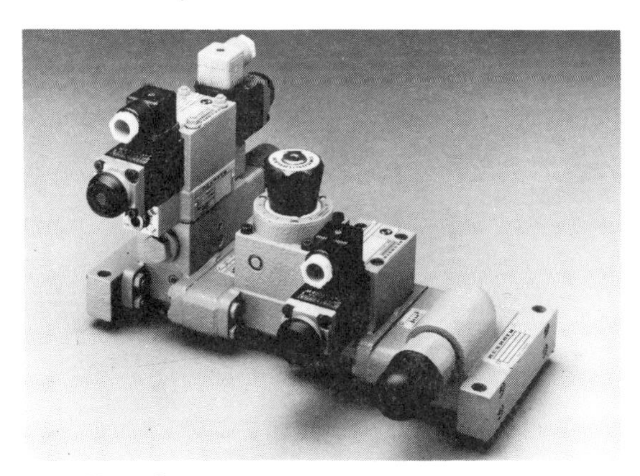

*Figure 2*
*A modular manifolding system.*

Stacked assemblies can normally be used on standard circuits with separate legs to each actuator but have the flexibility of being altered on site without remaking pipe connections. Correct sealing of each stack and torque tightening of the holding bolts must be observed.

Circuit diagrams can be drawn using circuit symbols drawn onto modular blocks and this can reduce draughting work required to produce circuits. Adhesive symbols are available from manufacturers. When the stacking assembly circuit is drawn, the mounting dimensions of the assembly are fixed, the maximum overall dimensions being determined by the type and method of stacking assembly (see Figure 3).

**Manifold Blocks**
Manifold blocks can be used in place of modular subplates and consist of a solid block of aluminium alloy, cast iron or steel, with drilled oil ways. They can be standard multi-station manifold blocks with one face drilled with valve interfaces, to meet CETOP and ISO standards, and other faces tapped with piped threads for the service ports, and the ends tapped for through pressure and tank return ports. Alternatively, they can be custom-built and drilled to fit special valves and applications.

The advantages of manifold blocks are that there is not the sealing or bolting up problems encountered with modular subplates and the length of the assembly can be kept to an absolute minimum. Vertical assemblies can be built up on multi-station manifold blocks in very compact assemblies especially with miniature valves.

The disadvantages are that a degree of flexibility is lost in not being able to change valve positions, as with modular subplates, and internal leakages, due to incorrect cross drilling in the block, are very difficult to diagnose. Deep bore or 'gun' drilling techniques have to be used to maintain the accuracy and correct depth of the drilled holes.

Orthographic views of the drilled oil ways are difficult to visualize and computer techniques are now being used to show the drillings in three dimensional graphics, which can be rotated on a screen, with pressure, return and service oil ways shown in different colours.

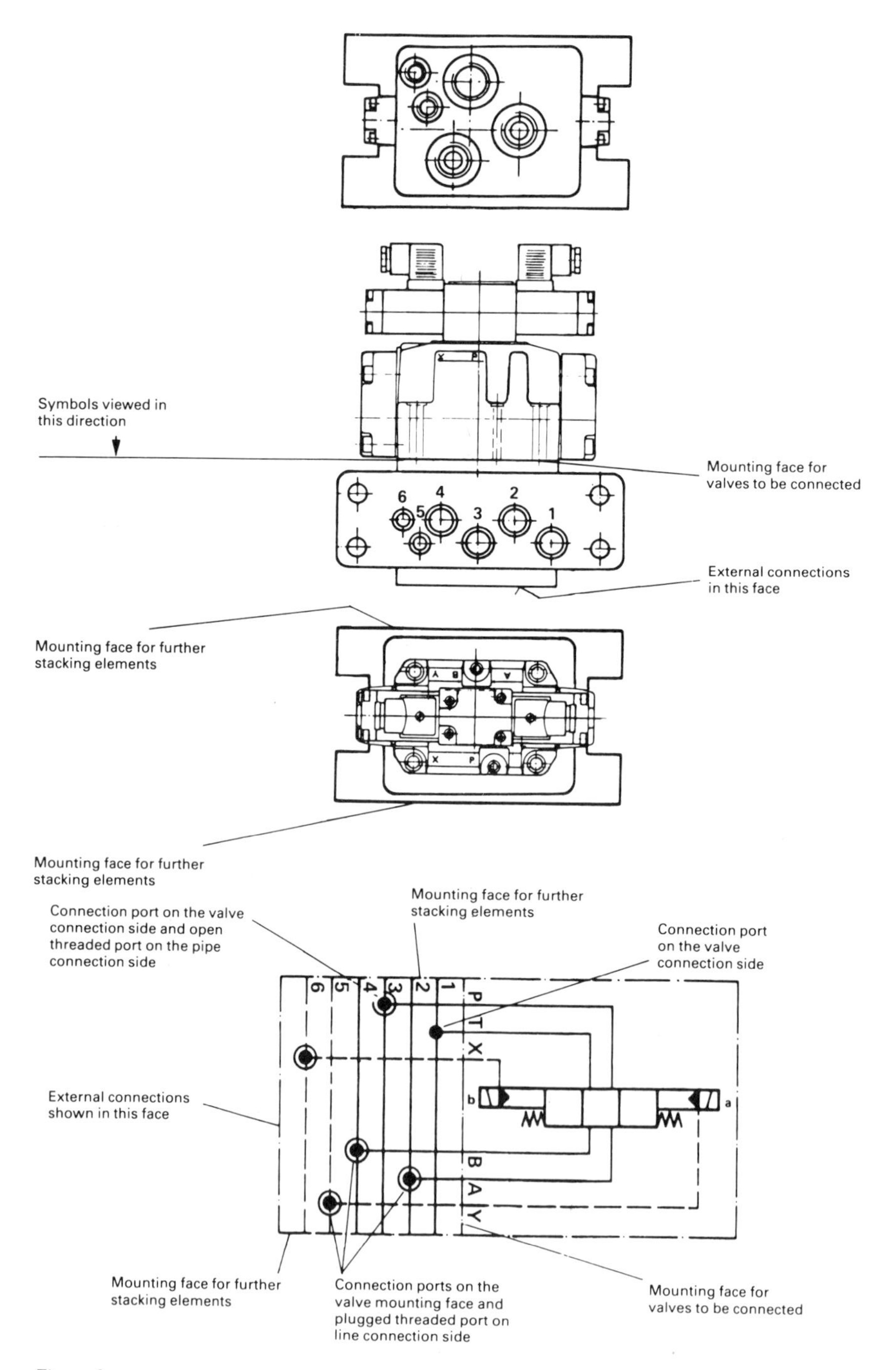


*Figure 3*
*Drawn sections of stacking assemblies.*

Problems can also be experienced in making sure that all the drilling swarf has been removed from the block after drilling, as this could contaminate a hydraulic system at a later date.

Due to the lack of flexibility of changing valve positions and the problems of fault diagnosis, manifold blocks should only be used on proven systems rather than a prototype or trial systems where changes may have to be made.

Manifold blocks are not very suitable for high flow systems either, as the pressure drops are usually higher than through pipework, due to the small bore drillings meeting at right angles in the manifold block.

**Cartridge valves**

Following the acceptance of manifold blocks, it was found that space could be saved by mounting check and relief valves in the drilled passage ways in the block, in the form of screw-in cartridges. The poppets or balls of the valves can either seat directly onto a face machined in the block or onto a hardened and ground seating face which is fitted into a recess machined in the block. The latter course is preferred for aluminium alloy blocks and for applications where high wear of the valve seats is anticipated.

*Examples of valve cartridges.*

As the cartridge valves became more popular manufacturers were able to produce a range of valves for pressure, flow and directional control in cartridge form. These can be fitted into recesses in the users own manifold block which has to be drilled to specific dimensions and tolerances.

Alternatively, manufacturers are able to supply the cartridges already mounted in compact rectangular valve blocks machined from high strength aluminium alloy or steel.

The advantage of this type of construction is that valve blocks can be

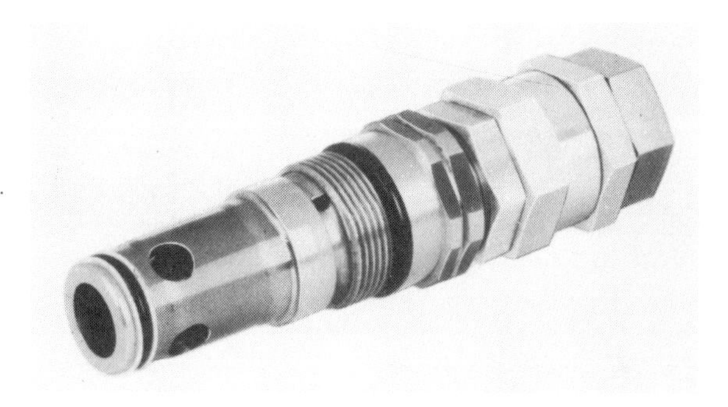

*Cartridge pressure relief valve.*

manufactured for standard circuit functions combining the use of cartridges for pressure, flow and directional control requirements. Numerous examples of this are found in hydrostatic transmission valve blocks, for use in closed loop circuits with a charge pump.

The block can contain cross-line dual relief cartridges set at the maximum allowable system pressure, a charge pump relief cartridge to control the boost pressure and shuttle valve and check valve cartridges to relieve the hot oil from the low pressure side of the loop. The valve blocks can be specially-manufactured to fit directly on either the pump or the hydraulic motor, or standard blocks are produced for mounting at an intermediate position, as shown in Figure 4.

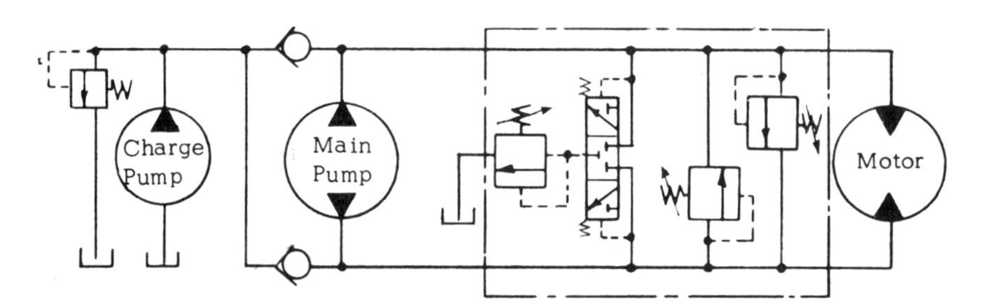


*Figure 4*
*Hydrostatic transmission valve manifold block.*

Modern cartridge valves usually contain their own hardened valve seats and O-ring seals and are inserted in the drilled passage ways and retained by threaded and sealed valve caps.

As with manifold blocks, pressure drops through the cartridge valve blocks can be higher, than through *cast pipe-mounted* or surface-mounted valves, where the internal passages can be shaped with smooth contours.

# Valve Operation

IT HAS been shown in Section 2, in the chapter on *Control Methods,* that the operation of valves can be manual and mechanical or operated by hydraulic pilot, electrical solenoid, servo-feedback signals and other remote controls.

**Manual Valves**

Manually-operated valves come into the category of valves which are either lever-operated, to change the valve position or opening, or are manually-adjusted, by means of a handwheel or screw, to set a pressure level or orifice size.

The most common lever-operated valves in use today are the *multiple-banked mobile* valves, which have already been described, and which have a separate lever for each spool section. The levers are often splayed out so that an operator can control up to six levers with one hand.

*Manually-operated valve with sealed lever linkage.*

On mobile valves used in arduous conditions the spools are protected from corrosion by chrome-plating.

Manually-operated valves used for marine or offshore applications should have the linkage between the spool and the lever totally enclosed and sealed from the corrosive salt water elements.

Manual override buttons are often built into electric solenoid valves, to enable the valve to be operated while commissioning the system, before the electric supply has been connected. They can also be used for emergency operation of the valve if the electric supply has failed. For marine and exposed valves the manual overrides should also be sealed to prevent ingress of moisture or dirt to the valve spool.

Mechanically-operated valves are similar to manual valves but are usually fitted with a cam following roller at the end of the spool or at the end of a lever. Similar precautions must be taken to seal the exposed end of the spool if used in arduous conditions, although mechanically-operated valves are usually found on machine tools with lubricated cams and rollers.

**Pilot-Operated Valves**

Pilot-operated valves are either single- or two-stage valves where a pilot pressure is generated within the valve to operate the valve spool. Two-stage pressure control valves are used so that quick response can be obtained from a large spool by using pilot pressure generated from a smaller spring-loaded poppet valve.

Alternatively, pilot pressure can be externally supplied to the valve from a pilot pressure source which is maintained at a constant level.

Large pilot-operated directional valves remotely-operated by manual or mechanical pilot valves have mainly been superseded by two-stage solenoid-operated pilot valves. This is due to the ease of running electrical cables from switches compared with the complexity of running small bore hydraulic pilot lines from the manual or mechanical operated selector valves.

Even in hazardous areas, flameproof solenoids or intrinsically safe solenoids are often preferred, due to the risk of damaging vulnerable hydraulic pilot lines. Long distance pilot lines have the additional problems of a pressure drop build up for medium pilot pressures and the occurrence of shock waves in the pilot fluid giving erratic main valve operation. The use of solenoid-operated pilot valves also allows the modern electrical or electronic computer generated signals to be used.

**Solenoid Valves**

Solenoids may be of the 'dry' or 'wet' type. In general, wet solenoids can be smaller for the same duty because of their lower static and dynamic friction. They also have the advantage that all moving parts are enclosed and lubricated and seals between the solenoid and valve body are eliminated.

They are described as *glandless* valves because, by arranging the solenoid armature to work in a sealed tube with the solenoid coil enveloping it, the sealing glands can be dispensed with, so simplifying the construction and eliminating one possible point of leakage. This principle has been applied extensively to the smaller valves, an example of which is shown in Figure 1.

This is a typical double solenoid directional valve, which can operate at pressures up to 315 bar (4500 lbf/in$^2$) and with flow rates up to 38 l/min (8.3 gal/min) and mounts on an ISO size 03 mounting surface to ISO 4401. The body is made of ductile iron to provide for superior wear characteristics which is especially important with 95/5 high water-based fluids.

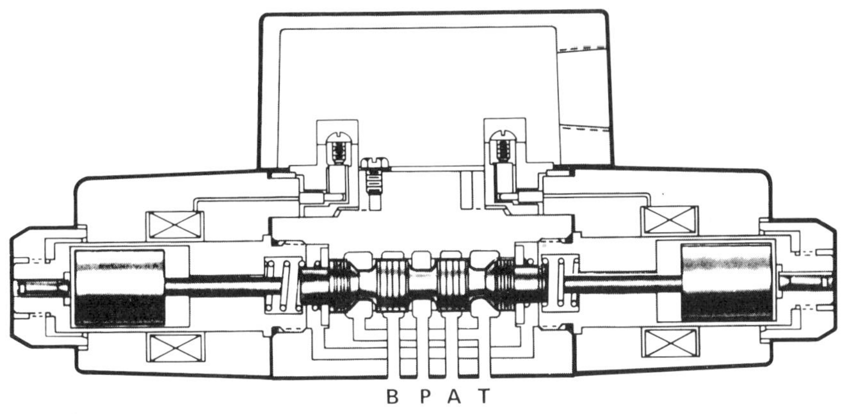


*Figure 1*
*Solenoid-operated valve.*

A standard body is machined with a range of different spool configurations which are interchangeable in the body. A four-land steel spool permits manufacturing of consistantly rounder spools for better balance in the bore and lower spool wear and leakage and U-shaped grooves keep the spool centred in the bore for longer spool life.

The solenoid coils can plug in over the armature tubes, for ease of servicing, and can be replaced without disturbing the hydraulic system or wiring cavity. Optional pin type or top-side plug-in electrical connectors can be supplied for easy valve replacement, with or without signal lights to indicate that the coil is energized. Dual frequency (50/60 Hz) two wire coils can be supplied to give the user a lower inventory of spare coils.

**Remote Control**
A remote 'joy stick' type control unit is used with either electro-hydraulic proportional solenoid valves or with pilot-operated 'load sensing' control valves.

In many mobile applications, such as mobile cranes, excavators and other similar equipment, the control valves are mounted on the machine with direct manual or solenoid proportional controls. By having an additional remote control station an operator can control the machine from an external position.

The control station can be sited in any position with no limitations between it and the position of the directional control valves and flexible electric cables can replace cumbersome pipe systems or mechanical linkages.

The joy stick control unit can even be mounted in a portable control box which can be strapped to the operator to enable him to move into different positions while operating the machine. The portable unit, which could weigh as little as 4 kg (8.8 lb) is easily plugged into different control stations with an electrical connector and cable.

The joy stick lever unit, shown in Figure 2, is a hand-operated co-ordinate lever which regulates the electric control signals. It can simultaneously control three proportional potentiometer-controlled signals. The lever is primarily intended for controlling electro-hydraulic converter valves which control mobile directional valves by means of hydraulic pilot pressure.

However, the lever unit can be used for any application where a lever movement

*Figure 2*
*Three proportional control, levers, electronic control units and banked directional control valves.*

is to be converted into electric signal. Linear lever and foot pedal control units are available to control one proportional potentiometer-controlled signal in each direction. The linear lever can be used alone or stacked to the required amount of functions.

The electro-hydraulic converter valve is an electrically-controlled pressure reducing valve designed for proportional remote control of different hydraulic functions. The output signal is a proportional hydraulic control pressure suitable for controlling directional valves, pump regulators and similar devices. The output control pressure is practically linearly proportional to the input current within the recommended control range and when fed with pulsating dc current the valve can have practically no hysteresis and a high pressure accuracy.

The joy stick lever unit, linear lever and foot pedal control units can be supplied as pilot pressure control valves which are piped directly to the pilot-operated proportional directional valve and a selection of these are shown in Figure 3.

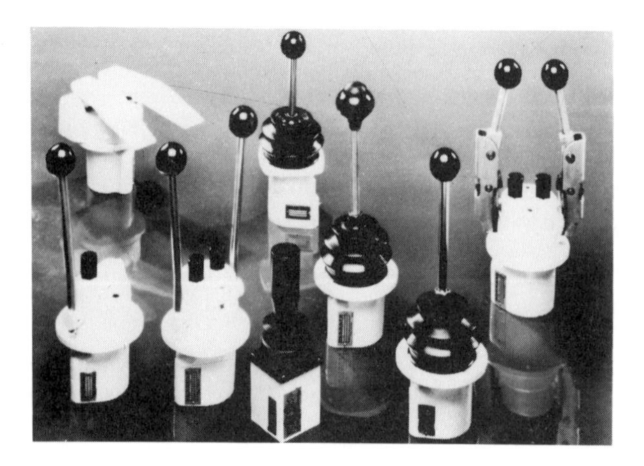

*Figure 3*
*Remote pilot control units.*

The largest stock
of both seamless and welded
steel tubes for cylinde
Honed or roller burnish
alta spa
Via Guevara, 3
21040 Abbiate Guazzone Tradate
Tel. 0331-849.377 (4 linee ric. aut.)
Telex 334233 ALTA I
Tubi per cilindri · Tubes for cylinders · Zylinderroh

**Servo valves**

The basic parameters of the hydraulic servo system are described in Section 2 *(Servo Control)*, which also describes the types of servo-valves and their first-stage configurations. Most servo-valves use the same construction for the second stage, *ie* a hardened spool operating in a hardened sleeve or bore. This makes them resistant to erosion of the control edges of the spool lands.

However, over long periods of operation, even in relatively clean fluid, erosion will occur. This erosion produces an effective underlap of the valve, increasing the flow gain and leakage around the null point. The effect of a given degree of erosion is dependent on the stroke of the valve: the longer the stroke of the spool, the smaller the effect of the erosion. This is because a longer stroke means a narrower port; hence less leakage and less change in flow gain. A longer stroke also reduces the proportion of the valve's operating range over which erosion effects occur.

The null stability of a servo-valve is controlled by the stability of its torque motor. Null shift can be caused by mechanical movement of components but, except in extremely high acceleration environments, this is very unusual.

The null stability of a servo-valve is controlled by the stability of its torque motor. Null shift can be caused by mechanical movement of components but, except in extremely high acceleration environments, this is very unusual.

Null shifts are usually caused by small internal mechanical movements in the torque motor from stress changes and by the stress changes themselves in the magnetic circuit. It is important, therefore, that a torque motor is chosen with a construction which minimizes these effects.

When the torque motor is constructed, the magnets may be attached to the frame by screws or by welding. Both methods impose initial stresses and those imparted by welding can be released by adequate temperature cycling.

*Miniature electro-hydraulic servo-valve.*

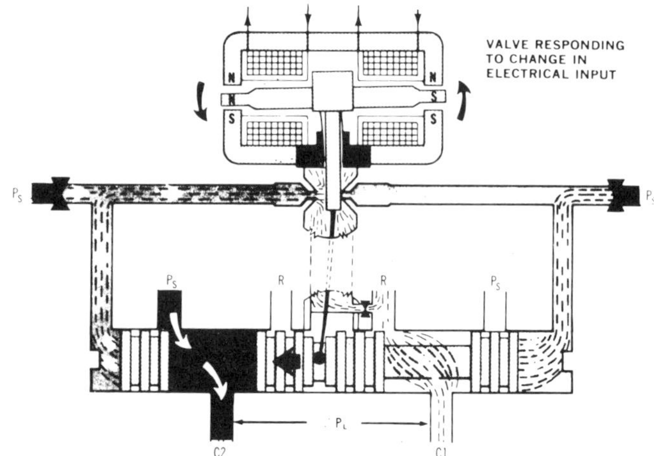


**OPERATION**

- electrical current in torque motor coils creates magnetic forces on ends of armature
- armature and flapper assembly rotates about flexure tube support
- flapper closes off one nozzle and diverts flow to that end of spool
- spool moves and opens $P_S$ to one control port; opens other control port to R

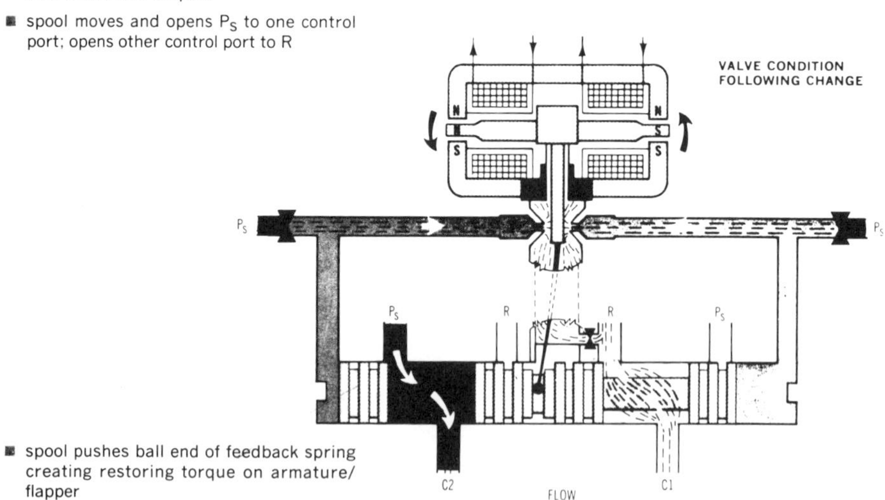


- spool pushes ball end of feedback spring creating restoring torque on armature/flapper
- as feedback torque becomes equal to torque from magnetic forces, armature/flapper moves back to centered position
- spool stops at a position where feedback spring torque equals torque due to input current
- therefore spool position is proportional to input current
- with constant pressures, flow to load is proportional to spool position

*Figure 4*
*Principles of operation of a flapper type servo-valve.*

The same cannot be done completely for screwed parts, because they rely on stress and strain for retention. Over long periods and through thermal cycles and other strains, the stresses in bolted parts will change, causing changes in the magnetic circuit and some small movements of components, thus causing null shift.

After thermal stabilization of a servo-valve, it is necessary to re-adjust it. It is important that the method of adjustment should not re-introduce the stresses just relieved. A biasing spring on the torque motor physically isolated from the magnetic circuit is a safe way of providing this feature.

There are two methods of closing the loop between torque motor and second stage spool. The two systems are: displacement follow up and feedback of a force proportional to spool position.

The follow-up system demands a torque motor or force motor displacement proportional to input current. This displacement moves a flapper device to a new fixed position. The resultant pressure unbalance moves the spool which, either directly or through a mechanism, moves the nozzles until the original null relationship is re-established.

There are two principle short-comings to this method. It is difficult to make an electro-magnetic device with a completely linear displacement/current characteristic and the internal-loop gain is very poor because it is limited by the 1:1 follow-up system. This tends to increase hysteresis and threshold response.

The force-feedback system avoids these difficulties. It consists of a spring between the second stage spool and the element driven by the armature – either jet-pipe or flapper. The receiver or nozzles are fixed, providing a first-stage null condition at fixed geometrical position of the armature. Thus, with a linear spring between spool and armature, a linear-force output electrical signal is required from the torque motor. This is relatively easy to provide at one fixed position.

*Figure 5*
*Flapper type servo-valve.*

The internal-loop gain can be readily adjusted by adjusting the change on the magnet. This has a large effect on the displacement-force gain of the torque motor, but only a small effect on the force/current characteristic. The principle is simple but requires some care in execution as the movements are small: a valve passing 22.5 l/min (5 gal/min) has a spool displacement of 1.5mm (0.06 in). One way is to attach one end of the feedback spring to the torque motor member (jet-pipe or flapper) with a ball on the end fitting into a slot in the spool. This requires an extremely close match between ball and slot.

If a clearance or backlash does develop, there is an area where the spool can drift to and fro with no possible control over it, because there is no feedback intelligence between second and first stage.

The forces on the ball can cause some friction which can degrade threshold, resolution and repeatability. Another way is to attach it rigidly at both ends and provide an elastic pivot to accommodate angular deflections.

The general principles of operation of a flapper type servo-valve are shown in Figure 4 and cut-way view of an actual valve is shown in Figure 5.

# Valve Selection

MANUFACTURERS' CATALOGUES will give values of maximum operating pressure, flow rate and temperature range for each valve, which obviously must not be exceeded without the manufacturers permission. The main parameter used for selecting a valve however, is the pressure drop through the valve.

**Pressure Drop**

Values of pressure drop or pressure loss through a valve are usually presented graphically as a curve of pressure drop against fluid flow for a particular fluid viscosity and temperature as shown in Figure 1.

If a valve is going to be used with a fluid of a different viscosity or at a much lower temperature, then this must be taken into consideration when assessing the allowable pressure drop through the valve.

With directional valves or sandwich-mounted valves, where the flow is in more than one direction through the valve, then the pressure drop must be considered in each direction and added together to give the total pressure drop.

Care should be taken when assessing the pressure drops in systems containing cylinders with large differential areas, as very large return flows can be generated from relatively small pump flows into a small annular volume.

The pressure drop must be calculated for each valve in the system and then added to the total pressure drop through the pipework to see if this value is within the allowable pressure drop for the system.

It must be remembered that, although modular-mounted valves and manifold blocks give a more compact installation, pressure drops through these assemblies are generally higher than through pipe-mounted valves. For this reason modular and manifold valves are usually associated with lower flow rates or miniature valve assemblies.

The *maximum operating pressure* stated relates to the maximum system working pressure as set by the main system relief valve or device. Most valves are tested at a pressure of 1.5 times the maximum operating pressure, with an additional safety factor before failing, so that transient pressure peaks are acceptable in a normal hydraulic system.

On some directional valves there is a limitation of maximum pressure on the tank return port, where this port drains the end caps and operating mechanism and this should be noted when applying the valve.

**Performance curves** (measured at v = 36 cSt and t = 50 °C)

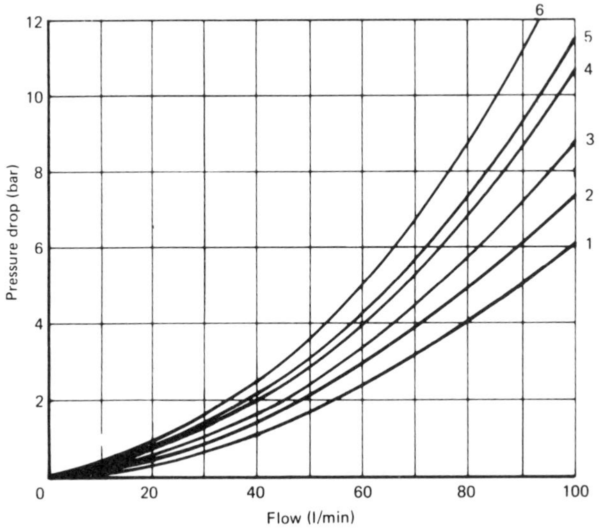

| Spool type | Flow direction P-A | P-B | A-T | B-T |
|---|---|---|---|---|
| A + B | 2 | 2 | – | – |
| C | 2 | 2 | 3 | 3 |
| D + Y | 2 | 2 | 3 | 4 |
| E | 1 | 1 | 2 | 2 |
| F | 1 | 2 | 2 | 2 |
| G + T | 4 | 4 | 5 | 6 |
| H | 1 | 1 | 3 | 4 |
| J | 2 | 2 | 1 | 2 |
| L | 2 | 2 | 1 | 3 |
| M | 1 | 1 | 2 | 3 |
| P | 1 | 1 | 3 | 3 |
| Q | 2 | 2 | 3 | 3 |
| R | 2 | 5 | 2 | – |
| U | 2 | 2 | 2 | 2 |
| V | 2 | 2 | 2 | 3 |
| W | 1 | 1 | 2 | 2 |
| Neutral position | | B-T | A-T | P-T |
| F | – | – | 1 | 3 |
| G + T | – | – | – | 4 |
| H | – | – | – | 1 |
| P | – | – | – | 4 |
| L | – | – | 5 | – |
| U | – | – | – | 6 |
| Switching position | P-A | B-A | A-T | P-T |
| R | – | 4 | – | – |

*Figure 1*
*Typical spool valve pressure drop curves.*

**Flow rate**

The maximum flow rate for a valve is again specified for a particular fluid viscocity which is usually taken as a normal hydraulic oil at an average operating temperature. If the maximum flow rate is exceeded, not only will the pressure drop become excessive but flow generated forces set up within the valve could cause it to malfunction or become unstable.

When selecting variable orifice valves for controlling flow rate it is best to choose a valve with the required flow rate at the middle of its flow range so that the pressure compensator is working approximately in mid-position.

The minimum pressure drop across the valve should not be less than about four times that across the throttle to maintain good response. If very low flows have to be controlled in a leg of a system where high flows are also required, a small flow control valve would have to be piped in parallel with a larger flow path to enable the low flows to be controlled.

In poppet type relief and check valves, high flows produce an high pressure override characteristic which can be overcome by using two-stage pilot-operated valves.

When selecting sliding spool valves, the *leakage* flow rate across the spool from the high pressure to the low pressure ports must also be considered. This information is usually available from the manufacturers data and some valves have selectively assembled spools and bodies so that minimum leakage flows can be specified.

Applications where minimum leakage flow is essential are where suspended loads have to be supported by counterbalance valves and where differential creep can take place in a horizontal cylinder.

The latter phenomenon is caused when pressure is maintained on the closed ports of a sliding spool directional valve controlling a single rod cylinder. As there is leakage across the blocked centre spool, pressure is generated in both ends of the cylinder and, due to the differential areas of the piston, a force is experienced equal to the pressure multiplied by the rod cross sectional area. This force tends to extend the cylinder rod and over a period of time the rod will creep forward at a rate determined by the leakage flow from the tank return port of the valve.

It must be remembered that, although sliding spool valves may be specified with minimum leakage values, the leakage flow will increase with time as the valve spool wears and clearances become larger. Also, although the use of low viscosity fluids will reduce the pressure drop, the leakage rate with these fluids will increase.

The leakage flow from spool type pressure control valves can either be internally drained to the low pressure port or externally drained through a separate drain line if pressure at the low pressure port will affect the operation of the valve.

**Temperature**

The operating temperature range specified for a valve is usually related to the temperature range of the fluid being used. High or low ambient temperatures will affect the valve as well as the operating temperature of the fluid.

Very large valves with close clearances, may be affected by thermal shock if subjected to sudden changes in temperature, as differential expansion or contraction of materials can cause changes in clearances. In this case the system should be designed so that the temperature is stabilized through out.

Valves which continuously bleed high pressure oil to tank, such as pressure reducing valves and pressure maintaining valves, are a source of temperature build up in a system, as the lost energy is converted directly into heat in the fluid.

In open loop hydraulic systems the reservoir is usually sized to dissipate the heat generated in the system, as well as allowing the fluid to settle, even if an additional cooler is added. The custom of mounting manifold blocks and modular assemblies on top of the reservoir can reduce heat disssipation from the tank.

Problems have been experienced when changing a proven system, with panel mounted sub-plate valves, piped with rigid piping, to a manifold modular assembly. It had not been appreciated how much heat was being dissipated from the pipework which was behind the vertical panel and was acting as a radiator.

When the system was converted to a modular manifold assembly, mounted on the reservoir, the extra pressure drops in the manifold and the inability of the solid steel assembly to dissipate the heat generated, meant that an extra cooler was required to stabilize the system temperature.

Conventional hydraulic fluids begin to break down at relatively moderate temperatures, and even below the break-down point will show accelerated deterioration. Mineral oils have a somewhat higher temperature rating than water-based fluids and can give satisfactory service at fluid temperatures of up to about 140 °C (280 °F). Phosphate ester fluids can be used up to about 150 °C (300 °F); but neither type is particularly suited for working at high temperatures. Hydrolytic stability is often suspect at temperatures above normal working figures. Many otherwise satisfactory fluids may break under such conditions; others become excessively acid and promote corrosive attack or show an excessive rate of oxidation.

All fluids, therefore, can be given maximum temperature ratings for continuous duty consistent with normal service life and a higher short-term rating for intermittent duty. The latter will result in reduced life, depending on the severity of the over-rating. Typical values are:

| | **Continuous** | **Short term maximum** |
|---|---|---|
| Water | 38 to 50 °C (100 to 120 °F) | 65 °C (190 °F) |
| Mineral oils | 50 to 65 °C (120 to 150 °F) | 120 to 140 °C (250 to 280 °F) |
| Water/oil emulsions | 50 to 65 °C (120 to 150 °F) | 65 °C (150 °F) |
| Water-glycol | 50 to 65 °C (120 to 150 °F) | 70 °C (160 °F) |
| Phosphate esters | 65 to 82 °C (150 to 180 °F) | 150 °C (300 °F) |
| Chlorinated aromatics | 95 °C (200 °F) | 150 °C (300 °F) |
| Silicones | up to 288 °C (550 °F) | 316 °C (600 °F) |

In general it is desirable to maintain fluid temperatures substantially below recommended maximum operating values as this will have a beneficial effect on fluid life. Also the use of synthetic fluids for higher temperature service may show unexpected limitations. It is a characteristic of synthetic lubricants that if they reach an excessively high temperature (as could occur at some localized point in the system) they can vaporize without leaving any residual surface film – *ie* leave surfaces dry and unlubricated.

**Fluid Compatibility**

Most modern industrial hydraulic valves are designed to work with a mineral oil fluid which lubricates and protects the internal steel components from corrosion.

If valves are selected to work with water or a high percentage water-based fluid then special internal corrosion protection must be included. The leakage rate and sliding action of spool valves is based on a required film thickness of lubricating fluid between the spool and body, so seated or poppet type valves are preferred for water or high water content systems.

The main question of fluid compatibility, which is usually considered when selecting valves, is the compatibility of the fluid with the seal material. This is dealt with in the chapter on *Properties of Hydraulic Fluids* in Section 1 and most manufacturers supply valves with alternative seals to match most hydraulic fluids which are commercially available.

The type of seal used is normally included in the valve code number on the name-plate, so that the correct seals are used if the valve has to be changed at a later date. Problems have been encountered where a mineral oil system was changed to a fire-resistant phosphate ester-based fluid at a later date, as not only did all the seals in the system have to be changed, but internal paintwork was also affected by the fluid.

**Mounting Style**

It has been shown how surface and manifold mounted valves, with interchangeable interfaces to ISO standards, are replacing pipe-mounted valves. When selecting valves, the advantages and disadvantages of the mounting method should always be borne in mind.

As well as the compactness of manifold and modular valves, for fairly standard circuits, safety can also be improved in hazardous applications by manifold mounting valves directly onto the actuator. This applies to counterbalance valves or pilot-operated check valves supporting crane jib cylinders.

Any failure of pipework, hoses, or the hydraulic system down-stream of the cylinder support valve will not cause the load to fall if the support valve is manifold mounted directly onto the cylinder head.

Servo-valves are also usually always manifold mounted onto a servo-actuator, as the trapped volume of fluid between the valve and actuator is an important parameter defining the valve's performance and stability.

Pipe-mounted valves can be compact on small bore systems and are retained on mobile hydraulic systems, where it is advantageous to group the hydraulic units in different positions on a vehicle, with hose connections between moving components.

The mounting style chosen should obtain the best compromise for mounting a valve, so that surface mounting, modular-, manifold- and pipe-mounted valves are combined in a system to the best advantage.

# Installation and Commissioning

TO ENSURE satisfactory operation of hydraulic valves, the manufacturer's installation and operating instructions must be observed. The following general instructions apply to most hydraulic valves, unless otherwise specified in certain cases by the manufacturers, and are subject to the regulations existing in various countries where valves are used. Certain local conditions under which the hydraulic valves are used may affect maintenance and operation.

The fluid used in the system must be of a type and degree of cleanliness acceptable to the valve manufacturer.

**Installation**

Hydraulic valves should be installed in accordance with the drawings and instructions issued by the manufacturer, particular attention being paid to the associated pipework and mounting of the valves. No stress from the fixing points or from the pipes must be allowed to cause distortion of the valves or their components.

Where sealing compounds are to be used, such as PTFE tape, they should only be of the type recommended by the valve manufacturer. Sealing compound such as hemp and putty are not permissable for hydraulic systems. The pipelines should be fastened in such a manner that they cannot vibrate or move unduly.

All equipment should be checked against the circuit diagram and any other specified requirements. With all electro-hydraulic equipment, voltage, current and frequency should be checked against the name-plates.

**Commissioning**

Before filling the system with the specified hydraulic fluid a careful check should be carried out to ensure that the interior of the system has been thoroughly cleaned. If this not the case then the system should be flushed with a flushing fluid at a flow rate of at least twice the maximum flow rate in the system, until the required cleanliness is obtained.

*Filling* should be carried out in accordance with manufacturers instructions. The system should be filled with a hydraulic fluid with specified properties and with a viscosity and temperature range suitable for the valves and equipment in the system. Wherever practicable the fluid should be pumped from the barrel into the reservoir through a filter of the correct micron size.

On completion of the filling operation valves should be shut or opened as appropriate and any air inclusion must be eliminated by bleeding, because it may endanger the proper function of the plant.

The pump or pumps should be started off-load, with an open flow path for the fluid to circulate or with relief valves set at their lowest settings, if possible. During the initial running period the system should be again thoroughly bled at fixed bleed points or other suitable connections.

Pressure adjustment of the valves should be made according to the instructions on the circuit diagram. Starting at the minimum possible pressure, each valve should be checked for operation and external leakage. This check should be continued until the valve operating pressure has been reached.

It is good practice to lock and seal the adjustment after setting, to avoid unauthorized re-adjustment. Where it would be dangerous to operate valves directing flow to actuators, the valves would be operated against shut isolating valves.

While the system is being operated the fluid level in the reservoir should be checked and maintained within the limits indicated on the reservoir or level guage. After starting, installed filters must be cleaned at intervals as dictated by experience. Contamination is being generated at a high rate during the running-in period and the system should get progressively cleaner, after a period of time, providing external contaminant is not introduced.

After an extended operating period under normal working conditions a check should be made that the recommended system temperature range is not being exceeded and that all connections and fixing bolts are correctly tightened.

**Maintenance**

The use of planned maintenance at fixed intervals using a log book is strongly recommended. After an operating period specified by the manufacturer all hydraulic valves must be checked for external leakage and satisfactory operation. Periodical maintenance should then be carried out at regular intervals. During overhaul work the manufacturers recommendations should be observed.

The fluid should be checked for deterioration and contamination and be replaced or cleaned whenever necessary. All the necessary steps should be taken to prevent any dirt from penetrating into the equipment during overhauls. Seals, packings and damaged or worn parts must be replaced. Pipe connections and fastenings should be checked and, if necessary, replaced or re-tightened.

Overhauls are best effected by using replacement valves and returning the original equipment to the manufacturer for servicing. The use of an operating instructions booklet is recommended and the external cleaning of the valve before it is removed prevents dirt from entering the system and helps the manufacturer to service the valve when it is returned. This is where the benefits of surface-mounted valves can be appreciated.

Correctly fitted hydraulic valves will operate without trouble for many years. In the event of any fault being detected, the cause should be determined and eliminated as soon as possible. This will be facilitated if a circuit diagram of the plant is available and one should be kept with the operating instructions booklet or the maintenance log book.

For short *shut-down* periods ( approximately one to two months) the hydraulic fluid may be left in the system. For more prolonged periods it may be more advisable to drain off the hydraulic fluid and re-fill and flush with a suitable temporary protective fluid in order to preserve the system.

Alternative measures can be taken on stopping the system for a period, depending on environmental conditions, hydraulic fluid and seals, *etc*. When hydraulic oil is kept in the system it may be useful to run the system from time to time to wet the interior as protection against corrosion.

When re-starting a system after a long period of inactivity the following precedure should be followed. Any preservatives and contaminants should be removed and a sample of fluid taken and examined. Water or sludge deposits should be removed also or the fluid changed and re-filled if necessary.

The strainers, filters and magnetic separators should be examined and cleaned or replaced if necessary. The system should be re-started as previously described with a check for external leaks and the functioning of the complete system.

# SECTION 5

## Pneumatic Principles

BASIC THEORY
TERMS AND DEFINITIONS
PROPERTIES OF AIR AND GASES
COMPRESSIBLE GAS FLOW

# Basic Theory

THE MAIN difference between the application of pneumatic and hydraulic power is that pneumatics uses atmospheric air, which is a compressible gas, as the fluid medium instead of an hydraulic liquid, which is assumed to be basically incompressible.

In an hydraulic power system the liquid is pumped around the circuit and pressure is generated wherever there is a resistance to the fluid flow. This pressure may be modified by relieving the resistance flow where required, or the pressure may be quickly generated by creating a resistance to the fluid flow.

In a pneumatic power system, the air is compressed and stored in a receiver under pressure, in an analogous manner to the early dead weight hydraulic accumulator system. When a valve is opened to allow the compressed air to move an actuator, the air expands according to one of the gas laws, depending on the rate of expansion and the temperature conditions. When the movement has been completed by the cylinder, or actuator, the air is exhausted to the atmosphere, thus obviating the need for return pipelines and the attendant pressure losses associated with hydraulic systems.

**Pressure**

Although air is essentially a mixture of gases, for general purposes it is assumed as being a simple gas containing one type of molecule only. These molecules move rectilinearly in all directions at considerable speed, and are in continuous collision with one another, impinging on immersed surfaces or container walls.

This impingement gives rise to a force acting on the surface and the force per unit area is known as the pressure, usually expressed in units of gf/cm$^2$ or lbf/in$^2$ *etc.* Atmospheric pressure varies according to height above or below sea level, decreasing with increase in altitude and increasing progressively below sea level. The pressure of the atmosphere at sea level may be taken as 762 mmHg or 14.7 lbf/in$^2$. In SI units this is equal to 1.013 bar where 1 bar = 105 N/m$^2$ = 105 Pascals (Pa).

As Pascals' laws state that the atmosphere pressure is acting at any point and in all directions, pneumatic pressure is measured by a pressure gauge as a pressure above atmospheric with atmospheric pressure equal to zero. This is known as *gauge pressure* and absolute pressure is guage pressure plus atmospheric pressure.

It is sometimes convenient to express compressed air pressure as being a multiple of atmospheric pressure and pressure is then quoted as being equal to the requisite number of 'atmospheres' (atm), *eg:*

2 atm absolute pressure = 2 x 1.013 = 2.026 bar.

It must be noted that 2 atm absolute pressure = 2 − 1 = 1 atm guage pressure or 1.013 bar.

In the case of pneumatic systems, the maximum pressure available from a compressed air supply is of the order of 7 bar (100 lbf/in$^2$). Compressed air is seldom produced and worked at higher pressures, except for specialized applications, because of the practical difficulties involved both in compressing and utilizing an essentially 'elastic' fluid at high compression ratios.

**Humidity**

Water vapour is constantly evaporating from lakes, rivers and seas and is absorbed by the atmosphere and carried by winds, until finally being condensed in the form of rain, mist, *etc.* Atmospheric air, therefore, is used by nature to transmit large quantities of water all over the earth.

The maximum quantity of water vapour which can be absorbed by air is dependent upon its condition of temperature and pressure. This maximum increases considerably with increasing temperature, but decreases as the pressure increases.

For given conditions of temperature and pressure, air will therefore sustain a maximum water vapour pressure and under these conditions it is said to be saturated.

*Relative humidity* is the ratio of the amount of water present in a given quantity of air, to the maximum possible amount which it can contain under the same conditions of pressure and temperature, and the ratio is usually expressed as a percentage. If the atmospheric humidity is stated as being 50%, this means that the amount of water in the atmosphere is half of the water contained in saturated air at the same temperature and pressure.

The water vapour in the air has to be removed from a compressed air system by various methods. The large surface area of the receiver cools the air from the compressor, thus, a portion of the moisture in the air is separated as water directly from the receiver.

While coarse separation of condensate is effected in a separator after the re-cooler, fine separation filtering and other subsequent treatment of the compressed air is dealt with at the point of usage. The compressed air can be dried by *absorption* drying, *adsorption drying* or low temperature drying.

**Cylinder Performance**

The fact that a pneumatic cylinder utilizes a relatively low pressure fluid has the advantage that cylinder construction can be simplified, reducing cost. The two great advantages of the air cylinder, however, are its speed response and the ease of control of both force and speed. Pneumatics is very much faster than hydraulics and independent of changes in fluid viscosity.

On the other hand, the elastic nature of air sets further problems where precise control is required, so that the actual performance achieved is very much dependent on the nature of the output load. This makes simple air cylinders generally unsuitable for powering movements where steady motions are required, or close control of movement.

It is also difficult to stop an air cylinder at any part of its stroke, except by direct mechanical braking and latching. These can be considered relatively minor limitations as to the general usefulness of pneumatic cylinders, however, because of the many advantages they can offer for a vast number of applications not requiring 'rigid' movements.

All pneumatic systems are often suitable for applications under conditions for which hydraulics are basically unsuited, *eg* at high ambient temperatures.

Pneumatics can worked at relatively high ambient temperatures with no particular modifications, except to ensure that suitable seals and constructional materials are employed.

Pneumatics is widely associated with low cost automation, where the three other forms of power available are air, hydraulics and electricity. Pneumatics generally offers the lowest initial and operating costs with simple, flexible control systems where pulling, pushing, lifting, positioning, clamping and lever motions are concerned.

To obtain the greatest advantage from the introduction of automation by pneumatics, however, it is necessary to pay particular attention to achieving maximum economy of air and maximum operating efficiency of the pneumatic components.

Hydraulic systems are self-lubricating, whereas pneumatic systems normally require the introduction of lubricant, in the form of oil mist or oil spray, and this, together with an air filter/regulator, can be a vital parameter in determining the over-all efficiency of the system. Although the air itself is free, the cost of compressing air can be relatively high, hence the importance of maximum use of the processed medium.

**Performance Calculations**

The theoretical maximum thrust obtainable from an air cylinder is equal to the product of the air pressure and the effective piston area *viz:*

(1) Single-acting cylinders or outward stroke of double-acting cylinders:

$$F = 0.7854 \times D^2 \times Pe$$

(2) return stroke of double-acting cylinder:

$$F = 0.7854 \times (D^2 - d^2) \times Pe$$

Where F = thrust force
D = cylinder bore diameter
d = rod diameter
Pe = effective air pressure in consistent units.

The theoretical maximum thrust is normally quoted for the static thrust rating of an air cylinder. This will be substantially correct provided that the effective air pressure rather than the nominal line pressure is used, *ie* an allowance should be made for the pressure drop in the line to the cylinder.

A typical figure for line losses in a simple circuit may range between 5% and 10% of the nominal pressure. More accurate values for static thrust rating, therefore, can be obtained from the following modified formulae.

Static thrust – case (1) $F = 0.7 \times D^2 \times P$
case (2) $F = 0.7 \times (D^2 - d^2) \times P$

Where P is the line pressure.

Typical values of static thrust are summarized in Tables 1 and 2. These may be used to estimate typical maximum 'break-out' forces available from air cylinders.

Under dynamic conditions, friction and other mechanical losses are introduced and, in practice, possible variations in the effective pressure. For cylinders in good condition, typical friction losses can be expected to lie between 5% and 15% or a mean of 10%, which is the figure commonly adopted as a working figure.

**TABLE 1 – TABLE OF FORCES FOR DOUBLE-ACTING AIR CYLINDERS – SI UNITS**

| Cylinder bore in mm | Piston rod direct-tion | Maximum theoretical force in N at effective air pressures of 1–10 bar (100–1000 kPa) | | | | | | | | | |
|---|---|---|---|---|---|---|---|---|---|---|---|
| | | 1 | 2 | 3 | 4 | 5 | 6 | 7 | 8 | 9 | 10 |
| 12 | + | 11 | 22 | 34 | 45 | 56 | 68 | 79 | 90 | 101 | 113 |
| | – | 10 | 21 | 32 | 42 | 51 | 62 | 72 | 82 | 92 | 103 |
| 16 | + | 20 | 40 | 60 | 80 | 100 | 140 | 160 | 180 | 200 | |
| | – | 19 | 36 | 52 | 70 | 87 | 105 | 122 | 140 | 168 | 175 |
| 20 | + | 31 | 62 | 94 | 126 | 154 | 188 | 220 | 251 | 285 | 314 |
| | – | 29 | 51 | 85 | 114 | 140 | 170 | 199 | 237 | 258 | 284 |
| 25 | + | 49 | 98 | 147 | 196 | 245 | 294 | 343 | 392 | 441 | 490 |
| | – | 38 | 76 | 114 | 152 | 190 | 228 | 266 | 304 | 342 | 380 |
| 32 | + | 80 | 160 | 240 | 320 | 400 | 480 | 560 | 640 | 720 | 800 |
| | – | 69 | 138 | 207 | 276 | 345 | 414 | 483 | 552 | 621 | 690 |
| 40 | + | 126 | 252 | 378 | 504 | 630 | 756 | 882 | 1008 | 1134 | 1260 |
| | – | 106 | 212 | 318 | 424 | 530 | 636 | 742 | 848 | 954 | 1060 |
| 50 | + | 200 | 390 | 590 | 780 | 980 | 1180 | 1370 | 1570 | 1760 | 1960 |
| | – | 170 | 330 | 500 | 660 | 830 | 990 | 1160 | 1320 | 1490 | 1650 |
| 63 | + | 310 | 620 | 940 | 1250 | 1560 | 1870 | 2180 | 2500 | 2810 | 3120 |
| | – | 280 | 560 | 840 | 1120 | 1400 | 1680 | 1960 | 2240 | 2520 | 2300 |
| 80 | + | 500 | 1000 | 1510 | 2010 | 2510 | 3010 | 3510 | 4020 | 4520 | 5020 |
| | – | 450 | 910 | 1360 | 1810 | 2270 | 2720 | 3170 | 3620 | 4080 | 4530 |
| 100 | + | 790 | 1570 | 2360 | 3140 | 3930 | 4710 | 5500 | 6280 | 7070 | 7850 |
| | – | 710 | 1410 | 2220 | 2820 | 3530 | 4230 | 4940 | 5640 | 6350 | 7050 |
| 125 | + | 1230 | 2450 | 3680 | 4910 | 6140 | 7360 | 8590 | 9820 | 11040 | 12270 |
| | – | 1150 | 2290 | 3440 | 4580 | 5730 | 6880 | 8020 | 9120 | 10310 | 11460 |
| 150 | + | 1760 | 3520 | 5280 | 7040 | 8800 | 10560 | 12320 | 14080 | 15840 | 17600 |
| | – | 1660 | 3320 | 4970 | 6730 | 8300 | 9950 | 11620 | 13250 | 14920 | 16600 |
| 160 | + | 2010 | 4020 | 6030 | 8040 | 10050 | 12060 | 14070 | 16080 | 18090 | 20100 |
| | – | 1890 | 3770 | 5650 | 7540 | 9420 | 11300 | 13190 | 15070 | 16960 | 18850 |
| 200 | + | 3140 | 6280 | 9430 | 2570 | 15710 | 18850 | 21990 | 25140 | 28280 | 31420 |
| | – | 3020 | 6030 | 9050 | 2060 | 15080 | 18100 | 21110 | 24130 | 27140 | 30160 |
| 250 | + | 4920 | 9840 | 14760 | 18550 | 24600 | 29520 | 34340 | 39360 | 44280 | 49200 |
| | – | 4640 | 9270 | 13900 | 19680 | 23200 | 27800 | 32500 | 37100 | 41750 | 46400 |
| 300 | + | 7080 | 14100 | 21240 | 28320 | 35400 | 42480 | 49560 | 56640 | 63720 | 70800 |
| | – | 6700 | 13400 | 20100 | 26800 | 33500 | 40200 | 47000 | 53700 | 60200 | 67000 |
| 350 | + | 9620 | 19240 | 28860 | 38480 | 48100 | 57720 | 67340 | 76960 | 86580 | 96200 |
| | – | 9120 | 18200 | 27300 | 36500 | 45500 | 54700 | 63700 | 72800 | 82000 | 91200 |
| 400 | + | 12580 | 25160 | 37740 | 50320 | 62900 | 75480 | 88060 | 100640 | 113220 | 125800 |
| | – | 11800 | 23600 | 35400 | 47200 | 59000 | 70080 | 82750 | 94500 | 106300 | 118000 |
| 450 | + | 15900 | 31800 | 47700 | 63600 | 79500 | 95400 | 111300 | 127200 | 143100 | 159000 |
| | – | 14950 | 29000 | 44800 | 59750 | 74750 | 89600 | 104500 | 118500 | 134200 | 149500 |
| 500 | + | 19650 | 38300 | 58950 | 78600 | 98250 | 117900 | 137550 | 157200 | 177750 | 196500 |
| | – | 18250 | 37000 | 55600 | 74200 | 92700 | 112000 | 130500 | 143200 | 167800 | 185200 |

1 N (Newton) = 0.1 kp (kilopond)
1 bar = 100 kPa
The values in the table give the forces for the plus and minus movements at air pressure of 1–10 bar (100–1000 kPa).

In fact, frictional losses may not exceed 5% in the case of horizontal cylinders in good condition, although an over-estimate rather than an under-estimate of losses is to be preferred in order to avoid selecting an undersized cylinder.

For this reason, air cylinders are commonly sized on the basis of a theoretical thrust rating of 25% to 50% in excess of the working force required, thus covering frictional losses, leakage losses and likely pressure drop through the inlet port.

## TABLE 2 – FORCES FOR DOUBLE-ACTING AIR CYLINDERS – US UNITS

| Cylinder bore in | Piston rod direction | Maximum theoretical force in pound-force (lbf) at effective air pressures of 20–140 lb/in$^2$ | | | | | | | | | | | | |
|---|---|---|---|---|---|---|---|---|---|---|---|---|---|---|
| | | 20 | 30 | 40 | 50 | 60 | 70 | 80 | 90 | 100 | 110 | 120 | 130 | 140 |
| 0.47 | + | 3.5 | 5.3 | 7.0 | 8.8 | 10.5 | 12.3 | 14.0 | 15.8 | 17.5 | 19.3 | 21.0 | 22.8 | 24.5 |
| | – | 3.1 | 4.7 | 6.2 | 7.8 | 9.4 | 10.9 | 12.5 | 14.0 | 15.6 | 17.2 | 18.7 | 20.3 | 21.8 |
| 0.63 | + | 6.2 | 9.4 | 12.5 | 15.6 | 18.7 | 21.8 | 25.0 | 28.1 | 31.2 | 34.3 | 37.4 | 40.6 | 43.7 |
| | – | 5.4 | 8.0 | 10.7 | 13.4 | 16.1 | 18.8 | 21.4 | 24.1 | 26.8 | 29.5 | 32.2 | 34.8 | 37.7 |
| 0.79 | + | 9.8 | 14.6 | 19.5 | 24.4 | 29.3 | 34.2 | 39.0 | 43.9 | 48.8 | 53.7 | 58.6 | 63.4 | 68.3 |
| | – | 7.8 | 11.7 | 15.6 | 19.6 | 23.5 | 27.4 | 31.3 | 35.2 | 39.1 | 43.0 | 46.9 | 50.8 | 54.7 |
| 0.98 | + | 15.2 | 22.8 | 30.4 | 38.0 | 45.6 | 53.2 | 60.8 | 68.4 | 76.0 | 83.6 | 91.2 | 98.8 | 106.4 |
| | – | 11.7 | 17.6 | 23.4 | 29.3 | 35.1 | 40.9 | 46.8 | 52.6 | 58.5 | 64.3 | 70.2 | 76.0 | 81.9 |
| 1.26 | + | 24.9 | 37.4 | 49.8 | 62.3 | 74.8 | 87.2 | 99.7 | 112.1 | 124.6 | 137.1 | 149.5 | 162.0 | 174.4 |
| | – | 21.4 | 32.2 | 42.8 | 53.5 | 64.3 | 74.9 | 85.7 | 96.3 | 107.1 | 117.1 | 128.5 | 139.2 | 149.9 |
| 1.57 | + | 38.9 | 58.4 | 77.8 | 97.3 | 116.7 | 136.2 | 155.6 | 175.1 | 194.5 | 214.0 | 233.4 | 252.9 | 272.3 |
| | – | 37.2 | 49.1 | 65.4 | 81.7 | 98.0 | 114.4 | 130.7 | 147.1 | 163.4 | 179.8 | 196.1 | 212.5 | 228.8 |
| 1.97 | + | 60.8 | 91.2 | 122 | 152 | 182 | 213 | 243 | 274 | 304 | 334 | 365 | 395 | 426 |
| | – | 51.1 | 76.6 | 102 | 128 | 153 | 179 | 204 | 230 | 255 | 281 | 306 | 332 | 357 |
| 2.48 | + | 96.6 | 145 | 193 | 242 | 290 | 338 | 386 | 435 | 483 | 531 | 580 | 628 | 676 |
| | – | 86.9 | 130 | 174 | 217 | 261 | 304 | 347 | 391 | 434 | 478 | 521 | 565 | 608 |
| 3.15 | + | 156 | 234 | 312 | 390 | 467 | 545 | 623 | 701 | 779 | 857 | 935 | 1013 | 1091 |
| | – | 141 | 211 | 281 | 351 | 422 | 492 | 562 | 633 | 703 | 773 | 844 | 914 | 984 |
| 3.94 | + | 244 | 366 | 488 | 610 | 732 | 854 | 976 | 1098 | 1220 | 1844 | 1466 | 1588 | 1710 |
| | – | 219 | 329 | 438 | 548 | 657 | 767 | 876 | 986 | 1095 | 1205 | 1815 | 1424 | 1534 |
| 4.92 | + | 380 | 570 | 760 | 950 | 1140 | 1330 | 1520 | 1710 | 1900 | 2090 | 2280 | 2470 | 2660 |
| | – | 355 | 533 | 710 | 888 | 1065 | 1243 | 1420 | 1598 | 1775 | 1953 | 2131 | 2308 | 2486 |
| 6.30 | + | 624 | 936 | 1248 | 1560 | 1872 | 2184 | 2496 | 2808 | 3120 | 3432 | 3744 | 4056 | 4368 |
| | – | 585 | 878 | 1170 | 1463 | 1755 | 2048 | 2340 | 2633 | 2926 | 3218 | 3511 | 3803 | 4006 |
| 7.87 | + | 975 | 1462 | 1948 | 2430 | 2924 | 3410 | 3896 | 4378 | 4860 | 5354 | 5848 | 6334 | 6820 |
| | – | 912.5 | 1370 | 1825 | 2290 | 2740 | 3195 | 3650 | 4115 | 4580 | 5030 | 5480 | 5935 | 6390 |
| 9.84 | + | 1525 | 2290 | 3060 | 3805 | 4580 | 5350 | 6120 | 6865 | 7610 | 8385 | 9160 | 9930 | 10700 |
| | – | 1438 | 2155 | 2880 | 3530 | 4310 | 5035 | 5760 | 6410 | 7060 | 7840 | 8620 | 9345 | 10070 |
| 11.81 | + | 2195 | 3310 | 4400 | 5480 | 6620 | 7710 | 8800 | 9880 | 10960 | 12100 | 13240 | 14330 | 15420 |
| | – | 2075 | 3120 | 4170 | 5280 | 6240 | 7290 | 8340 | 9460 | 10580 | 11530 | 12480 | 13530 | 14580 |
| 13.78 | + | 2980 | 4475 | 5970 | 7440 | 8950 | 10445 | 11940 | 13410 | 14880 | 16390 | 17900 | 19395 | 20890 |
| | + | 2825 | 4230 | 5660 | 7075 | 8460 | 9890 | 11320 | 12735 | 14150 | 15535 | 16920 | 18350 | 19780 |
| 15.75 | + | 3870 | 5840 | 7820 | 9760 | 11680 | 13660 | 15640 | 17580 | 19520 | 21440 | 23360 | 25340 | 27320 |
| | – | 3655 | 5480 | 7330 | 9150 | 10960 | 12810 | 14660 | 16480 | 18300 | 20110 | 21920 | 23770 | 25620 |
| 17.72 | + | 4925 | 7380 | 9850 | 12340 | 14760 | 17230 | 19700 | 22190 | 24680 | 27100 | 29520 | 31990 | 34460 |
| | – | 4630 | 6740 | 9275 | 11580 | 13480 | 16015 | 18550 | 20855 | 23160 | 25060 | 26960 | 29495 | 32030 |
| 19.69 | + | 6080 | 9130 | 12200 | 15240 | 18260 | 21330 | 24400 | 27440 | 30480 | 33500 | 36520 | 39590 | 42660 |
| | – | 5740 | 8610 | 11500 | 14480 | 17220 | 20110 | 23000 | 25980 | 28960 | 31700 | 34440 | 37330 | 40220 |

The values in the table give the forces for the plus and minus movements at air pressures of 20–140 lb/in$^2$

Adopting a middle figure, corresponding thrust formulae are:

case (1) $F = 0.55 \times D^2 \times P$
case (2) $F = 0.55 \times (D^2 - d^2) \times P$

where P is the line pressure as above.

Bearing in mind that selection would normally be made from standard production sizes of cylinders rather than theoretical sizes, Table 3 data is also useful as a rapid means of estimating the effect of varying cylinder diameter for a given line pressure. This has been calculated for 'production' sizes in imperial units, but can also be read for metric units and is based on a thrust factor of 1.0 for 1 in (25mm) diameter cylinder.

**TABLE 3 – CYLINDER THRUST FACTORS**

| BORE | | THRUST FACTOR |
|---|---|---|
| (inches) | (mm) | |
| 1 | 25 | 1.0 |
| 1¼ | 32 | 1.56 |
| 1½ | 38 | 2.25 |
| 1¾ | 45 | 3.06 |
| 2 | 50 | 4.0 |
| 2½ | 63.5 | 6.25 |
| 3 | 76 | 9.0 |
| 3½ | 90 | 12.5 |
| 4 | 100 | 16.0 |
| 4½ | 115 | 19.0 |
| 5 | 127 | 25.0 |
| 5½ | 140 | 30.25 |
| 6 | 152 | 36.0 |
| 7 | 178 | 49.0 |
| 8 | 203 | 64.0 |
| 9 | 230 | 81.0 |
| 10 | 254 | 100 |
| 11 | 280 | 121 |
| 12 | 305 | 144 |

The thrust factor for any other size of cylinder then gives the thrust of that cylinder in terms of the thrust of a 1 in cylinder, *eg* a 4 in cylinder would develop 16 times the thrust on the same line pressure. Equally, to compare the thrust of any two sizes of cylinders, their thrusts can be determined as the ratio of their thrust factors.

**Air Consumption**

Air consumption is determined directly from the geometry of the cylinder.

Air consumption = piston area x stroke x strokes per minute in consistent units. It is usual to express consumption in terms of a single stroke:

*ie*: Air consumption per stroke = piston area (A) x stroke (L)

$$\text{Air consumption (l)} = \frac{0.7854 \times D^2 \times L}{10^3} \quad \text{(1) outward stroke}$$

$$= \frac{0.7854 \times (D^2 - d) \times L}{10^3} \quad \text{(2) inward stroke}$$

where bore (D) rod diameter (d) and stroke (L) are in cm.

$$\text{Air consumption (ft}^3\text{)} = \frac{D^2 \times L}{2200} \quad \text{(1) outward stroke}$$

$$= \frac{(D^2 - d^2) \times L}{2200} \quad \text{(2) inward stroke}$$

where bore, rod diameter and stroke are in inches.

For practical purposes it is necessary to render air consumption in terms of free air delivered (FAD). This is determined by multiplying the volumetric displacement given by the above formulae by the compression ratio,

$$\text{Where compression ratio} = \frac{P + Pa}{Pa}$$

Where P = line pressure
Pa = atmospheric pressure

$$\text{Compression ratio} = \frac{P + 1.013}{1.013} \quad \text{where P is in bar}$$

For most practical purposes this can be simplified to (P + 1).

$$\text{Compression ratio} = \frac{P + 14.7}{14.7} \quad \text{where P is lbf/in}^2$$

Corresponding working formulae:

Air consumption FAD in litres

$$= \frac{0.7854 \times L D^2 (P + 1)}{10^3} \quad \text{(1) outward stroke}$$

$$= \frac{0.7854 \times L (D^2 - d^2)(P + 1)}{10^3} \quad \text{(2) inward stroke}$$

where P is in bar.

Air consumption FAD in $ft^3$

$$= \frac{L D^2 (P + 14.7)}{32\ 300} \quad \text{(1) outward stroke}$$

$$= \frac{L (D^2 - d^2)(P + 14.7)}{32\ 300} \quad \text{(2) inward stroke}$$

where P is in $lbf/in^2$

In the case of a single-acting cylinder the effective FAD requirement per stroke is given by a single calculation (1), unless the cylinder is a through-rod type, when (2) applies.

In the case of a double-acting cylinder the effective FAD requirement per complete stroke is given by summing the requirements of (1) and (2), unless the cylinder is of the through-rod type, when the total consumption per stroke is twice that given by (2). The consumption in terms of litres/min (FAD) or $ft^3$/min (FAD) then follows by multiplying by the number of strokes completed per minute.

In the case of double-acting cylinders the reduction in air consumption on the inward stroke is commonly ignored and taken as twice that of the outward stroke.

It is often more convenient to render consumption directly in terms of the operating time (t) to complete a stroke, when the corresponding formulae are:

Consumption (FAD) in $m^3$/min

$$= \frac{D^2 \times L \times (P + 1)}{t \times 10^4}$$

where D and L are in mm,
t is in seconds and
P is in bar

Consumption (FAD) in $ft^3$/min

$$= \frac{0.00186\, D^2 \times L \times (P + 14.7)}{t}$$

where D and L are in inches
t is in seconds and
P is in $lbf/in^2$

It follows that the actual speed of operation has a significant effect on the economics of operation, both as regards to air consumption and optimum port and valve sizes. In practice, without throttling, piston velocity is finally restricted by the pressure drop through the inlet port, which in turn is determined by the size of the port. With small ports and light external loading, pressure drops of approximately 2 to 3 bar (30 to 45 lbf/in$^2$) may readily be realized, roughly equivalent to halving the output force available from normal line pressures.

Cylinders will, therefore, normally have a characteristic speed performance determined by their bore size and port size (*eg* see Figure 1). Throttling is usual to control cylinder operating speed to reduce wear and tear and air consumption.

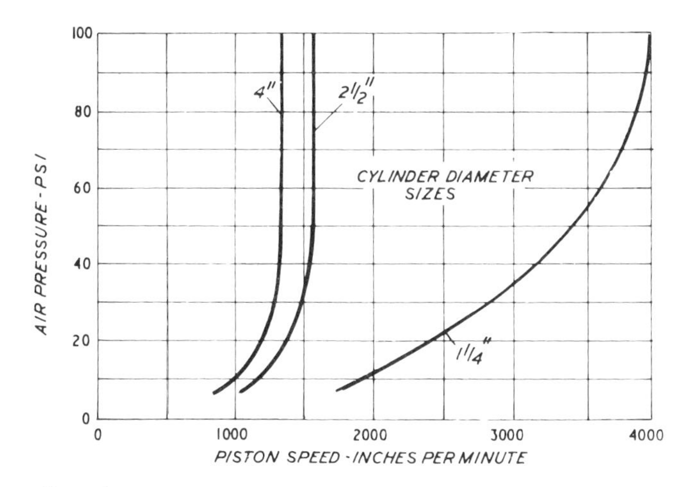


*Figure 1*

The simplest method of speed control is to reduce the supply pressure *via* a pressure regulator. This will, of course, also reduce the output force, and the amount of reduction needed to reduce speed could drastically reduce the force available. This again depends on the specific speed characteristics.

Thus, in Figure 1 the smaller cylinder woud be amenable to speed reduction *via* a pressure regulation over a fairly wide range. In the case of the two larger cylinders, however, no reasonable speed reduction is achieved until the pressure has been reduced to about a quarter of normal line pressure, reducing thrust available by a similar amount.

**Circuit Calculations**

Unlike hydraulics, circuit calculations for pneumatics tend to become extremely complicated, due both to the compressibility of the working fluid and the impossibility of establishing its exact expansion characteristics. Empirical determination of valve output curves and cylinder efficiencies provides only part of the data required.

The subject is an important one, for lack of exact knowledge of actual circuit performance normally leads to the use of oversize cylinders for the actual requirements and considerably oversized valves. The latter result partly from the use of an over-sized cylinder and partly from a lack of knowledge of piston speed, the two effects being additive.

The following method starts by analysing the behaviour of a double-acting cylinder controlled by a four-way valve. Valve and cylinder performance must, of course, be considered together as shown in Figure 2 and 3.

At T = O the valve is moved to reverse the cylinder. The pressure Pm in the head end rapidly changes from 0 to the mains pressure (Pd) because the cylinder volume is at its smallest; at the same time pressure in the rod end (Pr) falls slowly, depending on the size of the valve passage and the fact that the annular volume is at its largest.

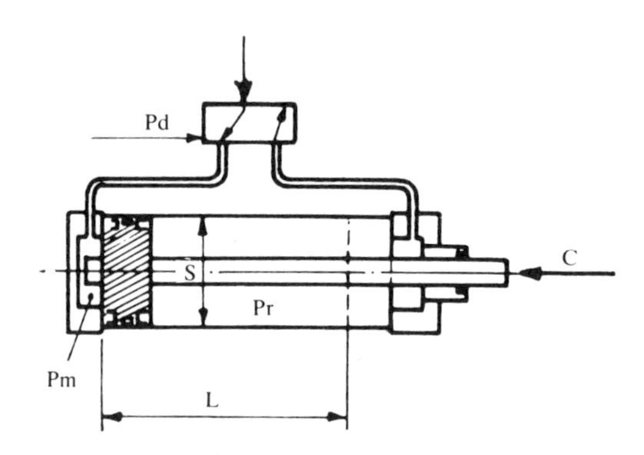


*Figure 2*

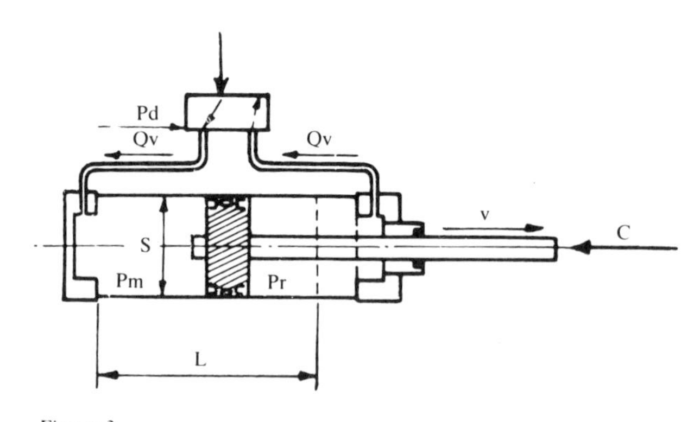


*Figure 3*

At time ($t_1$) when the difference between the pressure has reached $\frac{C + f}{S}$ the cylinder starts moving.

where Pd = mains pressure
Pm = pressure at head end
Pr = pressure at rod end
C = load
f = friction force
S = piston area.

At the end of the stroke, the piston may be stopped slowly by a cushion or suddenly, if not damped (see Figure 4).

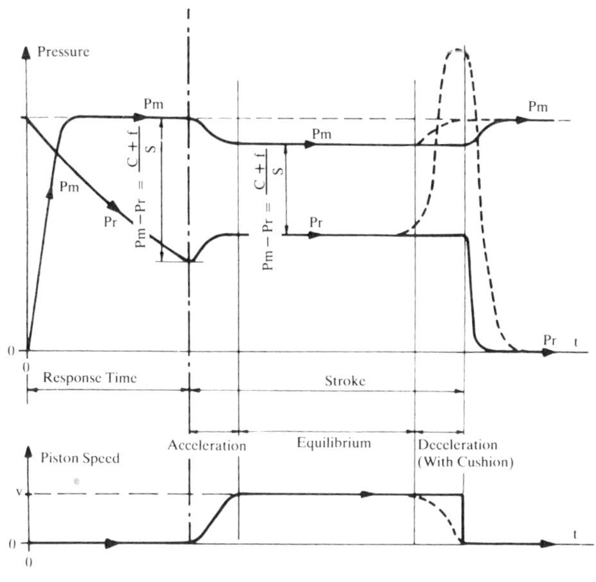


*Figure 4*

**Calculation of the Response Time**

According to the above it appears that the response time of a cylinder depends only on the rate of discharge of air from the mains pressure (Pd) to the pressure required to start the piston moving (Pr). From the exhaust time (t) for the valve (see Figure 5) it is easy to calculate the response time.

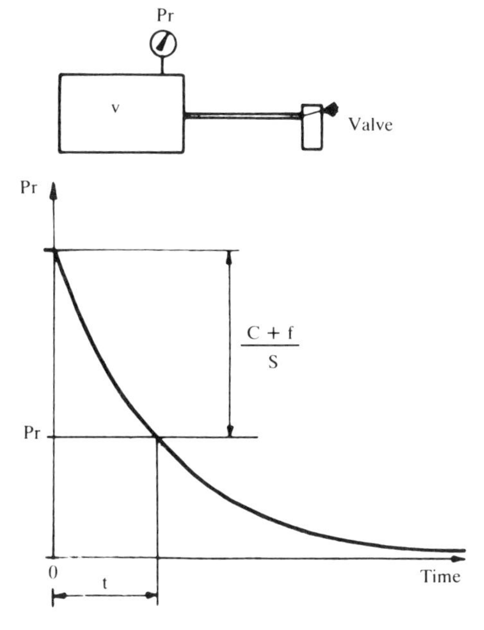


*Figure 5*

The most useful method entails drawing the discharge time curves for a given capacity and pressure. In Figure 6 this has been done for three sizes of valve connected to a capacity of 1 litre, assuming an inlet pressure of 5 bar (70 lbf/in$^2$).

These curves give the response time directly for a given piston load C + f. Knowing the discharge time between two cylinder inlet pressures, S is proportional to the volume of the capacity to be discharged; the time for the capacity under consideration is given by simple proportion.

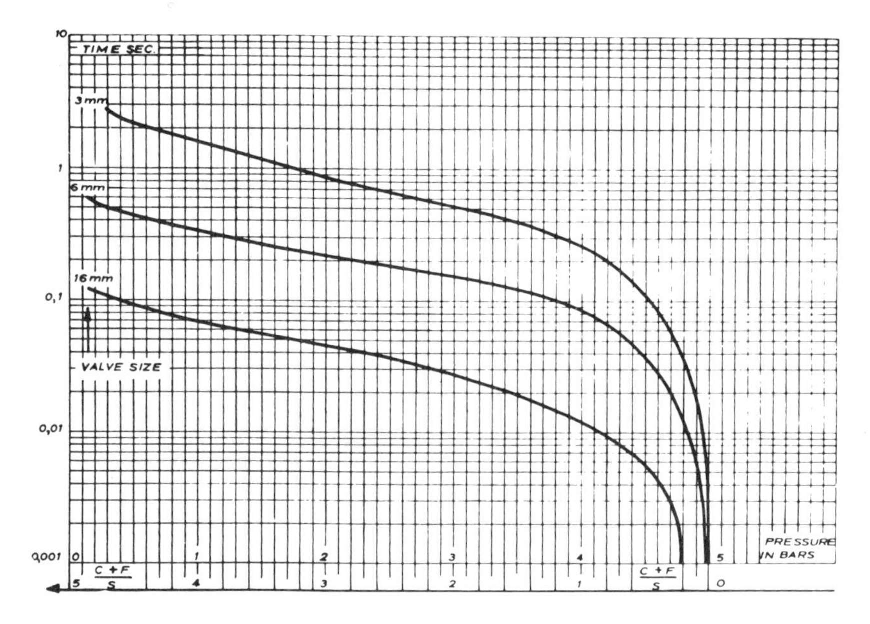


*Figure 6*

**Calculation of Time for One Stroke**

Figure 7(a) shows a double-acting cylinder with the piston in dynamic balance and moving at constant speed. Figures 7(b) and 7(c) show the circuits between valve and cylinder for the air entering and discharging, respectively.

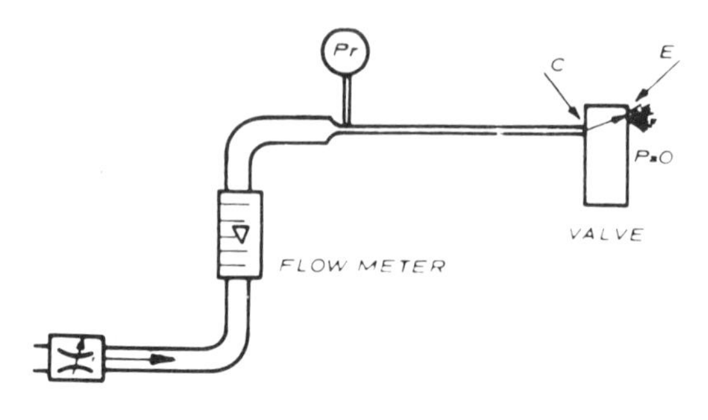


*Figure 7(a)*

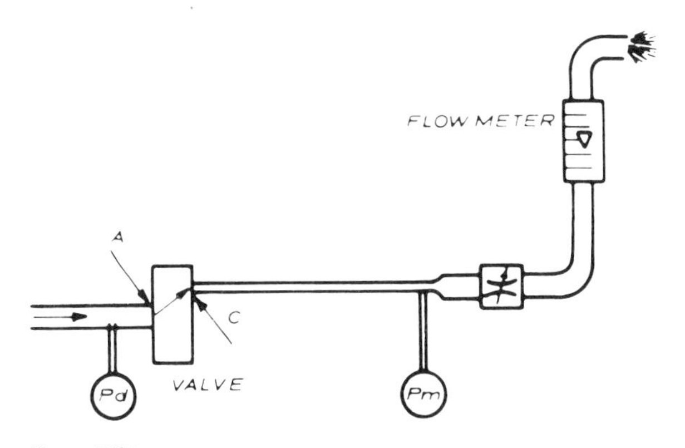


*Figure 7(b)*

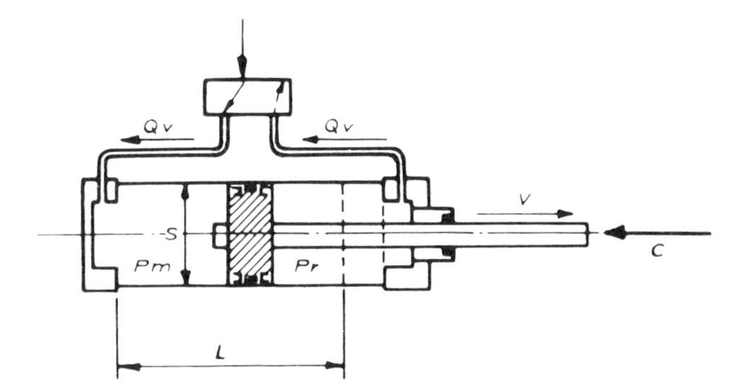


*Figure 7(c)*

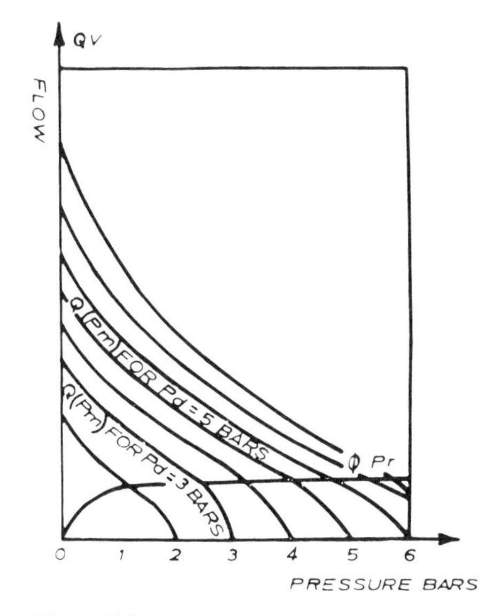


*Figure 7(d)*

Induction flow Q = f (pd.Pm) (a system of curves for varying Pd) Exhaust flow Q = f (Pr) where Q = volume flow of free air. Figure 7(d) is obtained by superimposing Q = f (Pd.Pm) on the curve Q = f (Pr).

The cylinder movement entails equal volumes of air on both sides of the piston (ignoring the effect of the rod):

Q Pm = Q Pr – see Figure 7(e).

For an unloaded and frictionless cylinder $\left(\frac{C + f}{S}\right) = 0$

Pm = Pr. Figure 7(e) has the same scale for Pm and Pr and will give the required volumetric output when Q Pr intersects Q Pm for the given inlet pressure (Pd). The piston speed is therefore given by:

$$V = \frac{Qv}{S}$$

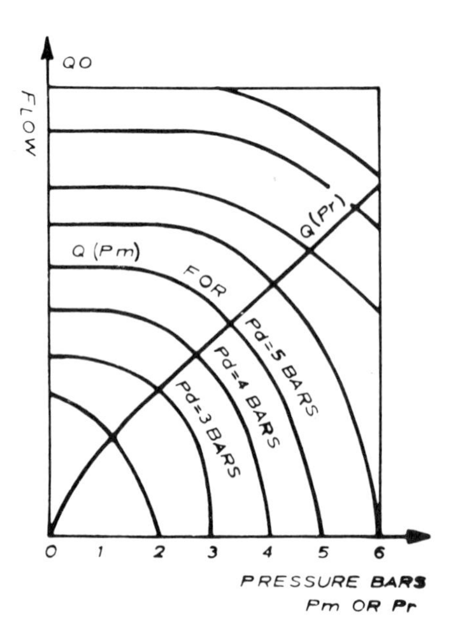


*Figure 7 (e)*

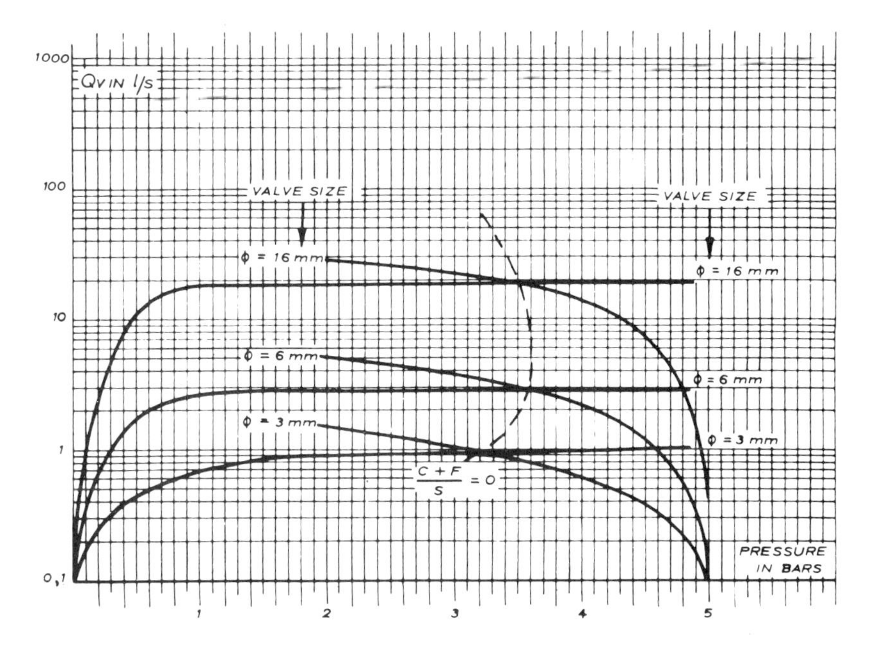


*Figure 8*

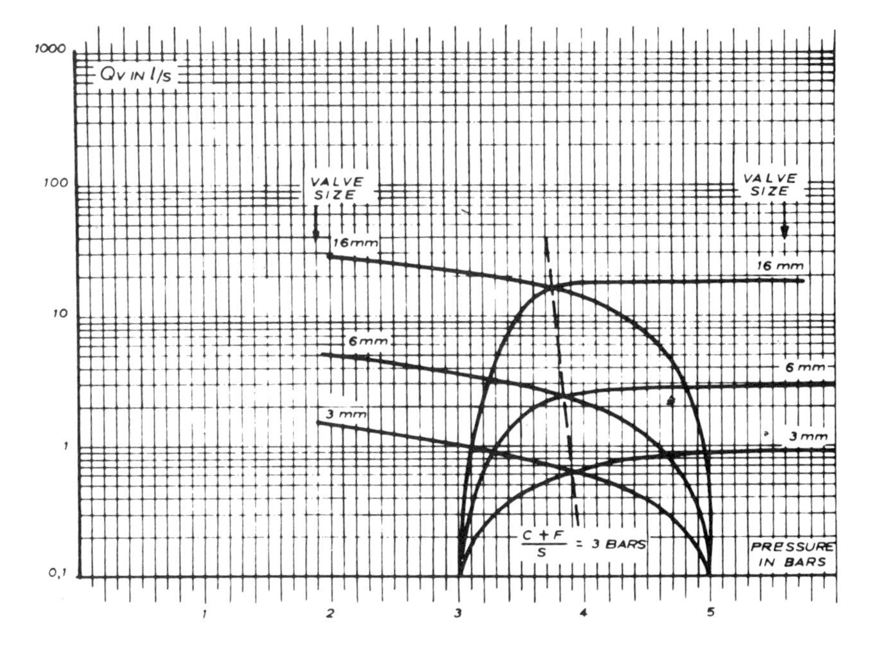


*Figure 9*

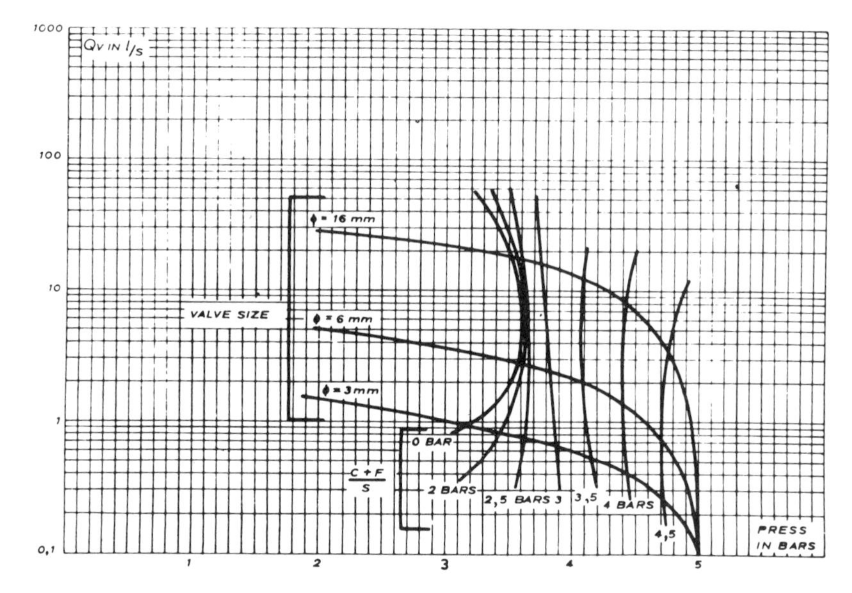


*Figure 10*

**Applying the Method**

When calculating the response time, the inlet pressure (Pd) is usually taken as fixed, but the valve size may vary.

The curves in Figure 8 are for 5 bar inlet pressure, piston unloaded $\left(\frac{C+f}{S}\right) = 0$ and valves of nominal 3, 6, and 16 mm size.

In Figure 9 similar curves are given for when $\frac{C+f}{S} = 3$ bar (45 lbf/in$^2$)

Figure 10 gives a family of similar curves for values:

$\frac{C+f}{S}$ between 0 and 5 bar (0 and 70 lbf/in$^2$).

# Terms and Definitions

THE FOLLOWING terms and definitions apply in the pneumatics industry, arranged in alphabetical order for convenience of reference (but with closely related subjects grouped so that their inter-relationship is clearly established).

**Absolute pressure** — see *Pressure*

**Absolute temperature** — see *Temperature*

**Adiabatic process** — compression of gas under conditions of no heat exchange with the surroundings.

**Aftercooling** — removal of heat from the air after compression is completed.

**Booster** — a machine or device for taking in air (or gas) and delivering it at a higher pressure.

**Capacity** — (compressor) actual volume rate of flow compressed and delivered at the discharge point at stated inlet conditions.

**Clearance volume** — the volume remaining in a compression space in a cylinder at the end of a compression stroke.

**Compressibility factor (z)** — a factor giving the deviation of the behaviour of a real gas compared with an ideal gas.

**Compression ratio** — ratio of final pressure to original pressure.

**Condensate** — liquid formed by condensation of water vapour in air due to a fall in temperature.

**Displacement** — actual swept volume, or volume displaced by a compressing device per stroke or per stage.

| | |
|---|---|
| **Displacement compressor** | compressor working on the principle of compressing air by displacement of a member (*eg* a piston in a cylinder). |
| **Dynamic compressor** | rotary machine converting kinetic energy into pressure energy |
| **Filter** | a device for removing contaminants from a fluid. |
| **Free air** | air at atmospheric condition at the inlet point of a compressor. |
| **Ideal compression** | conditions appertaining to the isentropic compression of an ideal gas. |
| **Ideal gas** | a gas which conforms exactly to the theoretical gas laws. |
| **Intercooling** | the cooling of compressed air (or gas) between compression stages. |
| **Isentropic efficiency** | ratio of real gas isentropic power consumption to shaft inlet power. |
| **Isentropic power consumption** | theoretical power required to compress a gas under constant entropy. |
| **Isothermal power consumption** | theoretical power required to compress a gas at constant temperature. |
| **Packaged compressor** | a complete, powered compressor unit ready to operate. |
| **Polytropic process** | compression or expansion of a gas where the relationship $P.V^n$ = a constant applies, the exponent $n$ having various values from 1 to 1.43. |
| **Pressure – absolute** | pressure measured from absolute zero. |
| **atmospheric** | absolute pressure of the atmosphere. |
| **dynamic** | velocity pressure or total pressure minus static pressure. |
| **guage** | pressure measured above atmospheric pressure. |
| **static** | pressure measured under static conditions for the air or gas (*ie* not influenced by velocity. |
| **total** | pressure measured at a stagnation point when a moving air-stream is brought to rest. With stationary air or gas, equals the static pressure. |
| **vacuum** | difference between atmospheric pressure and a (lower) value or absolute pressure. |

| | |
|---|---|
| **Pressure ratio** | ratio between absolute discharge pressure and absolute inlet pressure. |
| **Pressure regulator** | a valve or similar device for reducing line pressure to a lower (constant) level. |
| **Pulsation damper** | a chamber designed to eliminate pulsations present in air flow. |
| **Relative clearance volume** | ratio of clearance volume to volume swept by a compressing element. |
| **Relative humidity** | ratio of water vapour actually present in air to the saturated water vapour content, at a given temperature and pressure. |
| **Separator** | a device for removing liquids (*eg* water and oil) from compressed air. |
| **Specific energy requirement** | input energy requirement per unit of compressed air produced, or shaft input power per unit of compressor capacity. |
| **Specific power consumption** | specific energy requirement. |
| **Stage pressure ratio** | compression ratio for any particular stage in a multi-stage compressor. |
| **Surge limit** | the flow limit below which operation of a dynamic compressor becomes unstable. |
| **Temperature – absolute** | temperature above absolute zero in Kelvin. |
| **ambient** | temperature of the environment. |
| **discharge** | total temperature at the standard discharge point of a compressor. |
| **inlet** | total temperature at the standard inlet point of a compressor. |
| **total** | temperature at the stagnation point of an airstream. |
| **Volumetric efficiency** | ratio of capacity to displacement of a compressor or vacuum pump. |

## NOTES ON UNITS

The following units are now generally accepted as standard in the compressed air industry in Europe.

**Pressure** – bar (1 bar = $10^5$ N/m$^2$ = $10^5$ Pa).

**Linear** – metre (m) for dimensions of 1 metre or greater.

– millimetre (mm) for dimensions less than 1000 mm.

**Volume** – metre$^3$ (m$^3$) is the standard unit for volumes

or

**Capacity** – above 1000 litres

– litre (l) is the standard unit for volumes below 1000 litres.
Compressor and air tool manufacturers prefer l.
Pneumatic equipment manufacturers prefer $dm^3$.
Both have the same numerical value (1 litre = 1 $dm^3$).
The smaller volume unit $cm^3$ is used in fluidics.

**Velocity** – metres/second (m/s).

**Rotational speed** – r/min or rev/min (rpm) continues to be widely used, although rev/sec (r/s) is the more correct unit.

**Force-Newton** – (N).

**Mass** – kilogram (kg), or tonne (t).

**Lifting capacity** – tonne (t).

**Work, Energy or Heat** – joule (j).

**Power** – watt(W) or kilowatt (kW).
*Note:* brake horsepower (bhp) is now regarded as obsolete and if referred to is replaced by *brake power.*

**Torque** – Newton.metre (N.m).

**Specific power consumption** – kilowatt-seconds/$m^3$ (kW.s/$m^3$) or J/$dm^3$.

**Stress** – Newtons/$m^2$ (N/$m^2$), or N/$mm^2$.
*Note:* the Pascal (Pa) is also used by the high vacuum industry
(1/$mm^2$ = 1MN/$m^2$ = 1 MPa).

**Flow** – litres/second (l/s).
*Note:* for a rough approximation l/s divided by 2 = $ft^3$/min;
for a close approximation 1 l/s = 0.47 $ft^3$/min, and 1 $ft^3$/min = 2.1 l/s.

**Mass flow** – Kilograms/second (kg/s).

**Moisture content of air or gases** – kg/$m^3$, or g/$m^3$.

# Properties of Air and Gases

CLEAN, DRY air is a mechanical mixture of approximately 78% by volume nitrogen and 21% oxygen, the remaining 1% being up with minor quantities of some 14 other gases. The *composition* of air remains substantially the same between sea level and an altitude of about 20 km, but its *density* decreases with increasing altitude, and varies with pressure and temperature. At sea level, with a pressure of 1 bar and a temperature of 15 °C the density of air is 1.208 kg/m$^3$ (0.0765 lb/ft$^3$). Thus 1 kg of air has a volume of 0.0818 m$^3$ (1 lb of air a volume of 13.07 ft$^3$).

At a specific temperature and pressure, the number of molecules in a unit volume is constant for any gas or mixture of gases. For a temperature of 0 °C and a pressure of one atmosphere, this number is 2.705 x $10^{19}$ molecules per cm$^3$ (7.84 x $10^{23}$ molecules per ft$^3$).

At standard temperature and pressure, the mean velocity of gas molecules is of the order of 500 m/s with a mean free path between intermolecular collision, for air, of the order of 3 x $10^{-6}$ mm. The rate of collision under such conditions is responsible for the pressure exerted by air (or any gas) on a surface immersed in it, or on walls containing the gas. Specifically, therefore, the pressure of any gas depends on its mass density, the number of molecules present and their mean velocity.

The effect of change in temperature is to modify the value of the mean velocity. Hence, in the absence of any other change, the resultant pressure will vary with the temperature. Similarly, any change in volume will effectively modify the mass present and again affect the pressure. Thus pressure, temperature and volume are inter-related. This relationship can be expressed in the following form for a perfect gas (which assumes that the molecules are perfectly elastic, are negligible in size compared with the length of their mean free path, and exert no force on each other):

$$PV = RT$$

where P = pressure
V = volume
T = absolute temperature
R = gas constant.

In engineering units:

$$PV = RT \times 10^{-5}$$

where P is in bar
V is the specific volume of gas in $m^3/kg$
R is the gas constant in J/(kg.K) = 287.1 J/(kg.K) for air.

**Atmospheric Air**

Atmospheric air normally contains water vapour and the total pressure of the air is the sum of the partial pressures of the dry air and water vapour. The air is saturated when the partial pressure of the water vapour is equal to the saturation pressure of the water vapour at that temperature. The saturation pressure is dependent only on the temperature (see also Table 1 and Figure 1).

When air is cooled at constant pressure, the *dew point* is reached, when the partial pressure is equal to the saturation pressure. Any further cooling will then result in water separating by condensation (see also Table 2).

**TABLE 1 – VAPOUR PRESSURE OF MOIST AIR**

**(Pressure in millibars)**

| Temp. °C | Relative Humidity in Percent | | | | | | | | | |
|---|---|---|---|---|---|---|---|---|---|---|
| | 10 | 20 | 30 | 40 | 50 | 60 | 70 | 80 | 90 | 100 |
| –10 | 0 | 1 | 1 | 1 | 1 | 2 | 2 | 2 | 2 | 3 |
| –5 | 0 | 1 | 1 | 2 | 2 | 2 | 3 | 3 | 4 | 4 |
| 0 | 1 | 1 | 2 | 2 | 3 | 4 | 4 | 5 | 5 | 6 |
| 5 | 1 | 2 | 2 | 3 | 4 | 5 | 6 | 7 | 7 | 8 |
| 10 | 1 | 2 | 4 | 5 | 6 | 7 | 9 | 10 | 11 | 12 |
| 15 | 2 | 3 | 5 | 7 | 9 | 10 | 12 | 14 | 15 | 17 |
| 20 | 2 | 5 | 7 | 9 | 12 | 14 | 16 | 19 | 21 | 23 |
| 25 | 3 | 6 | 10 | 13 | 16 | 19 | 22 | 25 | 29 | 32 |
| 30 | 4 | 8 | 13 | 17 | 21 | 25 | 30 | 34 | 38 | 42 |
| 35 | 6 | 11 | 17 | 22 | 28 | 34 | 39 | 45 | 51 | 56 |
| 40 | 7 | 15 | 22 | 30 | 37 | 44 | 52 | 59 | 66 | 74 |
| 45 | 10 | 19 | 29 | 38 | 48 | 57 | 67 | 77 | 86 | 96 |
| 50 | 12 | 25 | 37 | 49 | 62 | 74 | 86 | 99 | 111 | 123 |

**Water Content of Moist Air**

The water vapour content of moist air can be determined from the following equation derived from the gas laws:

$$m_W = \frac{Q \times P_S \times V_t}{R_W \times T} \times 10^5$$

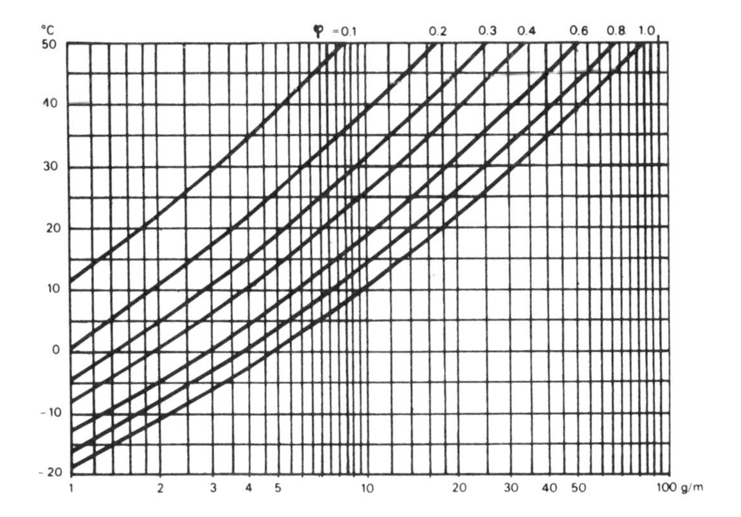


*Figure 1*
*Water content of air at different relative humidities.*

where $m_w$ = mass of water vapour in kg

Q = relative humidity (expressed as a decimal fraction)

$P_s$ = saturation pressure of water vapour in bar at given temperature (see Table 2).

$V_t$ = total volume or air/water vapour in $m^3$

$R_w$ = gas constant for water vapour = 461.3 J/(kg.K)

T = absolute temperature of air/water vapour mixture in K (= °C + 273.2).

The gas constant of the air/water vapour can also be determined as:

$$R_m = \frac{(287.1\ m_w) + (461.3\ m_w)}{(m_a + m_w)}$$

when $m_a$ = mass of dry air in kg.

From this density of the air/water vapour mixture follows as:

$$\text{density}\ (\rho) = 10^5 \times \frac{P_s}{R_m \times T}\ \text{kg/m}^3$$

**Compressibility**

For many engineering purposes the assumption that real gases conform to the perfect gas equation is satisfactory. The basic equation may, however, readily be extended to express the deviation of a real gas from a perfect gas when subject to compression, by introducing a compressibility factor (Z) when:

$$PV = ZRT$$

**TABLE 2 – SATURATION PRESSURE $\rho_s$ AND DENSITY ,$\rho_w$, OF WATER-VAPOUR SATURATION VALUES BELOW 0 °C REFER TO SATURATION ABOVE ICE**

| *t* °C | $\rho_s$ mbar | $\rho_w$ g/m³ | *t* °C | $\rho_s$ mbar | $\rho_w$ g/m³ |
|---|---|---|---|---|---|
| –40 | 0.128 | 0.119 | 5 | 8.72 | 6.80 |
| –38 | 0.161 | 0.148 | 6 | 9.35 | 7.26 |
| –36 | 0.200 | 0.183 | 7 | 10.01 | 7.75 |
| –34 | 0.249 | 0.225 | 8 | 10.72 | 8.27 |
| –32 | 0.308 | 0.277 | 9 | 11.47 | 8.82 |
| –30 | 0.380 | 0.339 | 10 | 12.27 | 9.40 |
| –29 | 0.421 | 0.374 | 11 | 13.12 | 10.01 |
| –28 | 0.467 | 0.413 | 12 | 14.02 | 10.66 |
| –27 | 0.517 | 0.455 | 13 | 14.97 | 11.35 |
| –26 | 0.572 | 0.502 | 14 | 15.98 | 12.07 |
| –25 | 0.632 | 0.552 | 15 | 17.04 | 12.83 |
| –24 | 0.699 | 0.608 | 16 | 18.17 | 13.63 |
| –23 | 0.771 | 0.668 | 17 | 19.37 | 14.48 |
| –22 | 0.850 | 0.734 | 18 | 20.63 | 15.37 |
| –21 | 0.937 | 0.805 | 19 | 21.96 | 16.31 |
| –20 | 1.03 | 0.884 | 20 | 23.37 | 17.30 |
| –19 | 1.14 | 0.968 | 21 | 24.86 | 18.34 |
| –18 | 1.25 | 1.06 | 22 | 26.43 | 19.43 |
| –17 | 1.37 | 1.16 | 23 | 28.09 | 20.58 |
| –16 | 1.51 | 1.27 | 24 | 29.83 | 21.78 |
| –15 | 1.65 | 1.39 | 25 | 31.67 | 23.05 |
| –14 | 1.81 | 1.52 | 26 | 33.61 | 24.38 |
| –13 | 1.98 | 1.65 | 27 | 35.65 | 25.78 |
| –12 | 2.17 | 1.80 | 28 | 37.80 | 27.24 |
| –11 | 2.38 | 1.96 | 29 | 40.06 | 28.78 |
| –10 | 2.60 | 2.14 | 30 | 42.43 | 30.38 |
| – 9 | 2.84 | 2.33 | 31 | 44.93 | 32.07 |
| – 8 | 3.10 | 2.53 | 32 | 47.55 | 33.83 |
| – 7 | 3.38 | 2.75 | 33 | 50.31 | 35.68 |
| – 6 | 3.69 | 2.99 | 34 | 53.20 | 37.61 |
| – 5 | 4.02 | 3.25 | 35 | 56.24 | 39.63 |
| – 4 | 4.37 | 3.52 | 36 | 59.42 | 41.75 |
| – 3 | 4.76 | 3.82 | 37 | 62.76 | 43.96 |
| – 2 | 5.17 | 4.14 | 38 | 66.26 | 46.26 |
| – 1 | 5.62 | 4.48 | 39 | 69.93 | 48.67 |
| 0 | 6.11 | 4.85 | 40 | 73.78 | 51.19 |
| 1 | 6.57 | 5.19 | 41 | 77.80 | 53.82 |
| 2 | 7.06 | 5.56 | 42 | 82.02 | 56.56 |
| 3 | 7.58 | 5.95 | 43 | 86.42 | 59.41 |
| 4 | 8.13 | 6.36 | 44 | 91.03 | 62.39 |

Compressibility charts may be used to determine specific values of Z for any particular gas or gas mixture. Alternatively, a generalized chart may be used, devised on the basis that the behaviour of all gasses is similar when their reduced pressures (pR) and reduced temperatures (tR) are the same, *ie*:

$$pR = \frac{P}{P_C} \quad \text{and} \quad t_R = \frac{T}{T_C}$$

where $P_C$ = critical pressure and
$T_C$ = critical temperature.

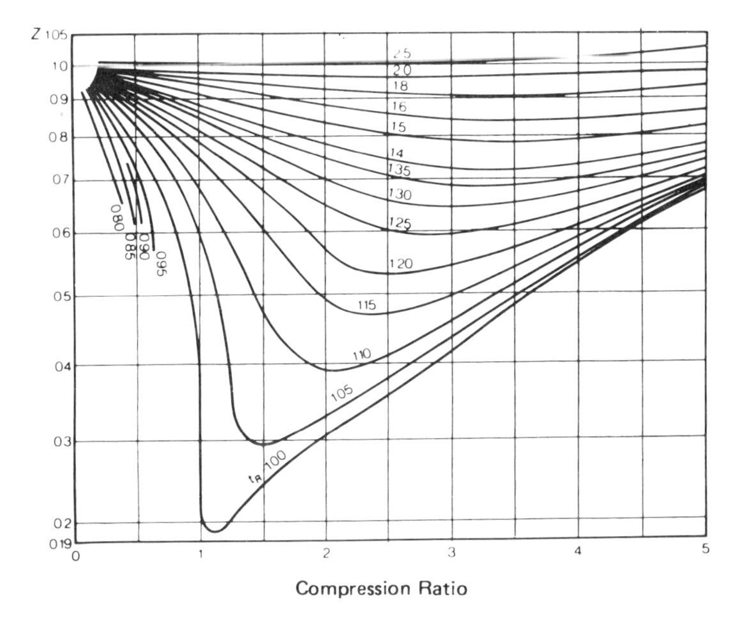


*Figure 2*
*Generalized compressibility factor Z.*

Such a chart is shown in Figure 2. To use, values of the critical constants pR and $t_R$ are determined on a proportionate basis and summed, *eg*

$$pR\ (\text{gas } 1) = \frac{P}{P_c\ (\text{gas } 1)} \times A\%$$

$$pR\ (\text{gas } 2) = \frac{P}{P_c\ (\text{gas } 2)} \times B\%$$

and so on, where A and B are the percentage proportions expressed as decimal fraction

Then, finally, pR = pR (gas 1) + pR (gas 2) + ......

The reduced temperature ($t_R$) of the mixture is determined in a similar way. The corresponding value of the compressibility factor can then be read off the chart. Table 3 gives critical pressure and critical temperature values for a wide range of gases.

**Air Compression**
The four possible ways of compressing air (or any gas) are shown in Figure 3. Of these, the first two involve either the removal or addition of heat and are not directly connected with pneumatic engineering which is primarily concerned with pressure/volume relationship and the general gas equation

$P \times V$ = a constant.

## TABLE 3 – PHYSICAL DATA FOR SOME GASES

| Gas | Formula | Molecular mass | Gas Constant J/(kg x K) | $P_{cr}$ bar (abs) | $t_{cr}$ °C | Normal boiling point °C | cp kJ/(kg x K) | $c_p/c_v$ |
|---|---|---|---|---|---|---|---|---|
| acetylene | $C_2H_2$ | 26.04 | 319.4 | 62 | 36 | –83 | 1.659 | 1.26 |
| air | – | 28.96 | 287.1 | 38 | –141 | –194 | 1.004 | 1.40 |
| ammonia | $NH_3$ | 17.03 | 488.2 | 113 | 132 | –33 | 2.093 | 1.30 |
| argon | Ar | 39.94 | 208.2 | 49 | –122 | –186 | 0.522 | 1.67 |
| benzene | $C_6H_6$ | 78.11 | 106.5 | 49 | 289 | 80 | 1.726 | 1.08 |
| butane, N- | $C_4H_{10}$ | 58.12 | 143.1 | 38 | 152 | –1 | 1.635 | 1.10 |
| butane, iso- | $C_4H_{10}$ | 58.12 | 143.1 | 36 | 135 | –12 | 1.620 | 1.10 |
| butylene | $C_4H_8$ | 56.11 | 148.3 | 39 | 144 | –7 | 1.531 | 1.11 |
| carbon dioxide | $CO_2$ | 44.01 | 188.9 | 74 | 31 | –78 | 0.833 | 1.30 |
| carbon disulfide | $CS_2$ | 76.13 | 109.2 | 75 | 278 | 46 | 0.657 | 1.20 |
| carbon monoxide | CO | 28.01 | 296.6 | 35 | –140 | –191 | 1.039 | 1.40 |
| carbon tetrachloride | $CCl_4$ | 153.84 | 54.0 | 44 | 283 | 77 | 0.833 | 1.18 |
| chlorine | $Cl_2$ | 70.91 | 117.3 | 77 | 144 | –34 | 0.498 | 1.35 |
| cyan | $(CN)_2$ | 52.04 | 159.8 | 61 | 128 | –21 | | 1.26 |
| dichloromethane | $CH_4Cl_2$ | 86.95 | 95.6 | 103 | 216 | 41 | | 1.18 |
| ethane | $C_2H_6$ | 30.07 | 276.5 | 49 | 32 | –89 | 1.714 | 1.19 |
| ethyl chloride | $C_2H_5Cl$ | 64.51 | 128.9 | 53 | 185 | –12 | | 1.13 |
| ethylene (ethene) | $C_2H_4$ | 28.05 | 296.4 | 51 | 9 | –104 | 1.515 | 1.24 |
| Freon 12 | $CCl_2F_2$ | 120.92 | 68.8 | 39 | 112 | –30 | 0.611 | 1.13 |
| Freon 13 | $CClF^3$ | 104.46 | 79.6 | 39 | 29 | –82 | 0.644 | 1.15 |
| Freon 14 | $CF_4$ | 88.01 | 94.5 | 37 | –46 | –128 | 0.694 | 1.16 |
| Freon 22 | $CHClF_2$ | 86.47 | 96.1 | 49 | 96 | –41 | 0.646 | 1.20 |
| helium | He | 4.00 | 2077.2 | 2 | –268 | –269 | 5.208 | 1.66 |
| hexane, N- | $C_6H_{14}$ | 86.17 | 96.5 | 30 | 234 | 69 | 1.617 | 1.08 |
| hydrogen | $H_2$ | 2.02 | 4124.5 | 13 | –240 | –253 | 14.270 | 1.41 |
| hydrogen bromide | HBr | 80.92 | 102.8 | 82 | 90 | –68 | | 1.42 |

| | | | | | | | | |
|---|---|---|---|---|---|---|---|---|
| hydrogen chloride | $HCl$ | 36.46 | 228.0 | 83 | 52 | –85 | 0.792 | 1.41 |
| hydrogen cyanide | $HCN$ | 27.03 | 307.6 | 49 | 184 | –89 | | 1.27 |
| hydrogen sulphide | $H_2S$ | 34.08 | 244.0 | 90 | 101 | –59 | 0.996 | 1.33 |
| hydrogen jodide | $HJ$ | 127.93 | 65.0 | – | 151 | –36 | | 1.40 |
| krypton | $Kr$ | 83.80 | 99.3 | 55 | –64 | –153 | 0.249 | 1.68 |
| methane | $CH_4$ | 16.04 | 518.3 | 46 | –83 | –162 | 2.205 | 1.31 |
| methyl ether | $(CH_3)_2O$ | 46.07 | 180.5 | 52 | 127 | –23 | | 1.11 |
| methylene chloride | $CH_2Cl_2$ | 84.94 | 97.9 | 63 | 238 | 40 | 0.826 | 1.18 |
| methyl chloride | $CH_2Cl$ | 50.49 | 164.7 | 67 | 143 | –24 | 0.826 | 1.25 |
| neon | $Ne$ | 20.18 | 412.0 | 27 | –229 | –246 | 1.032 | 1.66 |
| nitrogen | $N_2$ | 28.02 | 296.8 | 34 | –147 | –196 | 1.038 | 1.40 |
| nitric oxide | $NO$ | 30.01 | 277.1 | 65 | –94 | -152 | 0.990 | 1.39 |
| nitrous oxide | $N_20$ | 44.02 | 188.9 | 73 | 37 | –89 | 0.874 | 1.30 |
| oxygen | $0_2$ | 32.00 | 259.8 | 51 | –118 | –183 | 0.915 | 1.40 |
| ozone | $0_3$ | 48.00 | 173.2 | 69 | –5 | –112 | | 1.29 |
| pentane, N- | $C_5H_{12}$ | 72.14 | 115.3 | 34 | 197 | 36 | 1.624 | 1.06 |
| pentane, iso- | $C_5H_{12}$ | 72.14 | 115.3 | 33 | 187 | 28 | 1.601 | 1.06 |
| propane | $C_3H_8$ | 44.09 | 188.6 | 43 | 97 | –42 | 1.620 | 1.13 |
| propylene (propene) | $C_3H_6$ | 42.08 | 197.6 | 44 | 92 | –48 | 1.521 | 1.15 |
| sulphur dioxide | $SO_2$ | 64.07 | 129.8 | 79 | 158 | –10 | 0.607 | 1.25 |
| water (super-heated) | $H_20$ | 18.02 | 461.4 | 221 | 374 | 100 | 1.860 | *) |
| xenon | $Xe$ | 131.30 | 63.3 | 58 | 17 | –108 | | 1.66 |

*) for wet steam: $c_p/c_v \approx 1.035 + 0.1 \times X$
*where X is the steam content (X>0.4)*
*for super-heated steam (up to 10 bar and 800°C):*
$c_p/c_v \approx 1.332 + 0.112 \times 10^3 \times t$
*where t is the temperature in °C*

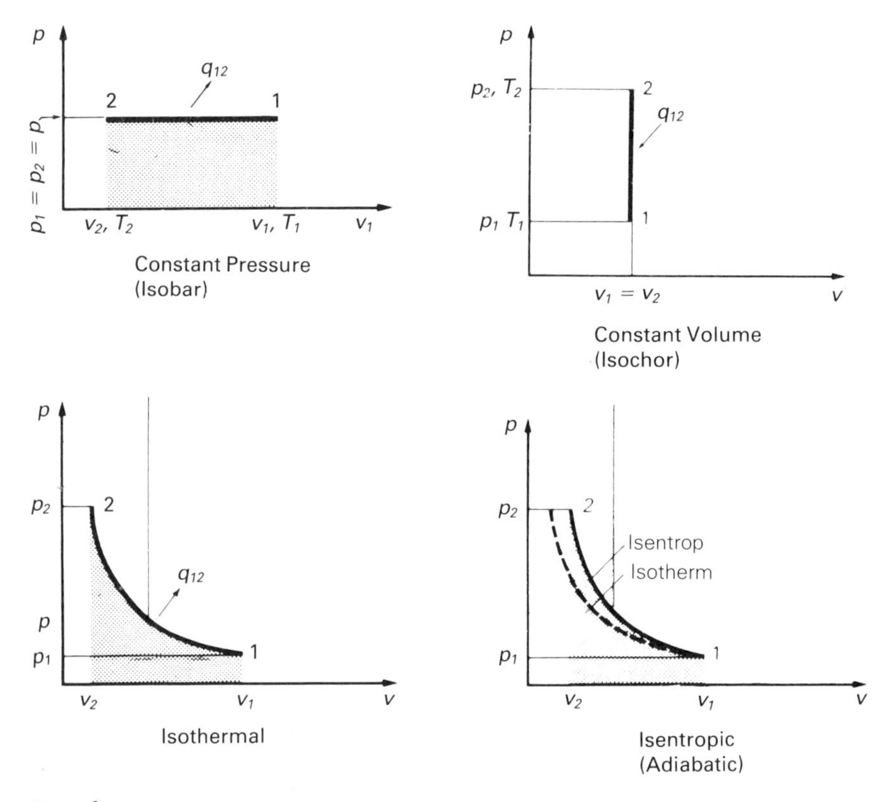


*Figure 3*

This, in fact, is consistent with *isothermal* compression where the process takes place at constant temperature, when pressure ratios and volume ratios have a direct reciprocal relationship

$$P_2/P_1 = V_1/V_2$$

To achieve this, cooling must be present to take away an amount of heat equal to the work of compression, *ie:*

heat to be lost $= m \text{ x } R \text{ x } T_1 \log_e (P_2/P_1)$ joules

or $= m \text{ x } P_1 V_1 \log_e (V_1/V_2) \text{ x } 10^5$ joules

where pressures $P_1$ and $P_2$ are absolute pressures in bars

m = mass of air (or gas) in kg

$V_1$ and $V_2$ are specific volumes in $m^3/kg$

$T_1$ is the initial (and constant) temperature in °K.

With *adiabatic* compression (also called *isentropic* compression) there is no heat exchange with the surroundings, *ie* no cooling present, and so the rise in pressure and reduction in volume are accompanied by a rise in temperature of the air (or gas). The pressure/volume relationship is:

$$P_2/P_1 = (V_1/V_2)^k$$

or $P \times V^k$ = constant

where k is the ratio of the specific heats.

The change in temperature from $T_1$ to $T_2$ is given by:

$$P_2/P_1 = (T_2/T_1)^{\frac{k}{k-1}}$$

or the temperature of the compressed air (or gas) $T_2$ is given directly by:

$$T_2 = T_1 (P_2/P_1)^{\frac{k-1}{k}}$$

In practice neither isothermal compression is likely to apply, but the process will be somewhere between the two or *polytropic*. The general equation then becomes:

$PV^n$ = constant

where the value of the exponential n lies between 1 and 1.405 for air

and compressed air temperature $T_2 = T_1 \ (P^2/P_1)^{\frac{n-1}{n}}$

**Compression Efficiency**

Compression efficiency can be defined as the ratio of the theoretical work required to compress a given amount of air to a specific volume, to the actual amount of work done by the compressor.

Because the units of pressure are $F/1^2$ and those of volume are $1^3$, work done can be expressed by the product of pressure and volume (units F x 1).

The mean effective pressure (mep) with isothermal compression is then:

$$\text{mep} = P_1 \log_e \frac{P_2}{P_1}$$

$$= P_1 \log_e r$$

where r is the compression ratio.

The isothermal indicated horsepower or *Isothermal Factor* is given by:

$$K \times P_1 \log_e r = K \times P_1 \log_e \frac{P_2}{P_1}$$

K being a constant factor, depending on the units employed.

More generally (or in metric units), the work for isothermal compression is given as:

$$\text{Work (joules)} = 10^5 \ P_1 \ x \ V_1 \ \log_e \ \frac{P_2}{P_1}$$

where P is the absolute pressure bar
V is the volume in $m^3$

For isothermal compression from atmospheric pressure:

$$\begin{aligned} \text{Work (joules)} &= 10^5 V_1 \log_e P_2 \\ &= 230.3 \, V_1 \log_e P_2 \\ &= 2.303 \, V_1 \log_{10} P_2 \end{aligned}$$

The power for isothermal compression follows by replacing volume flow in $m^3$ by $m^3$/sec, *viz*

$$\text{power (kW)} = 230.3 \ V_F \log_{10} /P_2$$

where $V_F$ = volume flow $m^3$/s
$P_2$ = final pressure in bar.

Power for adiabatic compression can be determined in a similar manner only in this case the mean effective pressure is:

$$\text{mep} = \frac{k}{k-1} \ x \ P_1 \left( r^{\left(\frac{k-1}{k}\right)} - 1 \right)$$

$$= 3.463P_1 \ (r^{0.29} - 1)$$

Thus power required for adiabatic compression is given by:

$$\text{power (kW)} = 2.303 \ V_1 \ \log_{10} \left( \left( \frac{p_1}{P_2} \right)^{0.29} - 1 \right)$$

**Equivalent Free Air Volume**

The relationship between free air volume and compressed air volume follows directly from the compression ratio, *viz:*

$$\frac{\text{free air volume}}{\text{compressed air volume}} = \text{compression ratio}$$

or

$$\frac{V_{FA}}{V_c} = r$$

whence

$$V_{FA} = \frac{P_2 + P_1}{P_1} \times V_c$$

where $P_2$ = final pressure, absolute
$P_1$ = atmospheric pressure

For $P_2$ in lb/in$^2$

$$V_{FA} = \frac{P_2 + 14.7}{14.7} \times V_c$$

The gas laws are based on the behaviour of perfect gases or mixtures of perfect gases. They are:

*Boyle's* law – which states that the volume (V) of a gas, at constant temperature, varies inversely with pressure (P).

$$V_2/V_1 = P_1/P_2$$

or

$$P_1 \times V_1 = P_2 \times V_2 = \text{a constant.}$$

*Clarke's* law – which states that the volume of a gas, at constant pressure, varies directly with absolute temperature (T).

$$V_1/V_2 = T_1/T_2$$

or

$$\frac{V_1}{T_1} = \frac{V_2}{T_2} = \text{a constant}$$

*Amaton's* law – which states that the pressure of a gas, at constant volume, varies directly with absolute temperature.

$$P_1/P_2 = T_1/T_2$$

or

$$\frac{P_1}{T_1} = \frac{P_2}{T_2} = \text{a constant.}$$

*Dalton's* law – which states that the total pressure of a mixture of gases is equal to the sum of the partial pressures of the gases present (partial pressure being the pressure each gas would exert if it alone occupied the same volume as the mixture).

$$P = \sum_{1}^{n} p_1$$

*Avagado's* law – which states that equal volumes of all gases under the same conditions of pressure and temperature contain the same number of molecules.

*Amagat's* law – which states that the volume of a mixture of gases is equal to the partial volumes which the constituent gases would occupy if each existed alone at the total pressure of the mixture.

*Poisson's* law – which states that for a process without any heat exchange with the surroundings the relationship between pressure and volume follows the mathematical relationship.

$$P_1 \times V_1{}^k = P_2 \times V_2{}^k$$

where k is the ratio of the specific heat of the gas at constant pressure to specific heat at constant volume.

The value of the exponential k varies temperature and pressure (see Figure 4) but for pressures of less than 1 bar (and all temperatures)

k = 1.40 for air
= 1.66 for mono-atomic gases
= 1.40 for two-atomic gases
= 1.30 for three-atomic gases.

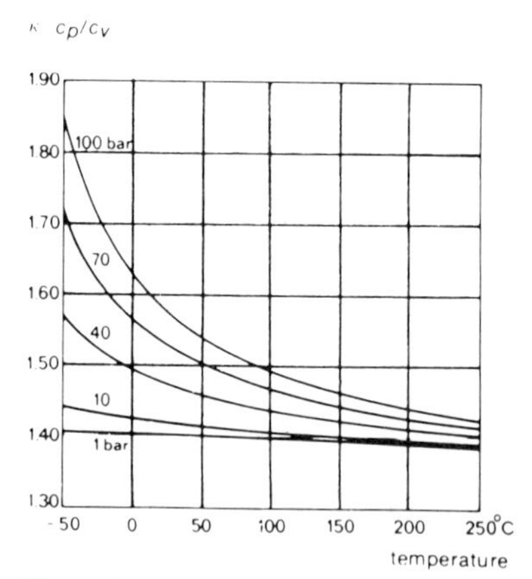


*Figure 4*

*General Gas* law – resulting from a combination of Boyle's law and Clarke's law.

P x V/T = constant.

*Laws of Thermodynamics*

(1) For any system of constant mass the amount of work done on or by the system is equal to the amount of energy transferred to or from the system.

(2) Energy exists at various temperature levels but is available for use only if it can pass from a higher to a lower temperature level.

# Compressible Gas Flow

AIR, STEAM and gases are all *compressible fluids* and the *d'Arcy* equation used for determining pressure and head losses with liquid flow is no longer applicable because the density of gases and vapours changes considerably with changes of pressure. For simplified general engineering calculations not requiring great accuracy, liquid (d'Arcy) flow formulae may be used if the pressure drop involved is less than 10% of the inlet pressure. Use of such formulae is also sometimes extended to pressure drops up to 40% of the inlet pressure, provided in this case the specific volume is taken as the average of the upstream and downstream conditions.

The real flow of a pressurized gas through pipes differs appreciably in a number of important characteristics from the flow of liquids in pipes. Pressure, for example, drops at an increasing rate along the pipe, rather than with a constant pressure gradient. At the same time velocity tends to increase up to a maximum defined by $V = \sqrt{kgRT}$ for air, but subject to a limiting or maximum length, which must correspond to the end of the pipe. At this point the pressure gradient is infinite, *ie* the pipe is effectively closed. In this equation k is the ratio of specific heats at constant pressure to constant volume, R the individual gas constant, and T the absolute temperature. An alternative formula is $M = 1/\sqrt{k}$ (= 0.845 for air), where M is the Mach number.

The general equation may be written in the same form as the d'Arcy equation for fluid flow, but with the addition of an extra term representing the pressure drop required to increase the flow momentum:

$$\frac{\Delta P}{L} = \frac{f}{D} \times \frac{\rho V^2}{2} + \beta \rho \bar{Q} \times \frac{dV}{dL}$$

where

| | |
|---|---|
| $\Delta P$ | is the pressure |
| L | is the pipe length |
| f | is a constant friction factor (but dependent on surface roughness) |
| D | is the pipe diameter |
| $\rho$ | is the gas density |

$\bar{Q}$ is the specific volume of gas
dV/dL is the velocity gradient
ß is a factor of the order of unity and normally taken as 1.0 (*ie* can be eliminated from the equation).

Actual flow conditions may range from adiabatic to isothermal. Adiabatic conditions are only likely to apply in short, well insulated pipes where no appreciable heat is transferred to or from the pipe. Isothermal flow or flow at constant temperature is commonly assumed as more consistent with normal practice, especially for long pipes. In fact, most practical pipelines will generate polytropic flow conditions, virtually impossible to analyse. The assumption of isothermal flow is thus a practical compromise.

**Isothermal Flow**
With isothermal flow, a formula developed from basic principles is

$$\frac{\Delta P}{L} = \frac{\frac{f}{D} \times \frac{\rho V^2}{2}}{kM^2 - 1}$$

The friction factor is dependent on the Reynolds number of flow and pipe roughness. It can be assumed independent of Mach number.

The Reynolds number will be constant for isothermal flow, but may vary with adiabatic or isentropic flow (and certainly with *diabatic* flow), in which case a mean value can be assumed.

The Reynolds number is a dimensionless quantity, given arithmetically by:

$$R_e = \frac{\rho VD}{\mu} = \frac{VD}{\nu}$$

where $\rho$ = is the mass density
$\mu$ = viscosity
$\nu$ = kinematic viscosity
} in units consistent with velocity (V) and pipe diameter (D)

For air at standard temperature $\nu = 1.45 \times 10^{-5}$ in$^2$/s

The friction factor (f) is common for all fluids (*ie* gases and liquids) and is normally determined from empirical charts. For *laminar* flow, the friction factor is dependent only on Reynolds number and is numerically equal to $64/R_e$, laminar flow being defined by the Reynolds number not exceeding 2000. For turbulent flow and smooth bore pipes, the Reynolds number can be calculated from the following empirical formula:

$$f = \frac{0.3164}{R_e^{0.25}}$$

As the Mach number (M) approaches 0, the denominator in this equation comes closer and closer to unity, reducing the equation to the same as that for liquid flow. There is thus some justification for using the d'Arcy formula for compressible flow calculations (as mentioned initially) as $M \rightarrow 0$ (*ie* consistent with short lengths of pipes with resulting low pressure drops). This also implies that at very low Mach numbers, compressible flow can be treated as incompressible. Pressure gradients ($\triangle P/L$) will, in fact, be within 5% of incompressible flow values at Mach numbers up to about 0.18 for air.

A complete working formula for isothermal flow is:

$$V_m^2 = \left[ \frac{144gA^2}{\bar{Q}\left(\frac{fL}{D} + 2\log\frac{P_1}{P_2}\right)} \right] \left[ \frac{P_1^2 - P_2^2}{P_1} \right]$$

where
$A$ = cross sectional area of pipe, $ft^2$
$g$ = acceleration of gravity = 32.2 $ft/s^2$
$L$ = length of pipe, ft
$f$ = friction factor
$D$ = pipe bore, ft
$P_1$ = absolute pressure (entry)
$P_2$ = absolute pressure (exit)
$V_m$ = mean velocity of flow, ft/s
$\bar{Q}$ = specific volume of gas, $ft^3/lb$.

**Limiting Values**

Maximum possible velocity ($V^*$) in a pipe is source velocity (M=1), given directly by

$$V^* = 11.32\sqrt{kg\,P\bar{Q}}$$

where
$V^*$ is in ft/sec
$k$ is the ratio of specific heats of the gas
$P$ is the absolute pressure in bar
$\bar{Q}$ is the specific volume of gas, $m^3/kg$

*Maximum Pressure* ($P^*$) can be determined on the basis of continuity, *viz:*

$$V_1 P_1 = VP = V^* P^*$$

Hence

$$\frac{P^*}{P_1} = \frac{V_1}{V^*} = \frac{M_1}{M^*} = M_1\sqrt{k}$$

where the suffix 1 refers to initial conditions.

Hence (because the velocity of sound is constant under isothermal conditions)

$$P^* = P_1 \times M_1 \times k$$

*Limiting Length* ($L^*$) can be determined from the general equation:

$$\frac{fL^*}{D} = \left(\frac{1}{kM_1^2} - 1\right) - \log_e \frac{1}{kM_1^2}$$

or

$$L^* = \frac{D}{f}\left(\left(\frac{1}{kM_1^2} - 1\right) - \log_e \frac{1}{kM_1^2}\right)$$

The implication of this is that velocity of gas flow can go on increasing in a pipe up to a maximum Mach number of $1/\sqrt{k}$. This increase ceases at a limiting length of pipe ($L^*$) which must be at the end of the pipe. If the actual length of pipe is less than, or equal to, $L^*$.

At the same time the limiting pressure ratio ($P^*/P_1$) and limiting length ($L^*$) are dependent only on initial velocity (Mach number) and the k value of the gas.

Pressures and lengths between two sections of a pipeline can thus be expressed as follows, knowing the Mach numbers at each section:

$$\frac{P_2}{P_1} = \frac{M_1}{M_2}$$

$$\frac{fL}{D} = \left(\frac{fL^*}{D}\right) M_1 - \left(\frac{fL^*}{D}\right) M_2$$

**Adiabatic Flow**

Similar treatment applies, although the corresponding formulae are more complicated and may require working as a series of approximations in order to reach real values in particular cases. In general, the pressure, temperature and velocity will always be slightly less than those for isothermal flow, but the limiting length will be similar. The differences are usually small enough to be negligable, except at higher Mach numbers, and thus, for simplification of calculations, isothermal formulae can be used for subsonic adiabatic flow.

Limiting Velocity:

This is given directly by:

$$\frac{V^*}{V_1} = \frac{1}{M_1}\sqrt{\frac{2\left(1 + \frac{k-2}{2} M_1^2\right)}{k+1}}$$

Limiting Temperature:

$$\frac{T^*}{T_1} = \frac{2(1 + \frac{k-1}{2}M_1^2)}{k+1}$$

Limiting Length:

$$L^* = \frac{D}{F}\left(\frac{1 - M_1^2}{kM_1^2}\right) + \frac{k+1}{2k} \log_e \left(\frac{(k+1)M_1^2}{2(1 + \frac{k-1}{2}M_1^2)}\right)$$

Limiting Pressure:

$$\frac{P^*}{P_1} = M_1 \left(\frac{2(1 + \frac{k-1}{2} M_1^2)}{k+1}\right)$$

**Stagnation state**
Flow is possible between two extremes. At one extreme, velocity is zero and temperature is a maximum because all the kinetic energy is converted to enthalpy. The speed of sound is also a maximum (stagnation point or stagnation state). At the other extreme, the velocity is a maximum and the temperature falls to absolute zero, all the enthalpy being converted into kinetic energy. The speed of sound is then zero (zero temperature state). Between these extremes the practical flow may be subsonic, trisonic, or supersonic Figure 1, although the zero temperature state can never be reached (*ie* is a hypothetical condition).

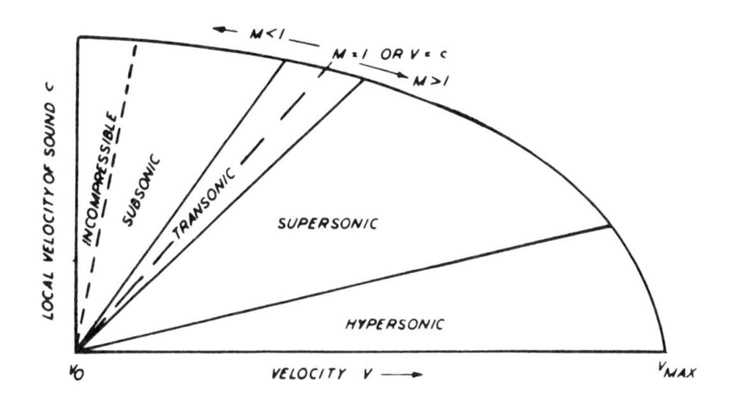


*Figure 1*

At the stagnation state:

$$\frac{0^2}{2} + h_O = \frac{V^2}{2} + h$$

or

$$h_O = \frac{V}{2} + h$$

From the general gas relationship it follows that the stagnation temperature ($T_O$) is given by:

$$T_O = T + \frac{V^2}{2sp}$$

where sp = specific heat at constant pressure.

Alternatively,

$$\frac{T_O}{T} = 1 + \frac{k-1}{2} \text{ x } M^2$$

where M = Mach number = $\frac{V}{c}$

The stagnation pressure can be derived as:

$$\frac{P_o}{P} = \left(\frac{T_o}{T}\right)^{\frac{k}{k-1}}$$

$$= \left(1 + \frac{k-1}{2} M^2\right)^{\frac{k}{k-1}}$$

This may be expanded in the form:

$$P_o = P + \frac{\rho V^2}{2} \left(1 + \frac{M^2}{4} + \frac{2-k}{24} M^4 + \ldots\ldots\ldots\right)$$

This can be compared with equation for incompressible flow:

$$P_O = P + \frac{\rho V^2}{2}$$

The difference between these two equations represents the effect of increased gas density due to compressibility, generally termed the *compressibility factor*. Values of the compressibility factor range from unity at very low Mach numbers (where there are no compressibility effects), up to 1.276 as the Mach number approaches 1 (velocity approaches the speed of sound in the gas).

Besides increasing the dynamic pressure of compressible flow, compared with incompressible flow, the rising value of the compressibility factor can also affect the flow velocity through ducts with varying area. The relationship between area and velocity changes is, in fact, a function of the local Mach number, and can be rendered in the form:

$$\frac{dA}{dV} = \frac{A}{V}(M^2 - 1)$$

With subsonic flow, a decrease in area produces an increase in flow velocity and *vice versa* (similar to incompressible flow) *ie*, area and Mach number changes are opposite. The flow velocity may be sonic only at a constant section. With supersonic flow, a decrease in flow area produces a decrease in flow velocity, and *vice versa, ie,* area and Mach number changes are the same.

**Flow from Stagnation Conditions**

Gas compressed and stored in a reservoir is essentially under stagnation conditions, where velocity is zero and the pressure and temperature are known (or can be determined). Where the reservoir is used as a supply, the velocity, temperature and pressure at any other section of flow are determined basically from the following relationships:

*Velocity at any arbitrary section:*

$$V = \sqrt{2sp\,T_O \left(1 - \frac{P}{P_O}\right)^{\frac{k-1}{k}}}$$

Alternatively, for adiabatic flow, the velocity at any section can be determined from the temperature at that section:

$$V = \sqrt{2sp\,T_O\left(1 - \frac{T}{T_O}\right)}$$

**Flow Rate**

Flow rate can be determined as the mass flow, *ie* mass = VA*p* or directly as the product of V and A in numerical consistent units:

$$\text{Dimensions are } \frac{L}{T} \times L^2 = \text{flow rate} = \frac{L^3}{T}$$

*Pressure at any arbitrary section:*

$$P = \frac{P_O}{\left(1 + \frac{k-1}{2} M^2\right)^{\frac{k-1}{k}}}$$

*Temperature at any section:*

$$T = \frac{T_O}{1 + \frac{k-1}{2} \times M^2}$$

At any (constant) section where the flow is sonic the flow conditions are described as critical, yielding a critical temperature ($T^*$) and a critical pressure ($P^*$), where:

$$\frac{T^*}{T_O} = \frac{2}{k+1} \qquad \text{(Adiabatic or isentropic flow)}$$

$$\frac{P^*}{P_O} = \left(\frac{2}{k+1}\right)^{\frac{k}{k-1}} \qquad \text{(Isentropic flow only)}$$

*Note*: For air, where k = 1.4, the value of critical pressure is $\left(\frac{2}{2.4}\right)^{\frac{1.4}{0.4}} = 0.528$

That is, the critical pressure is 52.8% of $P_O$. Similarly the critical temperature can be calculated as 83.3% of $T_O$.

*Critical Area:*

The relationship between the critical area ($A^*$) or throat area where $\mu = 1$ and the area of any other section (A) is given by

$$\frac{A}{A^*} = \frac{1}{M} \left(\frac{1 + \frac{k-1}{2} \times M_2}{\frac{k+1}{2}}\right)^{\frac{k+1}{2(k-1)}}$$

$$= \frac{1}{M} \left(\frac{1 + 0.2\,M^2}{1.2}\right)^3 \qquad \text{for air}$$

**Nozzle Flow**

Flow at the throat of a nozzle, supplied by a reservoir or similar source under stagnation conditions, will be sonic if the critical pressure is greater than the receiver pressure see Figure 2(a). This means that the flow will be critical. The flow velocity follows by relating to the critical temperature, where:

$$\text{flow velocity} = 49\sqrt{T^*} \text{ ft/s}$$

(The velocity of sound in air = c = 49 $\sqrt{T}$ ft/s where T is the absolute temperature in degrees *Rankine*).

If the critical pressure is less than the receiver pressure, then the flow cannot be critical – Figure 2(b). In this case the flow will be subsonic and the exit pressure will equal the receiver pressure. The temperature can be calculated from the general formula, or from:

$$T_1 = T_0 \left( \frac{P_1}{P_0} \right)^{\frac{k-1}{k}}$$

The velocity is likewise calculated from the general formula.

Similar analysis applies where the nozzle is of convergent-divergent form (Figure 2(c)). In this case it is first necessary to establish whether the flow is critical or not (at the throat). The throat velocity can be determined accordingly; and from this the final exit velocity from the divergent section.

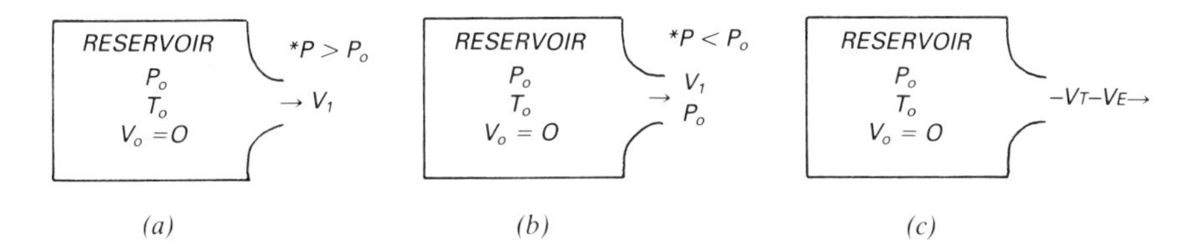


*Figure 2*

Flow through a nozzle can also be rendered directly in terms of flow rate and a discharge coefficient, this being a convention for engineering calculations. The complete nozzle formula is:

$$\text{mass flow} = A \times E\, c \times \delta \times d^2 \sqrt{h} \quad \sqrt{P2/T}$$

where A is a constant depending on the units employed
E = coefficient for the velocity of approach

$$= \frac{1}{\sqrt{1 - m^2}}$$

where $m = \dfrac{\text{cross-sectional area of nozzle}}{\text{cross-sectional area upstream}} = \dfrac{A2}{A1}$

$c$ = nozzle coefficient
$\delta$ = expansibility factor allowing for the change in air density which occurs during acceleration through the nozzle

$$= \frac{1 - 0.07h}{13.6\,P_2}$$ for values of m circa 0.16 (and h in in wg and $P_2$ in in of mercury)

$d$ = diameter of nozzle
$h$ = pressure drop across nozzle
$P_2$ = absolute pressure on downstream side of nozzle
$T$ = absolute temperature on downstream side of nozzle.

If T is °R, $P_2$ in in of mercury, h is in in wg and d is in in, a value of $A = 0.1148$ gives the mass flow in units of lb/s.

For a specific nozzle profile, the formula can be simplified by the use of a nozzle constant appropriate to that particular geometry and nozzle size. Rendered as a solution for conventional flow rate (Q)

$$Q = K\,(T_1/P_1)\,\sqrt{h}\,\sqrt{P_2/T_2}$$

where
$K$ = nozzle constant
$T_1$ = absolute temperature at specified inlet port
$T_2$ = absolute temperature at nozzle or specified point downstream
$P_1$ = absolute pressure at specified inlet point
$P_2$ = absolute pressure at nozzle or specified point downstream
$h$ = pressure drop across nozzle.

**Simplified Orifice Formula**

An orifice is a simple form of nozzle, formed by a circular hole cut in a thin flat plate. Flow can again be determined with reference to an empirical discharge coefficient or orifice coefficient. This will be much lower than for nozzles because of the less streamlined flow but, due to the simpler form of the nozzle, will be less subject to variation.

Thus nozzle coefficients may vary between 0.90 (or less) and 0.995 depending on size and geometry, whereas an orifice coefficient can be expected to be of the order of 0.61, regardless of size, and differing only if the orifice has a well rounded, as opposed to a sharp, entry.

Very much simplified formulae can therefore be applied to assess the discharge of air through orifices and the following are generally satisfactory for straightforward engineering calculations:

(1) For upstream pressures above 14.7 $lbf/in^2g$

$$Q\ (ft^3/min) \text{ for sharp-edged orifice} = \frac{218 \times A \times P_u}{\sqrt{460 + T}} \qquad (a)$$

$$\text{or} \quad = \frac{172 \times d^2 \times P_u}{\sqrt{460 + T}}$$

$$Q\ (ft^3/min) \text{ for rounded entrance orifice} \simeq \frac{417 \times A \times P_u}{\sqrt{460 + T}} \qquad (b)$$

$$\text{or} \quad \simeq \frac{327 \times d^2 \times P_u}{\sqrt{460 + T}}$$

(2) For upstream pressures below 14.7 $lbf/in^2g$

$$Q\ (ft^3/min) \text{ for sharp-edged orifice} = \frac{210 \times A \times P_u}{\sqrt{460 + T}} \qquad (c)$$

$$\text{or} \quad = \frac{166 \times d^2 \times P_u}{\sqrt{460 + T}}$$

$$Q\ (ft^3/min) \text{ for rounded entrance orifice} \simeq \frac{324 \times A \times P_u}{\sqrt{460 + T}} \qquad (d)$$

$$\text{or} \quad \simeq \frac{255 \times d^2 \times P_u}{\sqrt{460 + T}}$$

where $A$ = orifice area, $inches^2$
$P_u$ = upstream pressure, $lbf/in^2g$
$d$ = orifice diameter, inches
$T$ = upstream air temperature °F

Note that for an upstream air temperature of 60 °F these formulae further simplify to:

1(a) $Q\ (ft^3/min) = 11.9\ A \times P_u$

or $= 9.4\ d^2\ P_u$

(b) $Q\ (ft^3/min) \triangleq 18.3\ A \times P_u$

or $\triangleq 14.4\ d^2\ P_u$

2(c) $Q\ (ft^3/min) = 11.5\ A \times P_u$

or $= 9.05\ d^2\ P_u$

(d) $Q\ (ft^3/min) \triangleq 17.7\ A \times P_u$

or $\triangleq 13.92\ d^2\ P_u$

# SECTION 6

## Pneumatic Control

DIRECTIONAL CONTROL
FLOW CONTROL
PRESSURE CONTROL
LOGIC CONTROLS
ELECTRO–PNEUMATICS
PROGRAMMABLE CONTROLLERS

# Directional Control

IN A pneumatic system, compressed air flows from a compressor or receiver to an actuator or control device and then, after doing work, is exhausted to atmosphere. Directional control of the air is therefore only carried out by control valves.

**Directional Control Valves**
Directional control valves are categorized by the number of port openings or 'ways', *eg:* two-port (two-way) valve either opens or closes a flow path in a single line; three-port (three-way) valve either opens or closes alternative flow paths between one or other of two ports and a third port.

Other basic configurations are four-port(four-way) and five-port (five-way).

Such valves can also be described by number of *positions* provided and also whether the outlet is open or closed in the non-operated position, *ie* no-mally open or normally closed, respectively. Description may also be simplified by giving number of ports and positions as figures separated by a stroke, *eg* a 4/2 valve is a four-port two-position valve.

Figure 1, for example, shows a three-way two position directional control valve in simplified diagrammatic form.

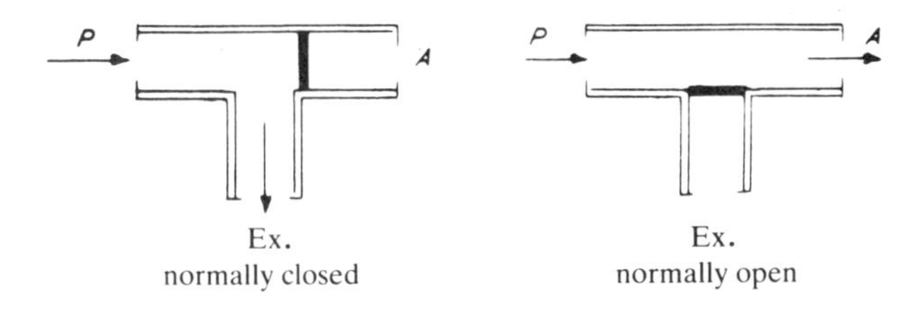


*Figure 1*

The three-way two position valve is thus a logical choice for a single-acting cylinder circuit control or for any other single circuit where downstream air has to be exhausted. It is not necessarily a complete answer to such circuits and there are cases where the use of two two-way two position (straightway) valves may be preferred.

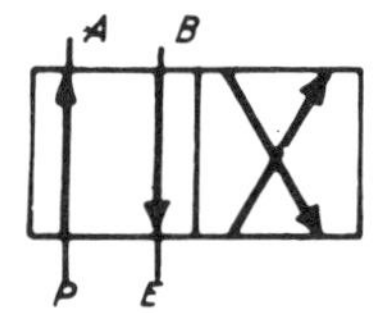


*Figure 2*

Increasing the number of ports, and if necessary the number of positions, extends the switching capability of a valve. Thus, the four-way two-position valve of Figure 2 has two through-connections. In one position (P) is connected to A and B can exhaust downstream air through E. In the second position, P is connected to B and A can exhaust downstream air through E. Thus a four-way two position valve can be used to operate a double-acting cylinder or any other device requiring alternate pressure and exhaust on two connecting lines.

While this is a logical control valve choice for a double-acting cylinder there are instances where the use of two three-way two position valves may be preferred, *eg* for close coupling or mounting the valves directly on the cylinder ports.

The five-way two-position valve in Figure 3 is similar to the four-way valve, except for the provision of an additional exhaust port. Thus, when either A or B is switched to exhaust, it operates through a separate exhaust. This can have specific advantages in particular applications.

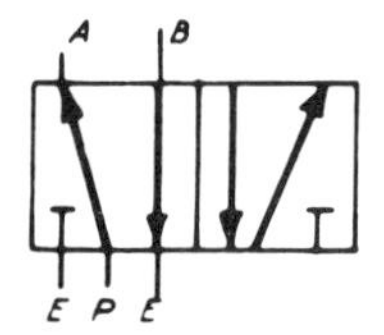


*Figure 3*

The other directional valve is the four-way three-position valve, which can have three possible modes Figure 4. In the first (left) the function is similar to a four-way two position valve except that in additional position is available with all ports blocked, *ie* no flow is possible through the valve in either direction. This is usually the normal position. A typical application is the control of a double-acting cylinder in which 'hold' facility is also available, *ie* pressurized air can be trapped on both sides of the piston.

In the second mode (centre), conventional four-way switching is available from the two extreme positions, but the mid-position shuts off the supply and connects both downstream lines to exhaust through the valve. Again this type of control is

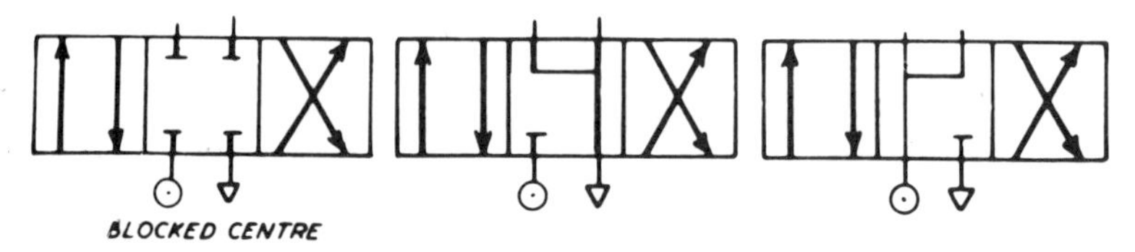


*Figure 4*

used with double-acting cylinders where it is required to free the piston by exhausting air from both sides of the piston.

In the third mode (right), normal switching is available from the two extreme positions but the centre position provides pressurized air to both downstream lines, with the exhaust closed. This provides an alternative hold facility for double-acting cylinders with both sides of the cylinder continuously pressurized. This will not, of course, give true hold on a single-rod cylinder, unless the pressure is dropped in the line to the blank end side of the piston to compensate for the loss of effective piston area on the rod side.

**Standard Valve Symbols**

Symbols used for designating valves on pneumatic circuits are conventionally in the form of adjacent squares, each square representing one position of the valve. Thus a two-position valve would have two adjoining squares and a three-position valve three adjoining squares. Interior connections between points in each position are then indicated by an arrow or arrows within the appropriate square, together with some indication of unconnected points.

Standard valve symbols adopted through Europe are shown in Figure 5. Additional symbols may also be appended to show the method of actuation Figure 6.

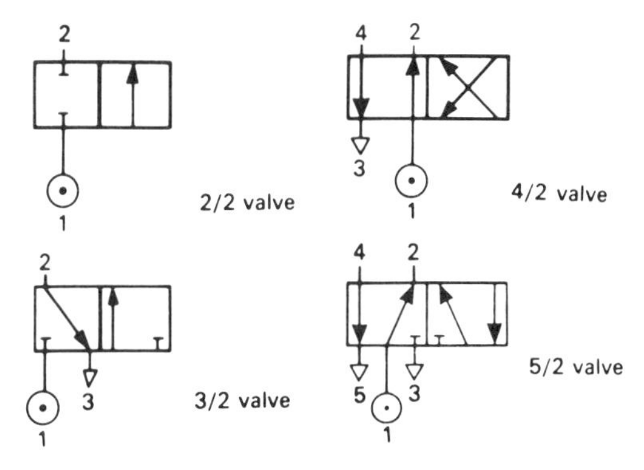


*Figure 5*

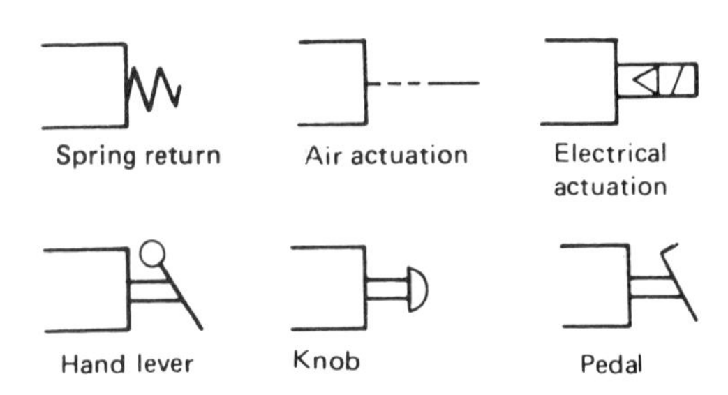


*Figure 6*

In addition main ports may also be indentified by numbers, *viz:*

1 = Normal inlet, *ie* main supply port.

2 = Normal outlet port on 3/2 and 3/3 valves.
2 & 4 Normal outlet ports on 5/2 and 5/3 valves.
3 Normal exhaust port on 3/2 and 3/3 valves.
3 & 5 Normal exhaust ports on 5/2 and 5/3 valves.

Port 1 is the main supply port, other odd numbers denote exhaust ports and even numbers outlet ports. Port 3 is always internally connected to port 2 when the valve is in one position. In the reverse position on a 5 port valve port 5 is always connected to port 4.

Pilot control ports are identified as follows: '10', '12' and '14'.

*Note:* Although this numbering is standardized it is not always used. Arbitrary numbering may be used for simplification or convenience in general descriptions.

**Valve Configurations**

The majority of valves used in the UK are of in-line configuration, *ie* designed for direct connection of inlet, outlet and exhaust lines. Continental Europe and USA favour the use of sub-base valves, *ie* with valve bodies in block form for mounting on a common baseplate or manifold. All lines are connected to the base.

With sub-base or manifold mounting the valve body normally has no tapped pipe connections. All internal ports are brought to the base of the valve and connection is completed by mounting on a matching sub-base or manifold carrying corresponding ports. Joints in such cases are normally sealed with gaskets or O-rings.

Sub-base mounts are designed to accommodate individual valves. Manifolds are designed to accommodate a number of valves on a common mount.

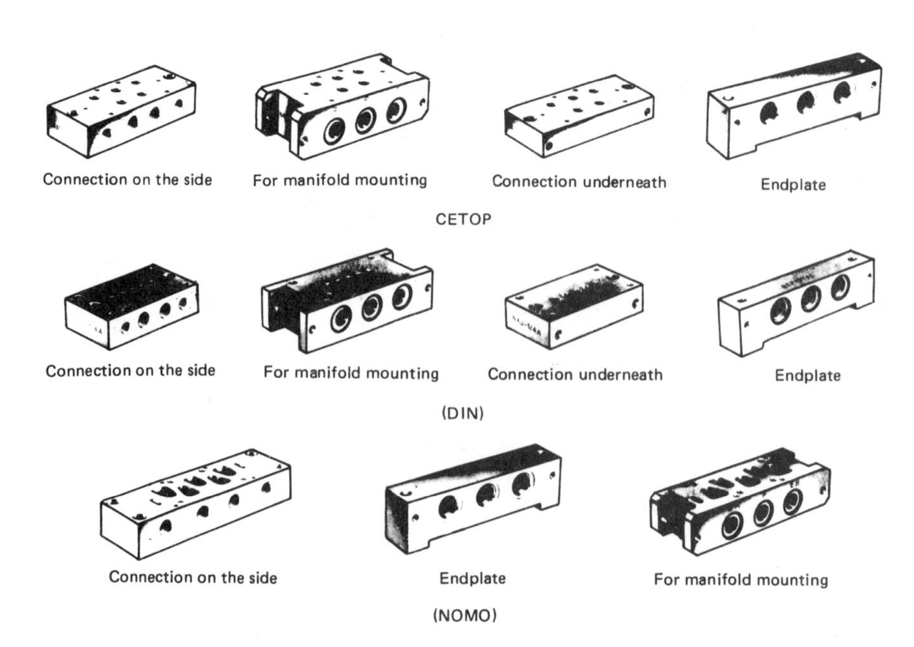


*Examples of sub-base patterns.*

**Shuttle Valves**

The shuttle valve is an 'automatic' type of directional control valve with three-way two-position characteristics. This is based on a simple three-port body with a moving element in the form of the free spool or shuttle or a ball (Figure 7).

A *shuttle* valve gives an outlet from either one of two pressurized inlets (whichever is the greater value), *ie* the higher pressure inlet moves the shuttle to a position where it blocks the other inlet. It is thus more of a special valve than a true directional control valve and is basically used for logic OR switching.

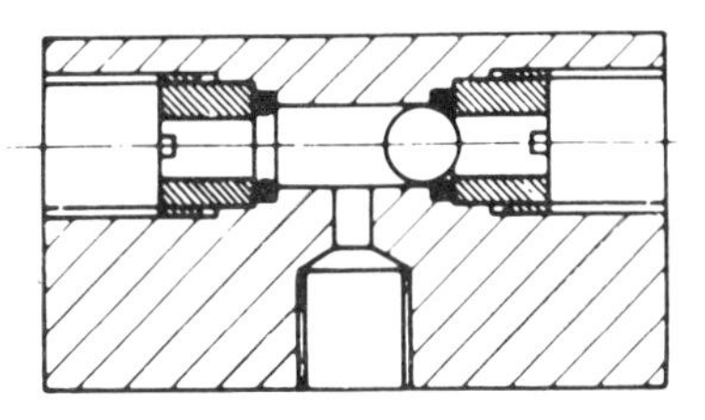

*Figure 7(a)*
*Shuttle valve with ball element.*

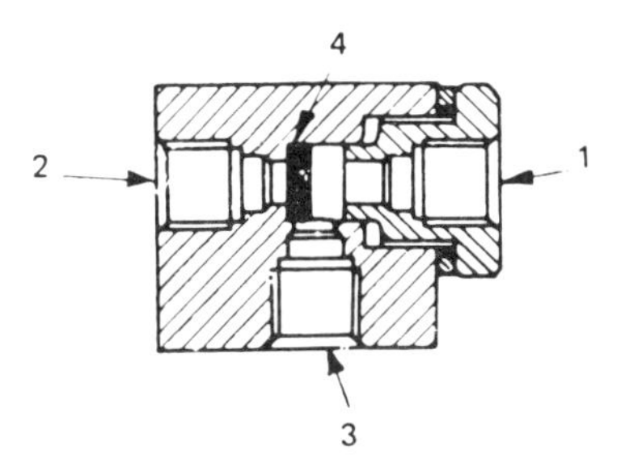


*Figure 7(b)*
*Shuttle valve allowing either one or two pressures to be applied to a single outlet without mutual interference.*

**Non-Return Valves**

Non-return valves allow flow of the air in one direction only, the other direction through the valve being at all times blocked to the airflow. The valves are usually designed so that the check is pressurized by the downstream air pressure to support the non-return action.

The simplest type of non-return valve is the *check* valve which completely blocks airflow in one direction and allows flow in the opposite direction with minimum pressure drop across the valve.

As soon as the inlet pressure in the free flow direction overcomes the internal spring force the check, which can be either a ball or poppet is lifted clear of the valve seat. Alternatively, the check may be lifted off the valve seat by some means, as in the case of a check valve within a quick-connect coupling.

Check valves are installed where different components need to be isolated or where restriction of flow through a component in one direction only is required for safety considerations.

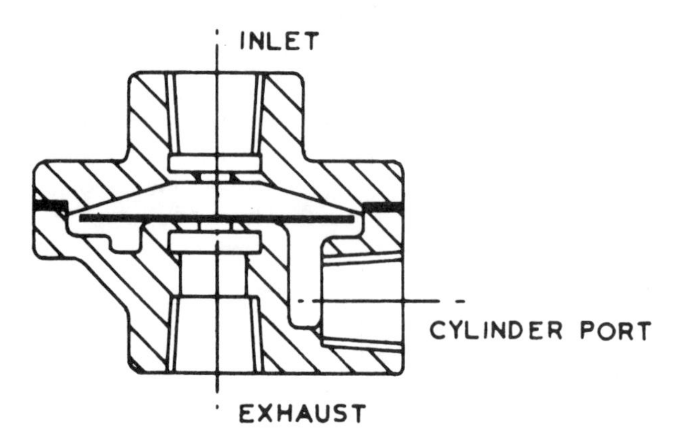


*Figure 8*

**Quick Exhaust Valves**

A quick exhaust valve, as shown in Figure 8, is designed to increase the speed of movement of a cylinder by allowing the exhaust air to vent directly to atmosphere. Ideally a quick exhaust valve should be screwed directly into the cylinder ports to obtain maximum effect.

Air pressure entering the inlet from a selector valve, forces the *cup* seal against the seat of the exhaust port and flows into the cylinder port. The cup seal keeps the exhaust port closed until the air connection from the selector valve is unloaded.

The exhaust air flowing from the cylinder port presses against the lips of the seal, forcing it against the inlet port to close it. The exhaust air then flows from the cylinder directly to atmosphere without having to go back through the *selector* valve.

# Flow Control

AIR FLOW is controlled by using throttling devices. Their geometry can range from that of a fixed orifice to adjustable needle valves or shaped grooves, *etc*, capable of providing progressive throttling. Their basic action may also be combined with a second function such as non-return in the opposite direction.

**Speed Control Valves**

Speed control valves for pneumatic applications are usually *restrictor check valves*. The throttling function is provided by a flow control orifice and the incorporation of a check function also makes them non-return valves when the orifice is closed. Usually the throttle is adjustable to permit regulation of air flow through the valve, with throttling in one direction of flow only. In the other direction free flow is provided through the check valve.

Figure 1 shows a *needle exhaust port flow regulator*. This type of valve is for use in conjunction with a five-port fully balanced spool valve for cylinder speed control.

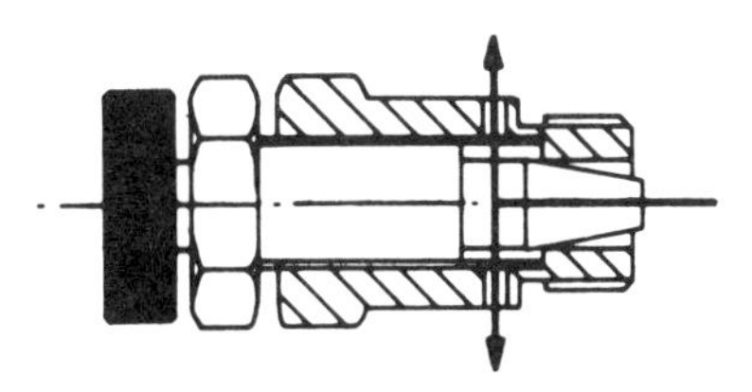

*Figure 1*

The setting of the needle regulates the rate of flow of air, to exhaust, through the control valve, thus regulating the speed of movement of the cylinder. A flow regulator is required in either of the two normal exhaust ports of a five-port valve to control the speed of movement of the cylinder in both directions.

A secondary effect of exhausting air directly to atmosphere from the valve is that it tends to produce a noisy operation. This can, to some extent, be overcome by the fitting of *exhaust silencers*. Exhaust flow regulators can be supplied as complete regulator/silencer assemblies comprising the regulator and a small sintered bronze silencer element (muffler) which is screwed to the regulator with a knurled coupling nut, (see Figure 2).

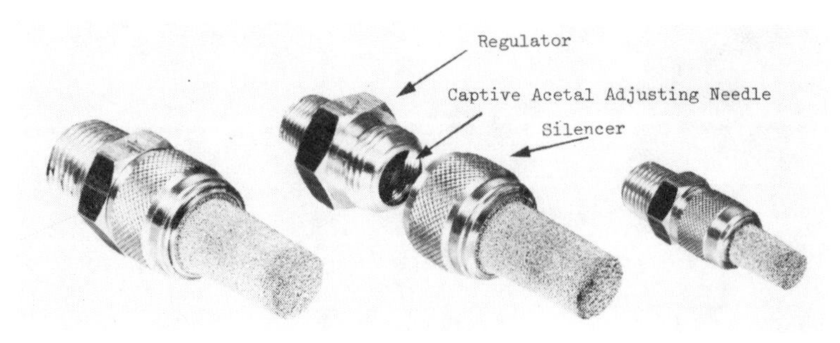


*Figure 2*
*Exhaust flow regulators/silencers.*

The regulator unit is a small neat assembly of brass body and inset captive acetal adjusting needle, which can adjust flow from zero to the maximum within six or seven turns, depending on the size selected. As the adjustment is partly hidden from view it is unlikely to be tampered with by unauthorized personnel. The addition of the silencer makes the flow setting even more tamper-proof.

A selection of in-line and *elbow flow* regulators, together with exhaust flow regulators are shown in Figure 3. The elbow flow regulators are designed to be directly screwed into the ports of double-acting cylinders, to give precise control of exhausting air flow to regulate stroke speed.

The regulators can also be used to control exhaust flow from valves, particularly when there is a requirement to pipe exhaust air away, as convenient 'push-in' fitting connections for Nylon tubing are included.

An alternative method of fitting a valve directly into the cylinder ports is with a *banjo* valve as shown in Figure 4. These provide an efficient method of simplifying speed control on cylinders where installation space is restricted.

*Figure 3*
*In-line and exhaust flow regulators.*

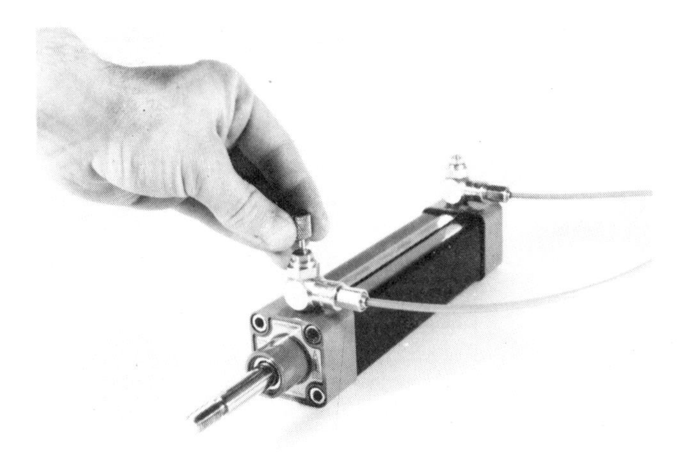

*Figure 4*
*Banjo flow control valves.*

Inlet air flow is unrestricted and flow out of the cylinder is controlled by an *Allan*-headed adjusting screw in the top of the valve, with a removable brass adjuster as shown. The banjo facility allows the valve to rotate freely about the centre of the cylinder port. This greatly simplifies piping up and the valve is locked to the required position as the thread is pulled down onto a fibre sealing washer.

**Speed Control**
The speed of operation of an air cylinder can be controlled by throttling either the inlet of outlet air or both. The degree of speed control available from such simple methods is suitable for a wide range of applications, although the actual response time will still be variable, if the load is variable, because of the varying compressibility of air against different loads.

Actual control of speed as such, however, is often less important than reducing the speed of operation to a satisfactory level or avoiding excessive operating speeds which increase wear on the cylinder and waste air.

Speed control accomplished by metering the inlet air will only produce a constant speed if the resisting force is constant (see Figure 5). Basically, pressure will first build up in the cylinder until the force available is greater that that of static friction (stiction) and resistive load, when the piston will start to move.

It will then continue to move forward at a more or less constant speed if the

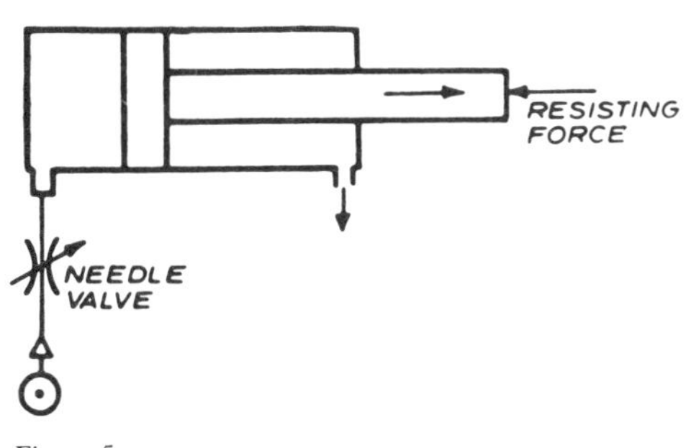


*Figure 5*

resistive load is constant, with speed governed by the rate of air admission. Any increase or decrease in resistive load will cause the speed of movement to slow down or increase, respectively.

Speed control by metered exhaust (Figure 6) will tend to give an initial rapid movement followed by a slowing down as air is compressed on the exhaust side of the piston. The movement is thus likely to be jerky. Much better control is provided if the modified circuit of Figure 7 is used where the piston is held in the retracted position under pressure.

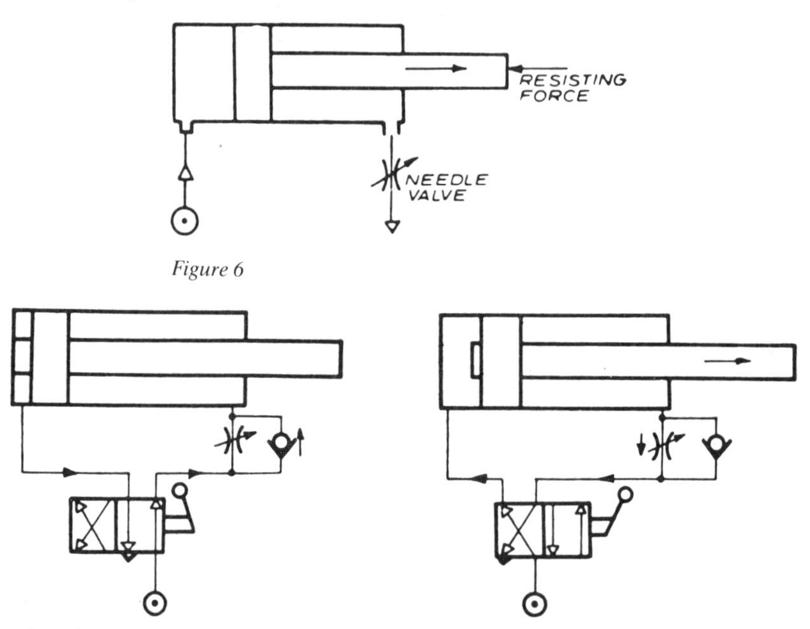


*Figure 6*

*Figure 7*

When the control valve is operated, the piston will start to move forward through pressure applied to the head before the air has had time to leak away from the rod end. The result is a freedom from initial jerk and smooth movement throughout. Also, tendency to jerk forward should the resisting force be removed is minimized.

The general recommendation is that speed control should preferably be applied by throttling the exhaust from a cylinder. A combination of both methods may, however, be used in specific applications.

Provided the load is reasonably constant, quite good accuracy can be obtained with such simple speed controls. With very low piston velocities, however, friction may become the most important parameter.

To achieve consistant results with speed control it may be necessary to use cylinders with very smooth and well-finished bores and low friction seals. This can be taken as applying when the piston velocity is of the order of 25 to 50 mm/s (1 to 2 in/s) or less.

For more precise speed control, and particularly to take care of fluctuations in load and/ or line pressure, surge damping may be applied to the system. This would normally demand the use of an 'oversize' system with a surge tank in the circuit, forming the equivalent of an accumulator in a hydraulic circuit.

# Pressure Control

THE INCLUSION of relieving devices into pressurized systems has been commonplace since the earliest days of steam power and has continued with the introduction of pneumatics as a power source.

Initially their use in pneumatic systems was restricted to limiting system pressure to a safe level in the event of a failure of control devices. As the industry developed, relieving devices were included to control the pressure as part of the normal working operation of the system.

**Pressure Control Valves**

Pressure regulating valves are known by specific terms in the pneumatics industry depending on the function they perform and the way they are applied to the system.

A *Back Pressure* Regulator is a valve with its inlet port connected to the system so that by automatic adjustment of its outlet flow to waste, the system pressure remains substantially constant. System pressure must, however, remain equal to or higher than the pre-set opening pressure of the valve. Under normal operating conditions the valve may be continuously flowing to waste.

A *Pressure Reducing* valve has its outlet port connected to the system so that with varying inlet pressure or outlet flow the outlet pressure from the valve remains substantially constant. Inlet pressure must, however, remain higher than the selected outlet pressure.

A *Pressure Relief* Valve has its outlet port connected to the system and limits maximum pressure by exhausting fluid when the system pressure exceeds the pre-set pressure of the valve. Under normal operating conditions the valve remains closed, only opening when unusual system conditions prevail.

A *Safety* Valve is the name given to a special type of pressure relief valve which protects a system from over-pressurization in the event of malfunction of a component forming part of the system. It opens and exhausts fluid when the system pressure exceeds its design pressure by a pre-selected amount. The valve is required to have a certified flow capacity sufficient to prevent the system pressure exceeding its design pressure by more than a stipulated percentage. Under normal operating conditions a safety valve remains closed and in the event of it functioning, the cause should be established before the system is allowed to be put back into operation.

A *Safety Lock-Up Valve* has its inlet port connected to one part of the system and

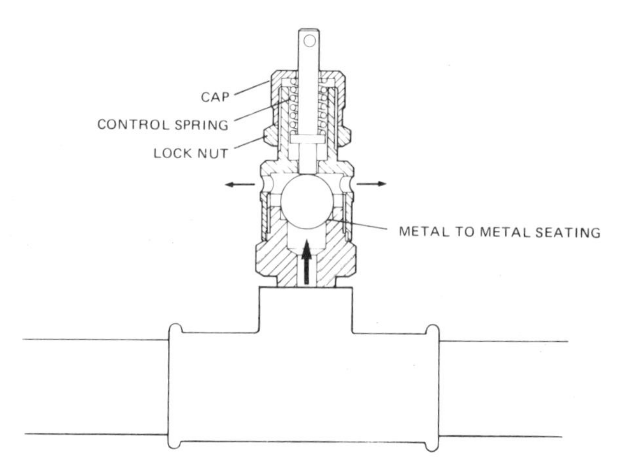


*Figure 1*
*Direct loaded relief valve.*

its outlet port to another part of the system so that should the inlet pressure fall below a preset amount the valve will close to lock up a minimum pressure in that part of the system connected to the outlet port.

A Sequence Valve has its inlet port connected to one part of the system and its outlet port connected to another part of the system and set so that until the inlet pressure rises above the preset opening pressure the fluid medium is prevented from flowing to the outlet port.

Pressure relieving valves can be either 'directly' loaded, 'pilot-controlled' or 'pilot-operated', depending on the type of design.

A *Direct Loaded Relief* valve is a relief valve in which the loading due to the fluid pressure underneath the valve disc is opposed by direct mechanical loading such as a spring. Direct loaded valves are divided into two main classes: a 'pop' type valve in which the sealing member is directly loaded onto a seat by means of a spring so that the system pressure acts directly on the sealing member, (Figure 1); and a

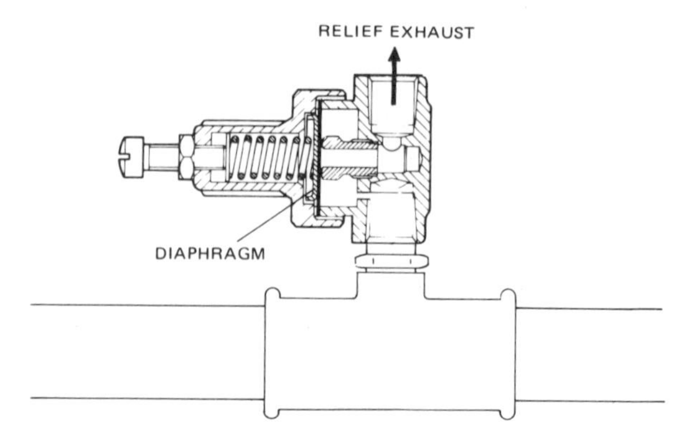


*Figure 2*
*Indirectly mounted diaphragm relief valve showing metal to metal seat.*

'diaphragm' valve in which the spring acts against a diaphragm which is exposed to system pressure (Figure 2).

If a metal ball is used to seal off the orifice the metal to metal seating of the ball on the body has the advantage of virtually no stiction between the ball and the seat. However, the problem with metal-to-metal seat designs is that it is difficult to guarantee a leak tight seal when producing units.

To overcome this short-coming, elastomeric material is introduced at the sealing area, as shown in Figure 3. The disadvantage of this type is that the 'bedding-in' of the valve creates stiction, which causes variations in the lift pressure, especially if the unit is not operated for long periods.

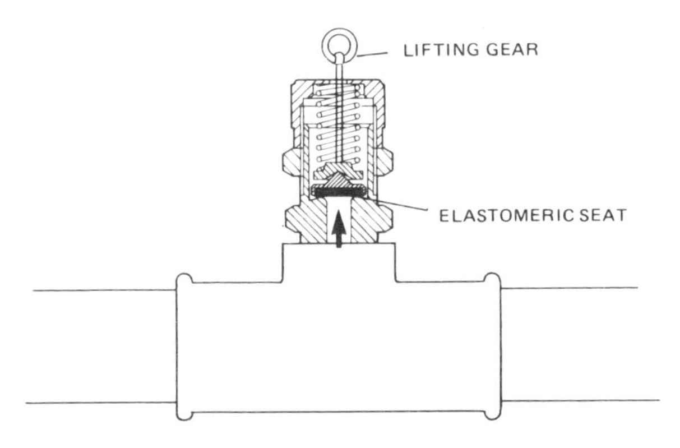


*Figure 3*
*Direct loaded relief valve showing elastomeric seat.*

Diaphragm valves help to eliminate some of the short-comings of the pop type valves, because pressure is sensed on a much larger area that the limited area of the seat. With this design the control spring exerts a force onto the diaphragm to close the orifice.

Provided that the applied fluid pressure beneath the diaphragm is lower than the spring force, the valve remains closed. Stiction can occur with this type, if the diaphragm is pressed directly onto the orifice. This is overcome by fitting a metal disc to the underside of the diaphragm and O-ring seal onto the orifice.

Proper design allows only a small amount of compression of the O-ring: sufficient to prevent leakage, but not to incur significant stiction, (see Figure 4).

A *Pilot-Controlled* Relief Valve is usually of the diaphragm type, in which the loading spring is replaced by fluid pressure from a controlling pilot. The pilot is normally a pressure regulator mounted remote from the relief valve and having its own fluid supply which may or may not be the same as the system fluid which the relief valve controls.

These valves provide superior operational characteristics to those obtained from direct spring controlled systems. Pressure is applied to the control side of the diaphragm from a pilot regulator, which can either be directly or remotely connected to the control port, (see Figure 5).

It is essential to incorporate a small bleed in the pilot control line or use a constant bleed pilot regulator, to prevent control variations due to temperature

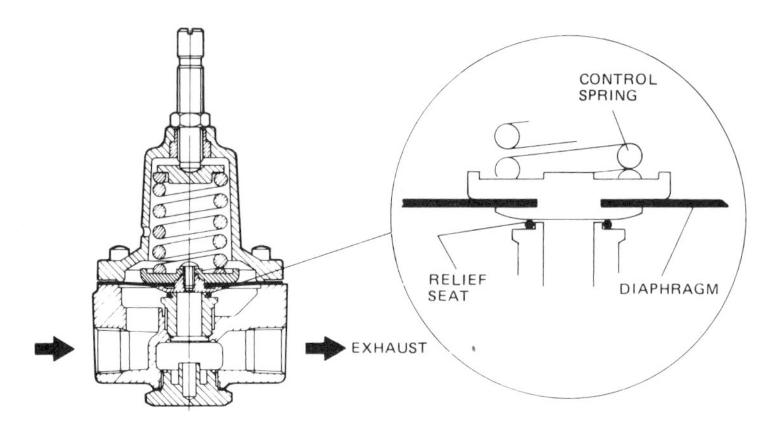


*Figure 4*
*Diaphragm relief valve seat.*

fluctuations and to improve accuracy by having a small continuous flow of air through the pilot regulator.

The controlling of relief valves by pilot regulators has short-comings. Malfunction of the pilot regulator may cause failure of the relief valve, because one cannot guarantee that regulators will fail safe. Also in some circumstances the need for a separate air supply to the controlling pilot is a disadvantage.

This problem is virtually eliminated by using integral Pilot-Operated Relief Valves. These are frequently of the diaphragm type, in which a small integral spring operated relief valve controls the opening and closing of the main valve seat. System pressure normally keeps the main valve closed and no external control fluid supply is required. They are sometimes referred to as 'spring-controlled integral pilot-operated valves', as shown in Figure 6. They are virtually two relief valves in a single body.

The actual relief pressure is set by means of the spring controlled pilot relief valve. Mains air passes through the unit and is in contact with the underside of the lower diaphragm assembly. It is also fed through a small drilling into the pilot

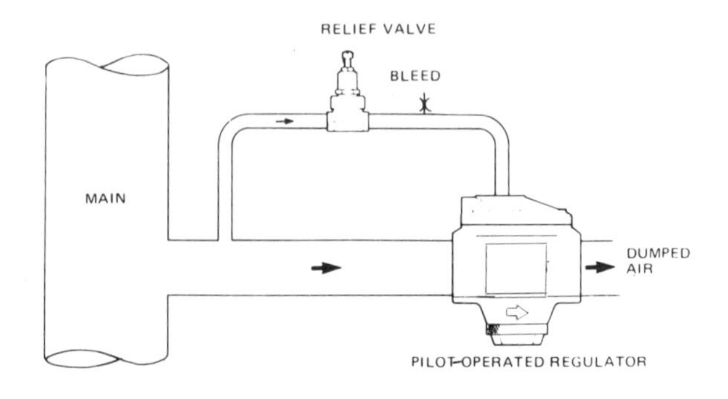


*Figure 5*
*Pilot-controlled relief valve.*

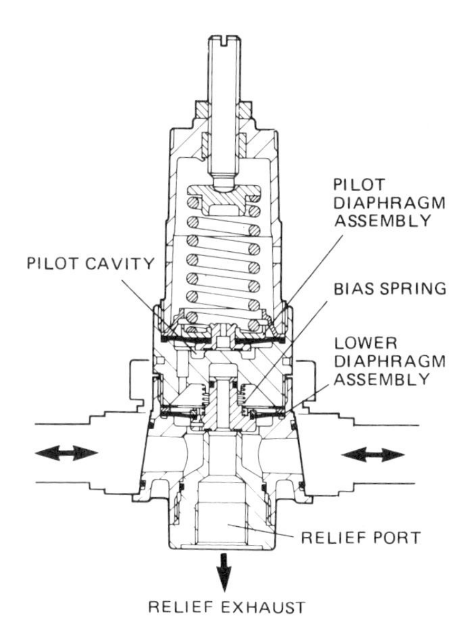


*Figure 6*
*In-line integral pilot-operated relief valve.*

section to pressurize the area above the main diaphragm so that equal pressure is obtained on either side.

The main relief seat remains closed due to the action of the bias spring situated above the assembly. Air also passes through the drilling into the cavity beneath the pilot diaphragm, which is counter-balanced by the force exerted by the control spring.

If the system pressure rises above the control spring setting, the upper diaphragm lifts, allowing the small volume of air within the unit to exhaust through the bonnet vent hole. This removes the pressure above the lower diaphragm. The system pressure raises this diaphragm – against the low pressure exerted by the bias spring and opens the relief port situated directly beneath the lower diaphragm to allow the excessive pressure to be bled away to atmosphere through the exhaust port situated in the base of the valve.

# Logic Controls

LOGICALLY CONSTRUCTED control circuits have numerous advantages over empirically-designed circuits (which are often a combination of intuition and 'guesstimation', finally proven by trial and error). These advantages show up particularly in complex control circuits, where only design by logic can be relied upon to provide the optimal solution.

Logic functions can readily be performed by simple valves or combinations of valves. Interest in pneumatic logic circuits was originally stimulated by the development of pure fluidic devices capable of extreme miniaturization and also incorporating no moving parts. While such a system can be shown to work it has several limitations in practice, notably the fact that it is essentially a low pressure system operating at air pressures of less than 1 bar, consequently needing amplification to power an actuator. Also such a system is critically dependent on air regulation and fine filtering of the air supply. It thus remains mainly of academic interest only.

Its counterpart in providing digital working – or logic circuitry – derived from the miniaturization of more or less conventional valves, *ie* moving part logic (MPL) elements operating at normal system pressure and capable of operating at high speeds. These now form the basis of virtually all practical pneumatic logic circuits and have been adopted on a relatively large scale.

Properly designed MPL elements are reliable, positive in response and suitable for use with both dry and lubricated compressed air supplies. Also the question of providing interconnections compatible with the size of these elements has been solved by the development of modular construction and assemblies.

The basic problem then remaining is how to design control circuits in terms of logic. The basic 'tools' are *Boolean algebra, Karnaugh-Veitch maps* and *Cascade techniques* primarily (there are others). Alternatively, there are 'half-way' measures, such as the *Step-counter method* of electro-mechanical sequencing. All of the basic methods need considerable study to master and apply, and are skills not readily acquired by the intuitive designer.

For that reason, producers of modern MPL components have developed their own system methods to make logic design easier to understand and minimize circuit design time. This simply parallels computer technology where special and simpler languages have been devised to make programming easier. In that case the language is tied specifically to a particular system, which in turn is based on a particular choice of logic elements providing all the logic functions necessary. This, like the language, can vary with different systems.

**Basic Logic Functions**

The basic logic functions are NOT, AND and OR, but for logic control functions we also need to add YES (which is only a modified form of NOT). The NOT function can be performed by a two-way two-position valve, normally open. A signal applied then moves it to closed, *ie* a signal gives no output. There is output only when there is NOT an input.

Simple two-way valves can be used in combination to provide AND or OR logic response (Figure 1). Equally a two-way valve can invert an output to produce the NAND or NOR functions or invert an input to provide INHIBITION.

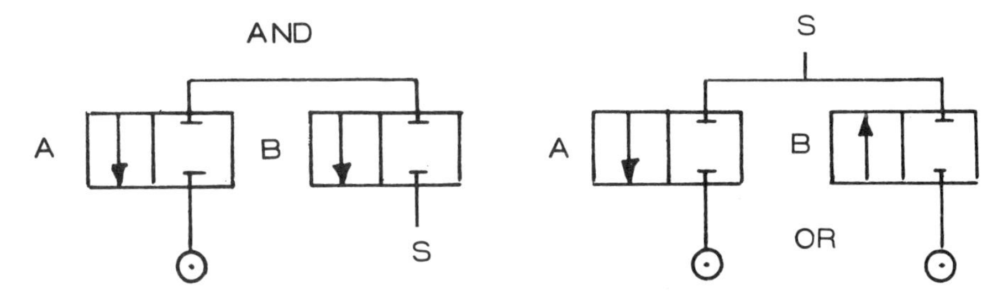


*Figure 1*

'Logically', in fact, a combination of on-off devices can provide all the logic responses required, except for MEMORY. This would require a combination of two-way valves. In practical pneumatic logic, however, it is far more realistic to devise single components to provide the logic functions required, which for full coverage of pnuematic logic circuit design requirements are: NOT, AND, OR, YES, NAND, NOR, INHIBITION and MEMORY.

Operating characteristics of these devices are detailed in Table 1. Again this listing represents a redundant requirement, *ie* not *all* these functions are necessary to cover all requirements in a logic circuit. The number of individual devices required can be reduced considerably by working with particular type of logic, *eg* AND logic, OR logic, NAND logic or NOR logic or combinations of these

Certain functions will, however, remain necessarily common with any type of logic chosen, *eg* NOT, INHIBIT and MEMORY. As far as pneumatic logic circuit components are concerned the choice of type of logic depends largely on the suitability of miniaturized valves to perform the various functions.

*Pneumatic logic units.*

**TABLE 1 – LOGIC CIRCUIT DEVICES**

| Logic | Pneumatic Symbol | Logic Symbol | Notes |
|---|---|---|---|
| NOT (three-way valve) | 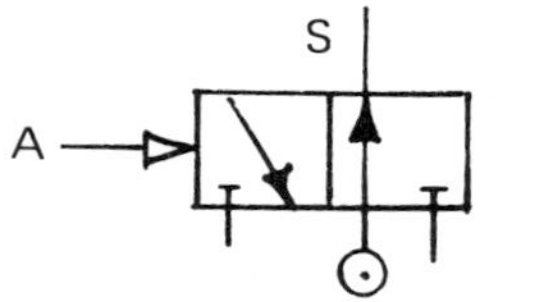 S, A | 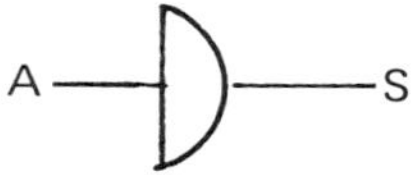 A, S | Output is present when there is NOT an input at A<br>NOT A = S or $\bar{A}$ = S<br>A = S |
| AND (tandem two-way valve) | 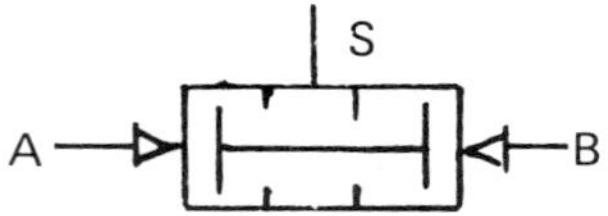 S, A, B | 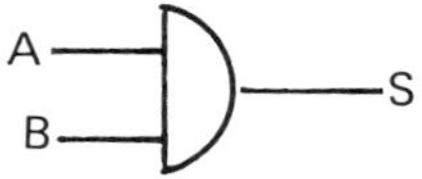 A, B, S | Output is present when there is input at A AND B<br>A *and* B = S or A.B = S |
| OR (shuttle valve) | 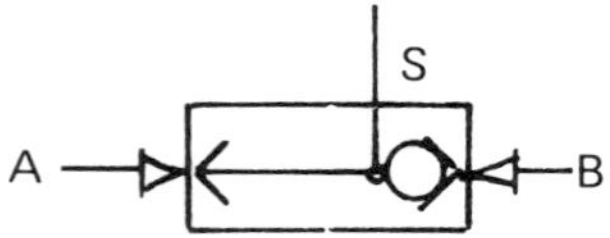 S, A, B | 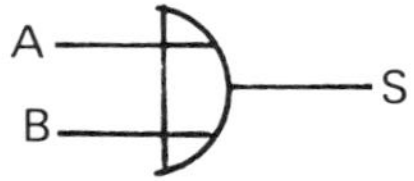 A, B, S | Output is present when there is input at A OR B.<br>A *or* B = S or A + B = S |
| INHIBITION (three-way valve) | 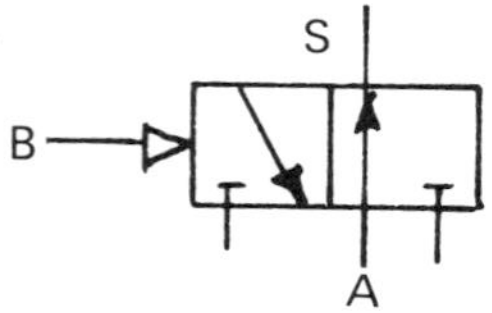 S, B, A | 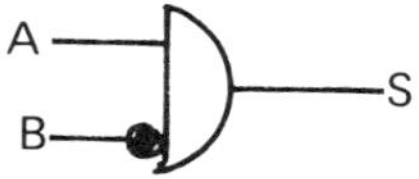 A, B, S | Output is present when there is input at B and there is no input at A. In other words, when a signal is applied to A it inhibits the signal path of B.<br>B *and not* A = S or B.$\bar{A}$ = S<br>B.A = O |

| | | | |
|---|---|---|---|
| NAND<br>(tandem two-way valve and three-way valve) | | | Output is present when there is *not* a signal, input at A and *not* a signal input B.<br>*not* A *and not* B = S, or<br>A.B. = S |
| NOR<br>(shuttle valve and three-way valve) | | | Output is present when there is *not* an input signal at A *or* an input signal at B.<br>*not* A *or not* B = S, or<br>A + B = S |
| YES<br>(three-way valve) | | | This is basically an amplifier. Line pressure is connected to B. This pressure is then available at the output when there is a signal input at A. (A = S)<br>In basic terms, this is an inverted NOT with a second input. |
| MEMORY<br>(four-way valve) | | | This is a flip-flop device. A signal input at A will give output state S1 and the memory will remain in this state, even when signal A is removed. A signal input at B then changes the output state to S2. It will remain in this state when signal B is removed, until receipt of the next signal A. In other words, the device 'remembers' the last input. |

If a multiplicity of elements is to be avoided, choice lies between NOR and NAND logic, or multi-functional devices. NOR and NAND can be made multiple-input devices when, by selecting either one or a number of inputs the other logic functions can be derived. This reduces the number of physical elements required, but each is more expensive in terms of size and cost; and as deployed throughout the circuit, many will be operating with redundant features. Multi-functional devices can provide the basic functions of OR, AND, NOT and MEMORY in smaller (individual) packages but using four-way spool valves can present problems in miniaturization.

It is virtually impossible to generalize on this subject, different manufacturers of systems having different methods of approach and component construction. Thus only one system, which has proved particularly acceptable in Europe, will be described in more detail.

**Pneumaid System**

This system adopts the multi-element approach and is based on five functions – OR, AND, NOT, YES and MEMORY. These devices are shown in Figure 2. It is also allied to a sophisticated mounting system (Polylog) made up of standard modules which can be built up to form a complete monobloc control centre without interconnecting piping.

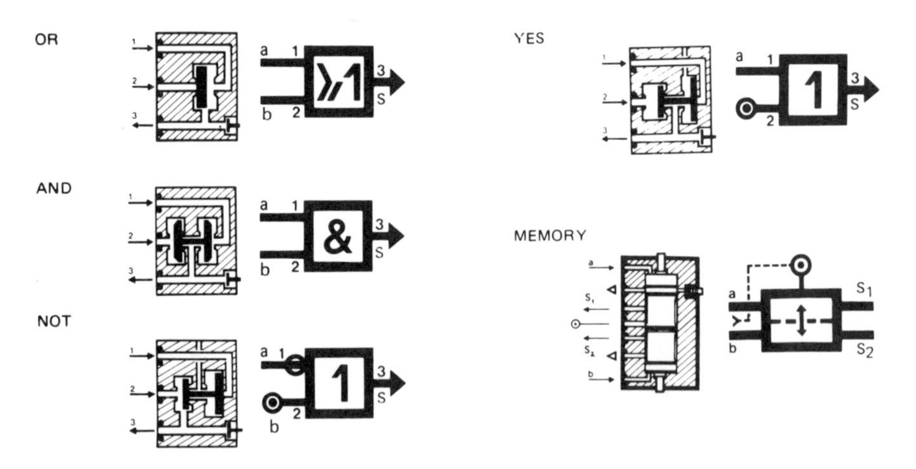


*Figure 2*

The devices can be divided in two categories: active and passive. The active devices are basically three-way diaphragm-operated, air-return poppet valves, normally open for the NOT function and normally closed for the YES function. These units have separate supply and exhaust ports in addition to the input and output ports.

Passive devices are simple poppet valves actuated by the input signals only, and can perform the AND and OR functions. The MEMORY function is normally supplied by a miniature four-way slide valve.

All MPL control systems required more than the five basic elements listed earlier. It is possible to produce a number of auxiliary devices from these basic gates.

With a fine restrictor valve having full flow in the reverse direction and a pressure reservoir (or even the capacity of the interconnecting tubing), it is possible to construct simple RC time delays with good repeatability up to 30 s. By combining these elements with the NOT and YES devices both normally open and normally closed delay functions can be achieved.

Similarly if a common signal is fed to the supply and input ports of a normally open delay arrangement, the result is a step-to-pulse converter which can be used to break the trapped signal in the pneumatic system. This is a problem with which most circuit designers are only too familiar. In a number of systems the logic device, reservoir and restrictor are stacked into integral time delay devices which occupy only one position of the mounting matrix.

By the addition of a simple diaphragm/nozzle pilot amplifier stage to the YES unit, a pressure amplifier can be developed which can accept signals from back pressure air sensing units.

The diaphragm-operated NOT unit is used in the multi-device logic system has been designed to carry out a function other than that of logic inversion. By setting the pilot pressure to reset at 10% (or less) of the supply pressure, the unit can be used to signal the end of the stroke of a pneumatic cylinder.

There is a refinement of this system in which the NOT unit supply pressure is derived from a port in the cylinder wall set just behind the piston when it has completed its stroke.

**Polylog System Design**

A circuit using the Polylog system can be designed and built after only one or two days' training. In fact it is possible to design and build the simpler circuit after only two or three hours' tuition. Therefore, no longer is it necessary to be a scholar of Boolean algebra and be able to calculate the resulting equations. Nor does one have the knowledge to find one's way round a Karnaugh map.

With the Polylog system there is a direct relationship between the machine cycle, the circuit and the way the elements are assembled. However, to make circuit design easier a simple form of diagram called *Grafset* is used. This is a European standard to show the graphical representation for expressions of automatic sequences. It is a step/command/transition control graph and a typical diagram is shown in Figure 3. This diagram relates to a circuit involving two cylinders A and B.

The right-hand squares represent the movements of the cylinders. A+ refers to cylinder A outstroking and A− the cylinder instroking *etc.* The squares on the left-hand side represent each step in sequence.

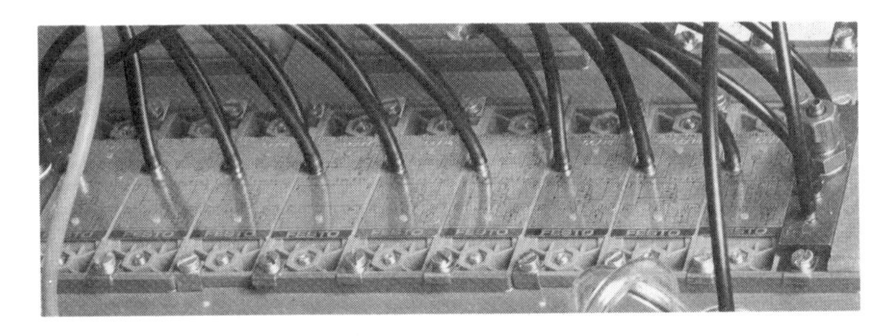

*Shift register chain base-mounted*

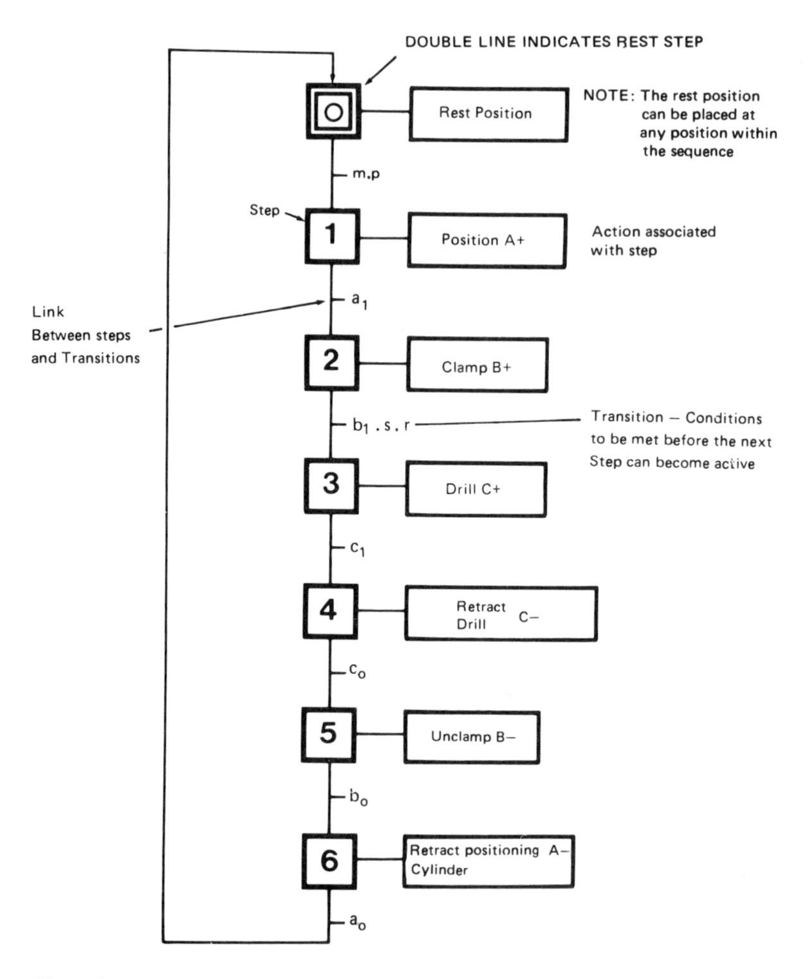


*Figure 3*

These are the principal steps. Each principal step requires a memory module – thus for this application we would require four memory modules. In between each step there are further letters; these represent interlocks confirming that the action of the principal step is complete or external signals setting the sequence in motion. These are called *intermediate steps*. This is basically all the circuit design that is necessary before beginning to assemble the system.

In order that the circuit can be quickly confirmed, special cards are available. When placed side by side these represent exactly the layout of the circuit. In this application the cards when put together would be as in Figure 4, shown together with the complete circuit. In this circuit four memory modules provide the four principal steps in the system. The example represents a very simple application.

**Pneumatic Logic versus Electronics**

Pneumatic logic control circuits work on exactly the same principles as programmed electronic controls – both work on a logic basis. Until comparatively

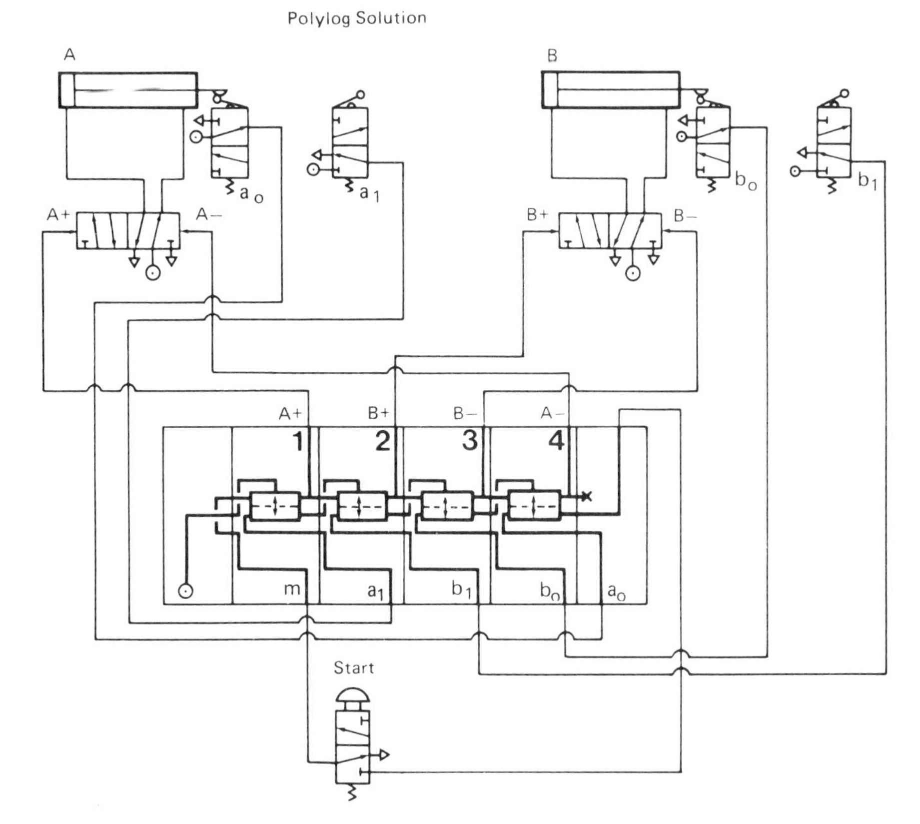


*Figure 4*

recently a pneumatic logic control circuit of medium complexity generally proved cheaper than its electronic equivalent.

With the advent of low cost microprocessors the position has changed considerably. Hybrid systems (electronic control of pneumatic systems) have many advantages to offer at comparable or even lower cost, particularly as the complexity of the control requirement increases. Apart from compactness, the big advantage with a microprocessor is that the programme can be re-written and sequences changed without re-piping the control system.

At the present state of the art, therefore, it would be reasonable to state that where only a relatively few functions are required (say a maximum of ten), pneumatic logic control circuits still have much to offer. Above that number, the Programmable Logic Controller microprocessor offers increasing overall advantages. Where pneumatic controllers can continue to remain unchallenged is in high-risk fire areas or hazardous areas, *eg* involving low and high flash-point materials in processing.

# Electro-Pneumatics

PRACTICALLY EVERYTHING that can be done with electrical controls can be done pneumatically. However, electrical controls may be preferred because of their compatibility with signals coming from electrical sources (*eg* timers), because wiring in neater than piping, or because of the convenience of microswitches to initiate control signals.

Using microswitches, proximity switches, *etc*, with appropriate ancillary devices, in conjunction with double-solenoid valves, very simple circuits can often be obtained. More complex control circuitry can be designed on intuitive lines or by logic.

The more complex the system, the greater the number of sequences required and the less competitive pneumatic control systems become compared with electronics, particularly with the introduction of low cost microprocessors and logic control units (LCUs).

**Solenoid Operation**

Solenoid operation is an electro-mechanical system. The solenoid can be mounted directly on the valve or directly connected to it by a lever, so that when an electrical switch is closed the solenoid is energized to apply direct force to the valve movement. This is *direct solenoid actuation*.

Double-solenoid actuation may be necessary with certain valves, *eg* three-position valves. If the valve is spring-controlled, one solenoid is energized to shift it to one position and the second to switch it to the other position.

Double-solenoids may also be used for *momentary* working where a valve position has to be held for an appreciable time. To reduce the current drain and solenoid heating, the valve is shifted by momentary solenoid actuation, the movement being mechanically latched. The valve then remains in this position until unlatched by momentary actuation of a second solenoid releasing the catch, when it returns under spring action.

An alternative form is *solenoid pilot actuation*. Here the solenoid movement is used to open a small *pilot* valve, which then directs air pressure onto the main valve piston to initiate main valve movement. With this system a relatively small solenoid can be used to actuate quite a large valve.

Another advantage is that the pilot (and its operating solenoid) does not necessarily have to be close to the main valve, but can be mounted at some remote

TABLE 1 – METHODS OF PROTECTION OF EQUIPMENT IN HAZARDOUS ATMOSPHERES

| TECHNIQUE | SYMBOL (Ex) |
|---|---|
| Oil immersion | O |
| Pressurization | P |
| Powder filling | q |
| Flameproof enclosure | d |
| Increased safety | e |
| Intrinsic safety | i |
| Non-incendive | N (n) |
| Encapsulation | m |
| Special protection | s |

and more convenient point to provide optimum location of both pilot and main valves. A solenoid and small pilot valve, when assembled as in integral unit, are generally referred to as *solenoid pilot*.

Although with pneumatics the option of using *air pilot* can be made, the use of solenoids to operate in hazardous areas is also required.

**Solenoid Protection**

The general term for all methods of protection of electrical equipment used in Europe is explosion-proof. In the USA the practice is to use this term for flameproof equipment. Table 1 lists the more usual methods of protection. Not all of these methods are applicable to solenoid protection, the more commonly used are listed.

(1) *Flameproof* protection entails enclosing the coil in a robust enclosure which will contain an internal explosion, should it occur, and prevent its transmission to the surrounding atmosphere.

*Flameproof solenoid enclosure.*

(2) *Non-incendive* (N-type) protection generally applies to non-sparking electrical equipment, such as a solenoid coil, which will not get abnormally hot even if the armature is locked out.

(3) *Intrinsically safe* (i-type) equipment uses a technique for ensuring that the amount of electrical energy, in a circuit, is too low to ignite the most ignitable mixture of gas and air, either spark energy or surface heat, under normal or probable fault conditions.

Intrinsic safety is essentially a low power technique and it is now possible to operate a solenoid within the constraints of the power restrictions. A typical i-type coil with a resistance figure of 370 ohms operates down to 12 V consuming, at this voltage, 33 mA (0.4 W). The coil is provided with diode protection, giving in practice, zero inductance and capacitance measurements.

The supply to the coil must be safe even under short circuit which is provided for by including a safety barrier in the supply line. The barrier consists of a network of zener diodes and resistors to restrict voltage and current to a safe level. The effect of such a barrier in the system is to drop the supply voltage (28 V maximum) by approximately 50% for the 'on load' voltage at the solenoid coil.

(4) *Special* protection is often a combination of one or more methods of protection and in the case of solenoids these are usually e and m types where the coil is potted and has over temperature protection.

**Miniature Reed Valves**

Miniature pneumatic valves have been developed which are sensitive enough to be actuated by the magnetic field of a small permanent magnet or a small coil. They are ideally suited to applications where low electrical power consumption is a requirement and space is limited.

The reed valve consists of two magnetic members, one fixed reed and one flexible reed separated by a small gap and mounted in a rigid plastic body which is then sealed (see Figure 1).

An external magnetic field applied in a direction parallel to the axis of the reed valve exerts across the gap an attraction sufficient to deflect the flexible reed towards the fixed reed, until the gap is closed. A small disc of elastomer bonded to the flexible reed opens or closes an orifice during the motion.

In the normally open configuration, the orifice is open in the absence of a magnetic field and conversely in the normally closed configuration, the orifice is closed in the absence of a magnetic field.

The magnetic field for actuation of the valve is provided by a small permanent magnet or a coil. Fluid flow through the valve is in the direction which tends to close the valve more tightly. Reversal of the high and low pressure connections results in the lifting of the seal from its seat and leakage of the valve.

In a printed circuit board-mounted version, the combination of low electrical power consumption, compact size and light weight makes it an ideal electrical-to-fluid interfacing device for today's microprocessor-based control systems.

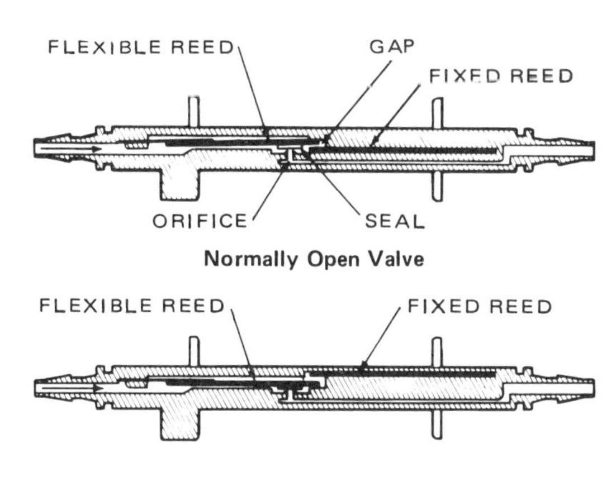


Normally Closed Valve

*Figure 1*
*Miniature reed valves.*

**Cylinder Position Monitoring**

Electrical reed contact switches are used in the opposite manner to reed valves to signal that a cylinder piston has reached a pre-determined position. The adjustable reed switches are mounted on the aluminium alloy barrel of small bore magnetic cylinders as shown in Figure 2.

The piston rod is made of stainless steel and small permanent magnet is fitted to the piston to actuate the normally open contacts of the reed switch when the piston reaches the desired position.

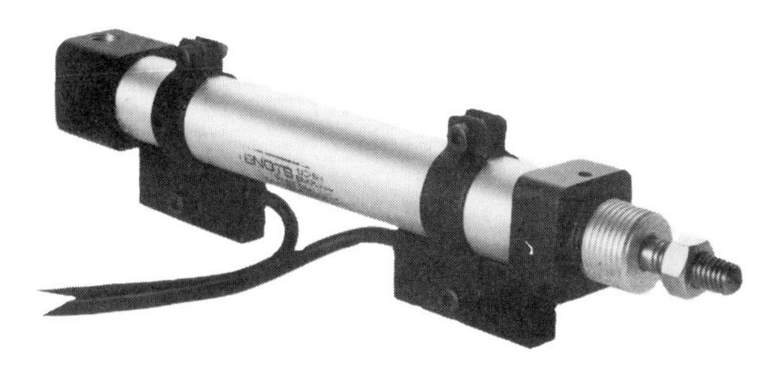

*Figure 2*
*Magnetic cylinder*
*with adjustable reed switches.*

On tie-rod cylinders, manufactured to ISO/DIS 6431 standard specifications with magnetic pistons, proximity sensors can be mounted with a stop screw in any position on the cylinders intergrated tie-rods. The piston's magnetic field actuates the reed contacts of the proximity sensor, which generates an electric output signal in any position along the entire stroke of the cylinder.

If several output signals are desired, several sensors can be mounted on the same cylinder (see Figure 3). The sensors can be connected directly to most control systems of relay or electronic type.

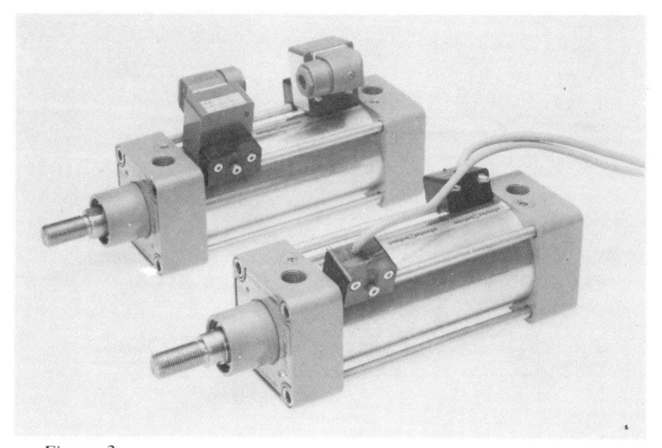

*Figure 3*
*Magnetic ISO cylinders with reed switch sensors mounted on tie-rods.*

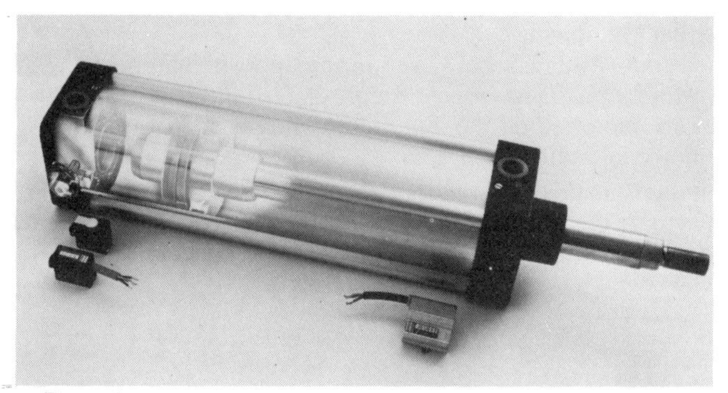

*Figure 4*
*Mechanical end position sensors with electric or pneumatic output signals.*

Mechanical end position sensing of pistons can be arranged to give either electrical or pneumatic output signals. Mechanical end position transmitters are built into the end covers. An output signal is generated *via* an electric or pneumatic detector, which is very easy to mount with a bayonet fitting as shown in Figure 4.

The fitting can be the same for both types of detectors. This makes it possible to change the type of output signal in simple manner without having to change the cylinder in conjuction with system changes, for example. Combinations with sensors are also possible if several output signals, in addition to the end positions, are required.

# Programmable Controllers

THE POINT at which 'mechanization' becomes 'automation' is where the machine operator's role is reduced to that of a supervisory nature, or even eliminated entirely. In other words, the essential feature of command is transferred to a timed operation of the system or semi- or fully-automatic control using a suitable detector generating feedback signals.

The simplest control method is a manually-operated control valve, which, even in an automated system, is normally retained for manual override or as an emergency control.

The manual control valve remains the usual solution when the control required is too simple to justify other control methods, although valves of this type, triggered by machine movements, may readily be adapted for automated working.

Today it is possible to solve control problems using various devices ranging from pneumatic sequence controllers to electronic Programmable Logic Controllers (PLC). An example is shown in Figure 1.

A typical pneumatic sequence controller provides 12 steps of control with clear indication of the step currently being executed. More than 12 steps are possible by cascading several of these controllers together. They can incorporate manual control for step by step operation. Safe operation is ensured since the next step

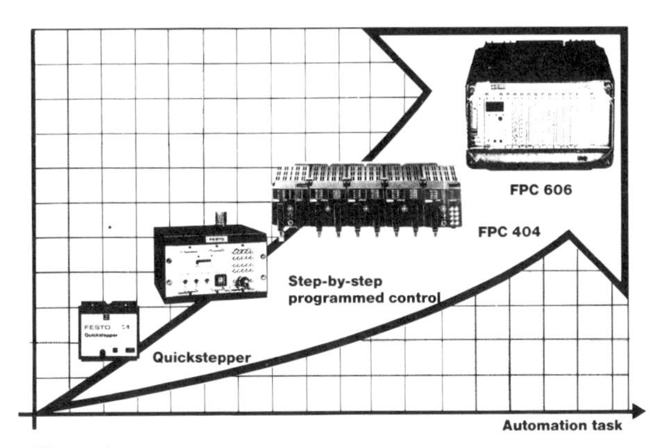


*Figure 1*

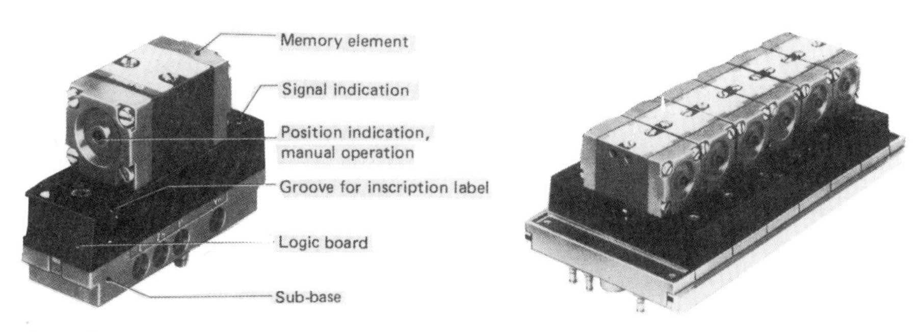


*Figure 2*
*Pneumatic shift-register module (left) and shift-register chain (right).*

cannot be executed until acknowledgement for the proceeding step has been received. Since these devices use only pneumatic techniques no specialized knowledge is required to program the controller. These pneumatic controllers continue to remain unchallenged for applications involving high risk fire areas, risk of explosions or hazardous areas, *eg* involving low and high flashpoint materials and where a relative few functions are to be controlled.

With the advent of low cost microprocessors, hybrid systems (electronic control of pneumatic systems) have many advantages to offer at comparable or even lower cost, particularly as the complexity of the task increases.

Programmable Logic Controllers (PLC's) are now readily available and range in size from those catering for as few as 8 inputs and outputs to very large units with thousands of inputs and outputs. Modular design allows optimum adaption to the task concerned by selecting the suitable system components. This flexibility of system configuration also enables a comprehensive range of facilities to be made available including such items as timers, counters, mathematical abilities, fast counting (for encoders *etc*.), multi-tasking, communications with other devices (*eg* printers, VDU's, other computers), networking several controllers together. This illustrates how easily and smoothly PLC's can be integrated into existing EDP procedures and hence higher level Company related information systems.

A major advantage of a PLC is that the logic for the controlled sequence is in the form of a written *program* and not in the form of tubing or cables. This means that the program can be easily modified to allow for varying requirements and that in-depth processing (*ie* complicated logic and/or sequences) can be achieved at low cost.

Maintenance of systems incorporating PLC's is simplified because of comprehensive fault/error diagnostic facilities provided within the PLC and the status of all inputs and outputs is clearly indicated at all times. Documentation of the program is readily available and this ensures the latest version of the program and hence all machine functions are retained in written form; all necessary information is available to service personnel at all times allowing malfunctions to be corrected rapidly. For the same reasons commissioning of installations with PLC is considerably more rapid than that of other technologies.

*Shift register modules* can provide a simplified approach to circuit design for pneumatic sequence controls. Basically each module is an in-line memory element (see Figure 2). Modules are then interconnected in such a way that a switching step is performed only when its turn comes up.

The number of individual movement steps indicate how many shift register modules are needed. The specific number of shift register modules are then lined up side by side to produce the required sequence or shift register chain. It is then only necessary to assign signals and instructions to the sequence chain for each step when the control section is more or less complete.

*Figure 3*
*Remote-controlled electro-pneumatic pressure regulator with closed loop integrated electronic control. The electronic module, shown separately, is contained with the regulator body. The output pressure is measured by an internal sensor and is compared and adjusted electronically to correspond to the control signal.*

**Electronic Control**

The introduction of electronic control to a pneumatic system is a natural progression particularly in complex production systems where a large number of steps may be needed in a given sequence. It also has the advantage over purely mechanical or pneumatic controllers where speed of operation can be a problem or where sequence changes must be introduced as the result of process changes.

Machine control applications fall into two categories: those using one or more fixed programs, which cover most special purpose machines, and those which need to be re-programmed by the user, such as machine tools and robots.

Simple fixed program requirements can be met by hard wire programmed sequencers. Most user programming requirements can be met by controllers which are very simply re-programmed by means of switches (see Figure 4).

More complex or multiple programs can be provided by the microprocessor-based controller where the program information is stored in an EPROM which is a permanent electronic memory.

In simple form, the microprocessor comprises three basic parts: the central processing (CPU), a memory and input/output (I/O) device. A microprocessor always contains a CPU and, in some instances, memory as well as I/O device. The CPU has the ability to send (address) information to either the memory or I/O device.

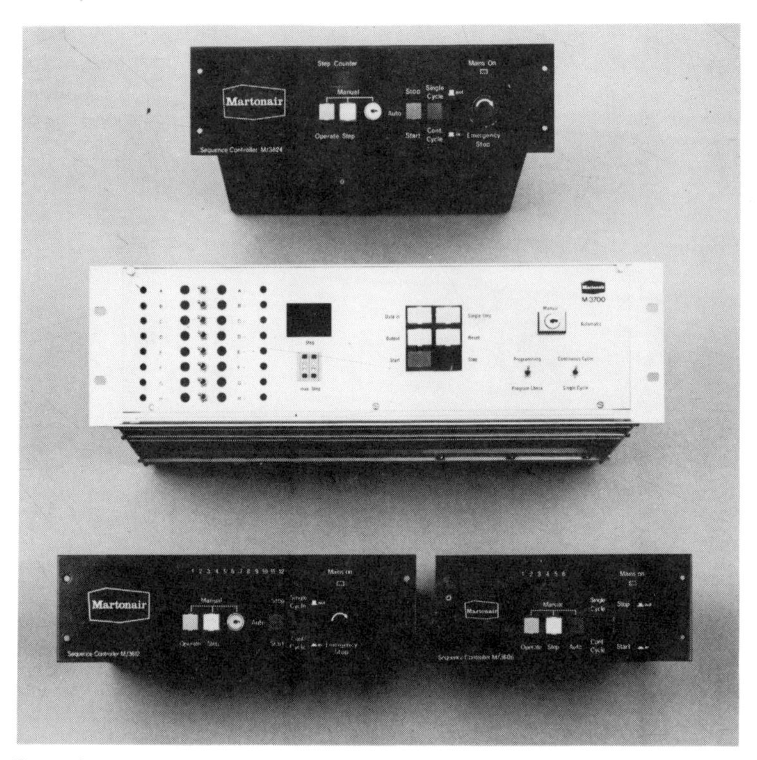


*Figure 4*
*Sequence controllers.*

Just as the memory must have an address before data may transferred to or from it, so must the I/O devices. Normally there is more than one input or output device on a system. Therefore, the CPU must decide which one it wishes to transfer data with and this is done by addressing.

To execute a particular program, the programmer may set aside a certain area of the memory for program storage. Other areas of memory may be set aside for data storage and these assignments may change for a different program.

Entries to the sequencer are made with a plug-in programmer. This can read out and modify all locations within the sequencers ladder element (memory) storage location area. When an entry is made, it is entered into and stored within this storage area. Timers and counters as well as series and parallel logic are contained therein.

The microprocessor/sequencer shown in Figure 5 offers simplicity in programming pneumatic cylinders and related components, in any sequence of operation. It has a capacity of 16 inputs, 16 outputs, plus eight dedicated inputs for start conditions or manual selections, one hand start, two hand start, latching start, emergency stop and program selection.

The programming is by handheld pendant which plugs into the control unit. It uses the simple combinations of letters normally used for sequencing pneumatic systems or if no mechanical parts are involved, then the outputs and inputs can be conditioned numerically.

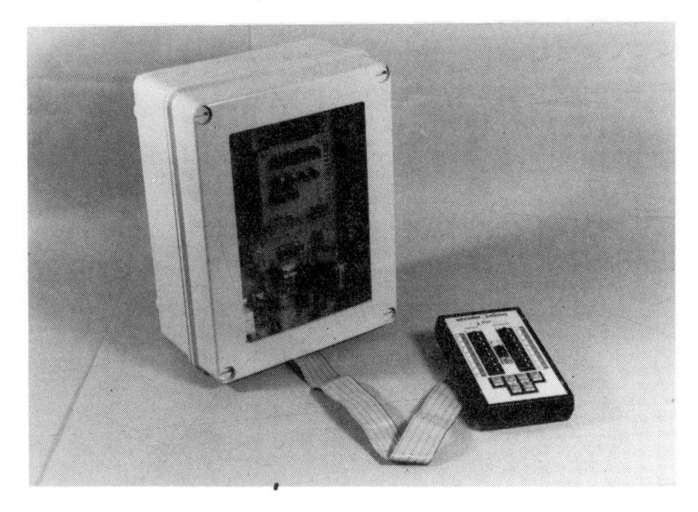

*Figure 5*
*Microprocessor/sequencer.*

This particular micro-processor/sequencer allows for four independent programs of 85 steps each, which can be married into one long program if required. A timer with a maximum of 99.9 s can be used at each stage and a value of up to 999 counts is available on each input or output.

Jumps can be made from each program line to any other program line, and sub-routines can be easily introduced. A plug-in module is available which contains a counter, a timer and a jump set plus the main information display. This can also be remotely mounted. A facility for remote signalling of the program number, is also included.

**Electro-Pneumatic Interfaces**
Where air valves are operated remotely by solenoids it is possible to signal them in sequence electrically by means of the solid state electronic sequence controllers. The series of inputs and outputs with relays can give sufficient power output to operate solenoid directional valves.

The microprocessor, whether in the form of computer, programmable controller or intergrated control circuit, allows almost infinite flexibility with the bonus of electronic output signals.

Designers have in the past faced problems of cost and complexity when converting these digital signals into physical movements of valves, switches or other controls. Air actuation has traditionally been the method preferred, by reason of its low cost and high reliability, though the air/electronic interlink was usually complex, with the associated circuitry bulky and potentially a maintenance hazard.

The difficulty has been overcome by using replaceable printed circuit boards, introducing the concept of the 'plug-in' electro-pneumatic card rack assembly as shown in Figure 6.

Miniaturized two and three-way NC and NO solenoid valves to a special patented design are mounted on a standard rack card which is linked *via* edge connectors to a computer or controller in the usual way by a ribbon cable.

*Figure 6*
*Electronic-pneumatic interface.*

On the pneumatic side, a standard card can accommodate up to eight valves, complete with light emitting diodes (LEDs) to indicate correct circuit performance, together with override switches for test purposes.

Output signals are transmitted *via* an eight-way tubing coupler and all essential accessories – pneumatic busbars,connectors, *etc* are available, to make the exchange of a 'pneumatic' card, in the event of test, malfunction or machine upgrade, as quick and simple as that of an electronic card.

The system has obvious advantages in terms of ease of maintenance when the techniques are applied to established machinery. An additional bonus is available to a designer, not totally familiar with the detailed operation of pneumatic systems and modern components because by using these techniques, he can concentrate his attention on system level solutions without any need to involve himself with problems related to the performance of individual components.

The pneumatic card may be regarded simply as a *black box*, converting electronic signals into their pneumatic equivalents which are fully susceptible to treatment by the Boolean logic with which the electronic engineer is already familiar. The electro/pneumatic printed circuit boards can convert electronic input signals into pneumatic output signals for:

(1) Air pilot supply for pneumatically operated valves.

(2) Direct actuation of low volume mini-cylinders.

(3) Direct air supply for instrumentation and general process control including explosion proof environments.

# SECTION 7

## Pneumatic Cylinders

TYPES OF CYLINDERS
CONSTRUCTION
PERFORMANCE AND SELECTION
APPLICATIONS

# Types of Cylinders

PNEUMATIC CYLINDERS can be categorized under the same type headings as those for hydraulic cylinders shown in Section 3, *ie* single-acting, double-acting, through-rod tandem and Duplex, cushioned, rotating and miniature. Pneumatic cylinders of these types can be manufactured on similar lines to the hydraulic cylinders except for different materials of construction and relative rod sizes, due to the much lower working pressure of compressed air. Special types of cylinders, however, are manufactured for working with compressed air only and these will be considered in this chapter.

**Diaphragm Cylinders (Thrusters)**

A *diaphragm* cylinder is basically a large diameter, short stroke cylinder fitted with a diaphragm instead of a piston. The construction usually takes the form of a pair of shallow convex housings, with a diaphragm sandwiched between them and a piston rod attached to the diaphragm. They are usually of single-rod type, either single-acting with spring return or double-acting as shown in Figure 1.

Diaphragm cylinders are capable of generating very high forces with very short strokes and are commonly called *thrusters*. Theoretical thrust available is equal to the product of the applied air pressure and effective diaphragm area. The stroke is normally limited to a maximum of about one third of the cylinder diameter.

Thrusters can be made from castings or from shallow steel pressings which are clamped or bolted together with the diaphragm as shown in Figure 2. The diaphragm can be moulded from Neoprene or fabric-reinforced Nitrile rubber and the piston rod is usually guided in lubrication-free nylon bearings.

**Cable Cylinders**

The cable cylinder is a specialized cylinder which employs a cable instead of a piston rod. The cable is connected to both sides of the piston and the continuous length passes over pulleys at each end of the cylinder. The load to be moved is attached to a plate connected to the external part of the drive cable and, because the cable runs parallel to the cylinder barrel, there is no increase in overall length as the stroke movement takes place.

This overcomes the problem with conventional long stroke cylinders, where there is a risk of the piston rod bending under load and where the overall length must always be twice as great as the maximum length of stroke required.

In some applications the cylinder can be designed for rotary motion, with a take

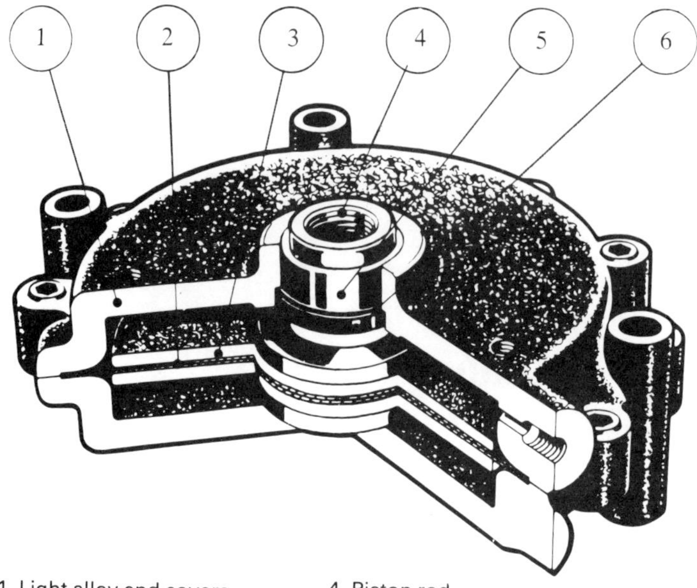


1. Light alloy end covers.
2. Neoprene diaphragm.
3. Support plates.
4. Piston rod.
5. Bearing.
6. Rod seal.

*Figure 1*

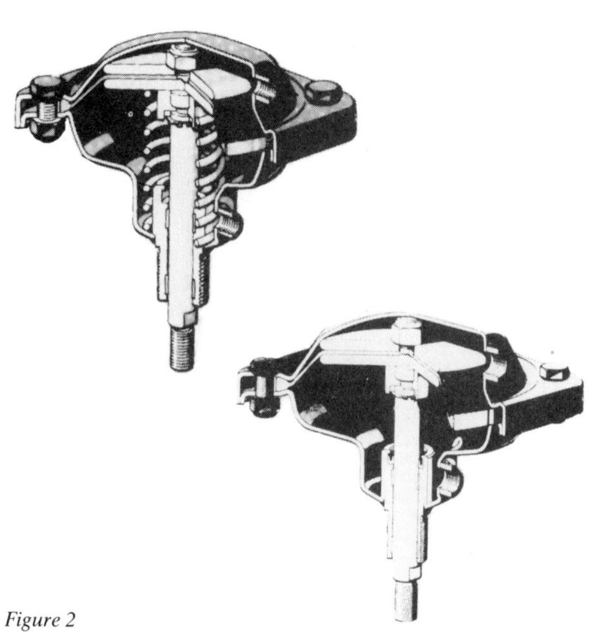

*Figure 2*
*Diaphragm cylinders.*

off point at one of the pulleys. Special designs enable the cable to round corners or the cable can be extended to wrap around a drum or can be threaded through a machine.

**Band Cylinders**

*Band* cylinders are similar to cable cylinders but employ a hardened spring steel band instead of a cable as shown in Figure 3. In terms of the stroke, bore and function of the cylinder, it is mainly the absence of a piston rod which produces a very compact unit. Because the band is thin it is possible to keep the diameter of the idler rollers small. The smooth surface of the band allows dirt to be wiped off easily and a low friction method of sealing in the cylinder heads is unaffected by working pressure or stroke to be used.

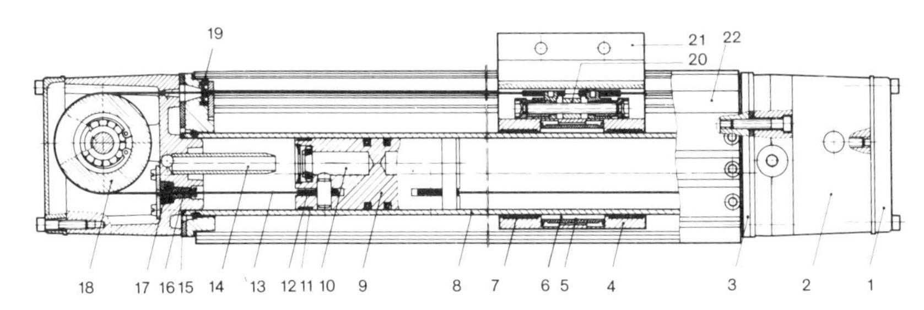


1 Cover
2 Cylinder head
3 Flange
4 Guide bush
5 Pneumatic brake
6 Brake lining (6-piece)
7 Slipper ring, guide bush
8 Cylinder barrel
9 Piston
10 Damping hole and seal
11 Slipper ring, piston
12 Reinforced belt end
13 Circulating steel belt
14 Air inlet and outlet with damping tube
15 O-ring barrel seal
16 Guide pads
17 Band Seal
18 Idler roller with ball bearings
19 Band wiper
20 Tension lock
21 Power takeoff T
22 Side cover

*Figure 3*
*Band cylinder.*

In some versions the outside surface of the barrel is used to guide the power 'take off' point with slipper rings enclosing the barrel completely. Additional guiding can be employed from the side covers enclosing the cylinder barrel so that a greater amount of side thrust can be tolerated.

The cylinder can be equipped with a pneumatic brake at the guide bush slipper ring as shown and this opens up a number of areas of application for the endless band cylinder.

The braking force and motive force of the cylinder are approximately equal when referred to the same pressure. The braking action is produced by a brake lining being pressed pneumatically onto the outside of the cylinder barrel by means of a seal ring operating at low pressure.

**Rodless Cylinders**

Recently-developed *rodless* cylinders provide an ideal solution for long stroke applications and supersede the cable and band cylinders in some respects. In the rodless design the cylinder barrel has a slit along the barrel and the piston in the barrel is connected directly through the slit to a mounting on the outside of the cylinder.

Thus, when the piston is moved by applying air on either side, the load connected to the external mounting is moved. The problem of providing a continuous and moving seal between the piston and the external mounting has now been achieved with a high degree of reliability. Not only can this design handle long strokes effectively but can also accept lateral loading at the mounting point.

An illustration of the construction of one type of rodless cylinder is shown in Figure 4. The drive from the piston to the carriage is *via* a substantial drive tongue which passes through the barrel slot. This drive tongue forms the centre part of a yoke which is incorporated in the carriage extrusion. The piston halves are pinned to the lower part of this yoke, thus joining the piston to the carriage.

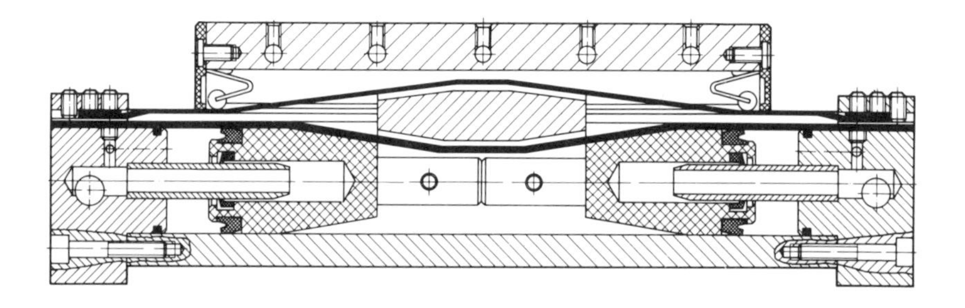

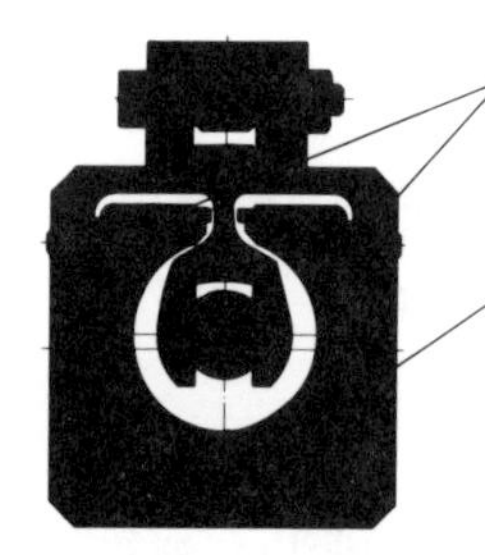

*Figure 4*
*Rodless cylinder.*

The volume between the piston seals, containing the yoke, is at atmospheric pressure. The pressure and dust slot seals are unclipped and parted by cam shapes in the yoke within this non-pressurized section and by the advancing movement of the piston carriage assembly along the cylinder.

After unclipping, the internal pressure seal slides under the lower part of the yoke adjacent to the piston halves, while the external dust seal slides over the upper part of the yoke within the carriage. The pressure and dust slot seals are then pressed together and re-clipped by the spring roller assembly in the carriage and the ramp shape of the piston and by the retreating movement of the piston carriage assembly along the cylinder body. Thus the drive is taken up from the piston to the carriage through the slot as this assembly traverses the cylinder.

A similar design of the rodless cylinder is shown in Figure 5 with integral control valves built into the end caps. The valve can be supplied for either pneumatic or electric operation complete with adjustable speed control and non-return valve. The main advantage is in having all of the control functions contained in the complete compact assembly.

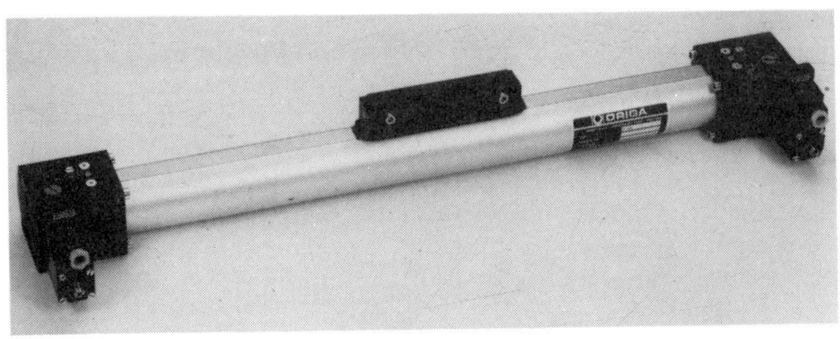


*Figure 5*
*Rodless cylinder with integral control valves.*

**Bellows Actuators**

A *bellows* actuator works in the reverse mode to an air spring and has similar geometry. Air springs connected to an external pressure source can be used as actuators, the most suitable form being the convoluted bellows with one, two, or possibly more convolutions depending on the stroke required (see Figure 6).

Nominally, and dependent on the detail design, the maximum stroke obtainable from a single convolution bellows is approximately a quarter of its overall diameter or less; approximately half the diameter with two convolutions; and rather more than half the diameter with three convolutions.

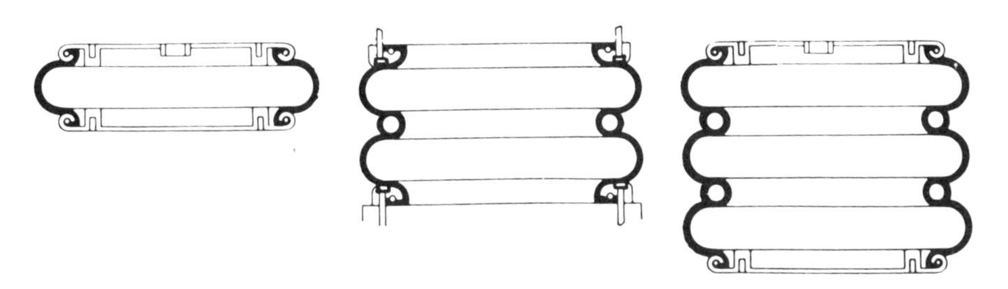

*Figure 6*
*1–, 2– and 3– convolution actuators.*

The bellows are formed from two plies of nylon reinforcing cord encased in synthetic rubber (Neoprene), the cord being designed to provide an interlocking effect as internal pressure is applied. Crimped head plates provide hermetically sealed ends as well as convenient mounting and accommodation for the entrance port.

With this form of bellows the effective area of the actuator varies with extension, thus the force characteristics vary through the stroke range. Design can, in fact, be varied to provide different force-stroke characteristics, even constant force if required.

A further feature of such actuator types is that they can accommodate considerable irregularity of alignment, thus dispensing with the need for a spherical bearing mounting with a conventional cylinder, in many applications where this working conditions exists.

**Buffer Cylinders**

Instead of incorporating an integral cushion a *buffer* cylinder may be used as deceleration control. This is a separate cylinder pre-pressurized and proportioned to absorb the excess momentum (Figure 7). The main requirement is that the buffer cylinder should have reserve capacity and that the maximum pressure realized in the buffer cylinder should not exceed the design rating (which can be very much higher than that for conventional air cylinders). With suitable design, cushioning of high speeds and forces can be provided by quite short buffer cylinders pre-pressurized to the order of 0.7 to 1 bar (10 to 15 lbf/in$^2$).

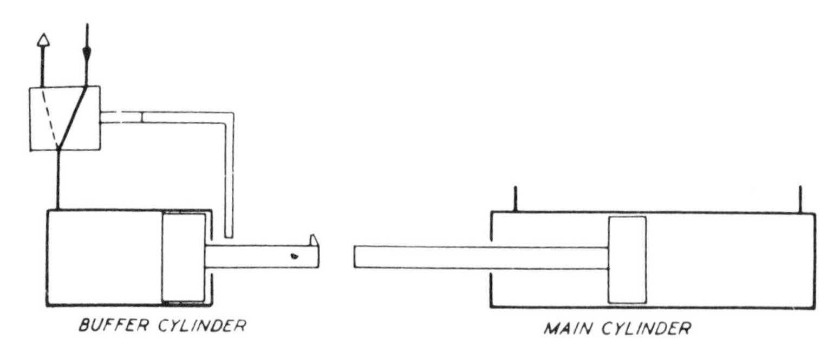


*Figure 7*

The valving circuit associated with a buffer cylinder must be capable of reversing the main control valve at the end of the buffer stroke, otherwise the high pressure generated in the buffer cylinder will tend to reverse the travel of the load.

Other limitations of integral cushions include the fact that they are generally ineffective where a cylinder is fitted with quick exhaust valves or is pre-exhausted, as the amount of air trapped in the cushion is likely to be inadequate for proper cushioning; there is also the practical problem that the bleed may be subject to clogging by contaminants unless periodically attended to. Further of course, integral cushions are only effective if the full stroke of the cylinder is used.

Nevertheless cylinders with integral cushions have a wide application and are generally capable of giving excellent performance in the majority of systems. Piston speed limitation is less likely to be a factor than may appear at first sight because air cylinders are seldom required to operate at their fastest speed, and the cylinder operating speed is frequently controlled.

**Impact Cylinders**

An *impact* cylinder is one arranged so that the speed developed is high enough for impact work, such as forging, piercing, *etc*.

It is possible to achieve this mode of working with a conventional cylinder fitted with extra large inlet and outlet ports and suitable valving. Such a system is shown in Figure 8. This provides for the air on the underside of the piston to be quickly exhausted, while simultaneously supplying air to the other side. The piston speed obtained can be varied by adjusting the appropriate valves. It is generally advisable – or even necessary – to use a reservoir of at least equivalent capacity to that of the cylinder displacement and which is connected close to the cylinder in order to avoid too high a pressure drop.

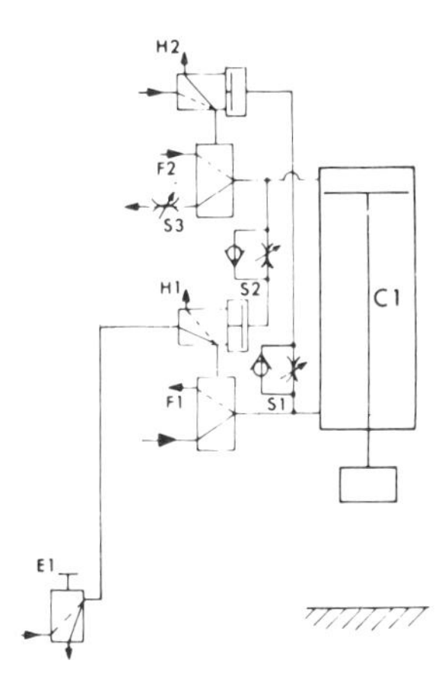


*Figure 8*
*Circuit for operating on air cylinder as an impact cylinder.*

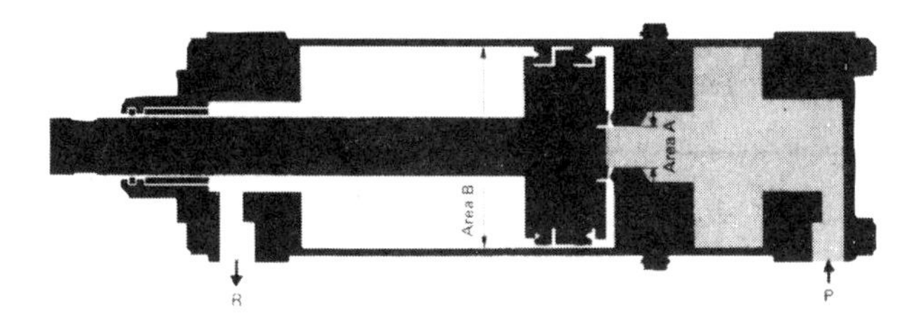


*Section through impact cylinder. Piston remains at rear end position (seat closed on pre-pressurizing chamber) until product of inlet pressure P times area A is greater than product of outlet pressure R times area B. Exhaust rate controlled by down-stream throttle valve.*

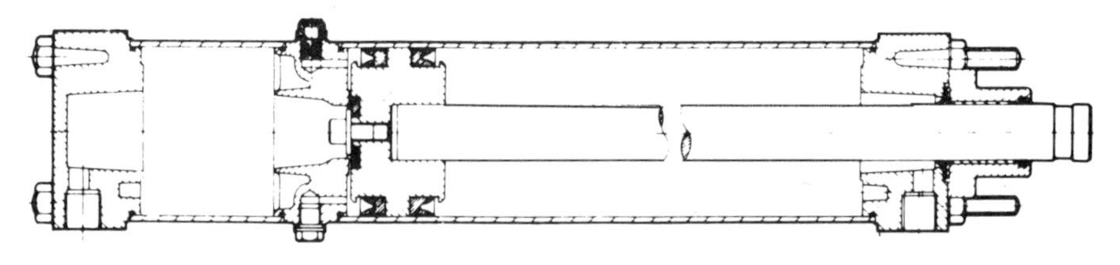

*Figure 9*
*Impact cylinder.*

An alternative approach is to incorporate a reservoir section in the cylinder, connecting directly to a large inlet port valve. Such a design is shown in Figure 9. Initially the piston rod end is charged with air at normal line pressure with the annulus between the piston seat and centre piece connected to exhaust. The control valve is then opened and pressure builds up in the reservoir. This acts on the piston seat, which is 1/n that of the piston area (*ie* A/n where n is the ratio of the piston to annulus area). The thrust force downwards is thus P x A/n.

The piston rod end of the cylinder is connected to exhaust when the pressure in this end falls to less than the supply pressure P. When it has fallen to just below P/n the piston moves away from poppet seat. Immediately this happens the whole of the piston is exposed to the reservoir pressure causing the piston to move rapidly downwards. Maximum piston speed is reached within the first 25 to 35 mm of the stroke and is of the order of 8 metres/s (25 ft/s) with supply pressure of 6 bar (90 lbf/$in^2$).

Because of the high shock loads developed, a necessary requirement for all impact cylinders is that they be mounted rigidly. For even higher energy outputs pairs of cylinders can be mounted in opposition and operated simultaneously. With proper alignment, shock loading can be negligible in such cases, all the energy being absorbed by the workpiece. Some impact cylinders are fitted with a cushion to prevent damage if they are used for other types of work.

Plastic cylinders may be all-plastic except for the piston rod, or mixed metal and plastic. For example, a plastic cylinder tube and end covers with a metal piston and stainless steel rod may have advantages when working in a corrosive ambience. Plastic materials used include nylon, polyacetal, polycarbonate, rigid PVC and reinforced epoxy resins.

Apart from corrosion resistance, plastic cylinders can also offer cost advantages. However the higher moduli associated with plastic materials can restrict the working temperature range of plastic cylinders and substantially reduce the maximum service temperature rating.

**Two Fluid Systems**

The basic limitation of a pneumatic cylinder is its general unsuitability for powering movements which require a constant output with varying load, particularly where precision feeds are required. This can be overcome by associating an air cylinder with a hydraulic cylinder, the latter acting as a rigid check unit. The two cylinders can be separate, or combined in the form of a single air-hydraulic cylinder, although in all cases the air and liquid circuits are quite separate.

Two-fluid systems involving separate but interconnected air and hydraulic (oil) cylinders fall into these categories opposite.

## TYPICAL STANDARD SIZE CYLINDER GEOMETRY

**In Size Cylinders**

| Cylinder Diameter in | Rod Diameter in | Typical Pcrt Size |
|---|---|---|
| ¾ | 5/16 | BSP 1/8 in |
| 1 | 3/8 | BSP 1/8 in |
| 1½ | ½ or 5/8 | BSP 1/8 in |
| 2 | 5/8 or ¾ | BSP ¼ in |
| 2½ | ¾ | BSP ¼ in |
| 3 | 1 | BSP 3/8 in |
| 4 | 1¼ | BSP 3/8 in |
| 5 | 1¼ or 1½ | BSP 3/8 in |
| 6 | 1.3/8, 1½ or 1¾ | BSP ½ in |
| 8 | 1½ or 1¾ | BSP ¾ in |
| 10 | 1¾, 2 or 2¼ | BSP 1 in |
| 12 | 2¼, 2½ or 3 | BSP 1¼ in |
| 14 | 2½ or 3 | BSP 1½ in |

**Metric Size Cylinders**

| Cylinder Diameter mm | Rod Diameter mm | Typical Port Sizes |
|---|---|---|
| 12 | 4 | M5 |
| 16 | 6 | |
| 20 | 8 | BSP 1/8 in |
| 25 | 12 | or M5 |
| 32 | 12 | |
| 40 | 16 | |
| 50 | 20 | BSP ¼ in |
| 63 | 20 | BSP ¼ in |
| 80 | 25 | BSP 3/8 in |
| 100 | 32 | BSP 3/8 in |
| 125 | 32 | BSP 3/8 in |
| 150 | 35 | BSP ½ in |
| 200 | 50 | BSP ¾ in |
| 250 | 60 | BSP 1 in |
| 300 | 70 | BSP 1¼ in |
| 350 | 80 | BSP 1.1/8 in |
| 400 | 100 | BSP 1½ in |
| 450 | 110 | BSP 2 in |
| 500 | 120 | BSP 2 in |

(1) *Check units* – where a small hydraulic cylinder (or cylinders) is mechanically connected to an air cylinder for accuracy of control of cylinder movement.

(2) *Air-hydraulic* cylinders – usually a back-to-back tandem arrangement of air and hydraulic cylinders with a common movement.

(3) *Intensifiers* – where a low pressure air cylinder is used to generate a much higher pressure in a smaller hydraulic cylinder.

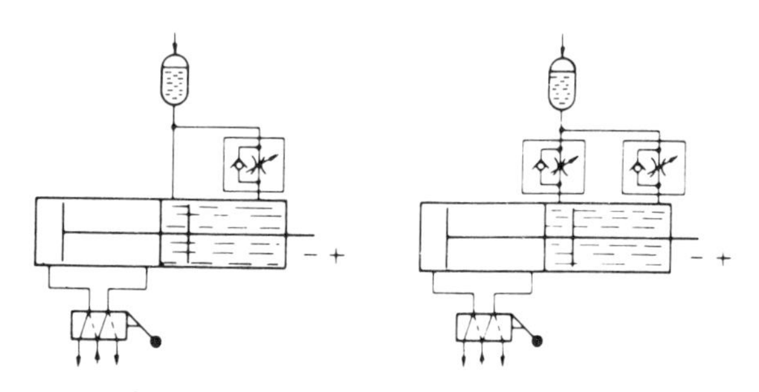


*Figure 10*

**Hydro-Check Cylinders**

Two examples of basic hydro-pneumatic cylinder circuitry are shown in Figure 10. In the left hand diagram speed is controlled in the + stroke of the cylinder. If the − stroke is to be controlled, the non-return restrictor valve is inserted in the other hydraulic cylinder line.

The second diagram shows the same circuit adapted to speed control in both directions. However, one throttle valve may substitute for both non-return restrictor valves where equal speed in either direction is called for, or can be accepted.

**Air-Hydraulic Cylinders**

Typical applications of air-hydraulic cylinders include feeds for drills, milling machines, circular saws, *etc* and similar operations where the power supply available is compressed air, output forces required are relatively light, but a rigid feed is required.

Hydro-pneumatic working is achieved by mechanically coupling a pneumatic cylinder and a low pressure hydraulic cylinder so that they have a common movement normally initiated by compressed air applied to the air cylinder, which is a double-acting type. The normal configuration is back-to-back mounting with a common piston rod (Figure 11). The hydraulic (oil) cylinder is also double-acting, but with a closed loop circuit. Control of the rate of flow of oil through this circuit governs the speed of operation of the tandem cylinder.

In practice, it is essential that air cannot leak into the hydraulic circuit, otherwise this would cause 'spongy' operation. Thus two sets of rod seals are employed, one on the air side and one on the oil side. Provision also has to be made on the hydraulic side to accommodate any changes in working volume of the oil, *eg* by the entry and withdrawal of the rod in the simple system of Figure 11 and also oil volume changes due to ambient temperature changes.

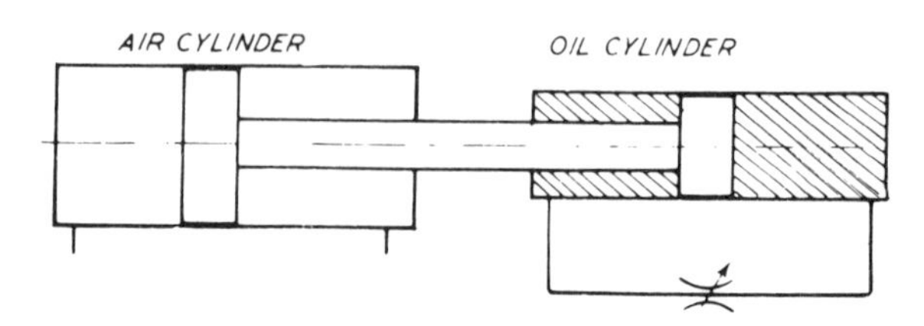


*Figure 11*

A logical design configuration is to make the air cylinder the blind end of the combination so that the hydraulic cylinder then has a through-rod giving equal displacements with either direction of motion. It is then possible to reduce the volume of the compensating oil reservoir to minimum proportions or even eliminate it entirely if the cylinder will be working at a substantially constant temperature.

**Speed control**

Further controls can be provided in the pneumatic circuit, which is the 'working' circuit. However, unloading or pressure relief valves cannot be used in the air circuit because these could be operated by the damping produced by the hydraulic circuit, causing loss of working pressure. If unloading is required on the air side, this would have to be done *via* microswitches operated by mechanical movement or suitable pilot valves.

Throttling may be provided on both sides of the hydraulic piston for adjustable speed control in both directions or on one side only with a flap valve for differential speed control with fast movement on the opposite stroke, as in Figure 12.

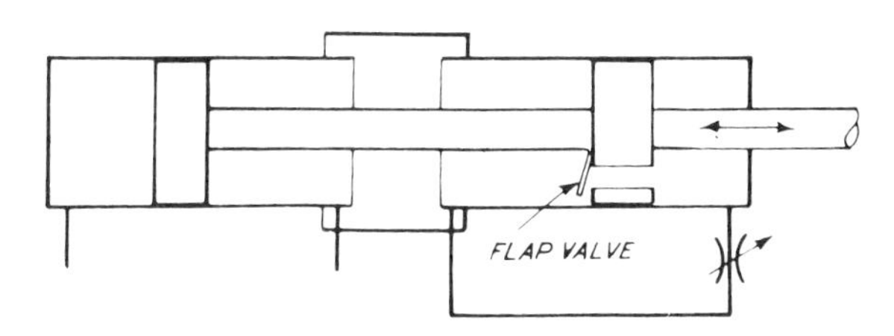


*Figure 12*

Basic movement controls possible with the control unit(s) located in the hydraulic circuit are:

*Speed, one direction* – one throttle valve or restrictor.

*Speed, both directions* – two throttle valves or restrictors.

*Fast Approach, with final controlled approach* –

(1) throttle and bypass valve in parallel;

(2) mechanical (applicable only in case of separate cylinders).

*Fast Return* – flap valve in piston.

*Fast Approach, controlled return* – flap valve in piston plus throttle valve in return flow line.

*Controlled Approach, fast return* – throttle valve in one line; flap valve in piston.

Various other combinations of the aforementioned are possible, *eg* fast approach with controlled final approach, followed by fast return. In that case the control units required would be those for both types of movements, on appropriate sides in the hydraulic circuit (see Figure 13).

In general:

*Throttle valve* (or *restrictor*) provides speed control (in appropriate line).

*Throttle valve with bypass valve* provides differential speed control (in appropriate line).

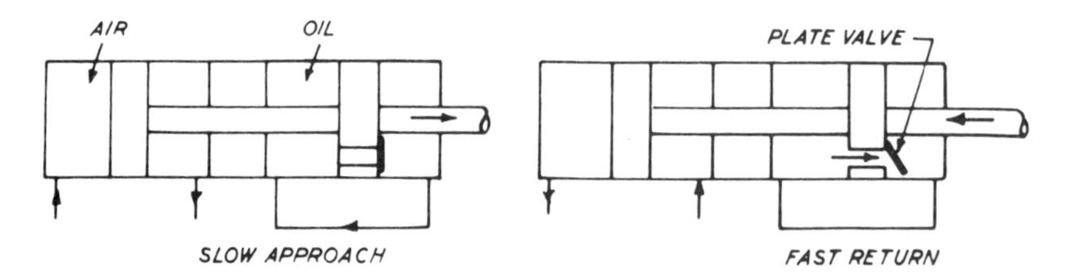


*Figure 13*

*Flap valve* in piston provides uncontrolled fast approach, or fast return (depending on which side of the piston it is fitted).

**Intensifiers**

Intensifiers are normally a special type of air-hydraulic cylinder whereby low pressure air can be used to generate a supply of high pressure fluid. The high pressure fluid may be air (or gas) or a liquid (*eg* hydraulic oil or water). Intensifiers may be used directly as a source of low volume high pressure fluid or as high pressure hold devices. In the latter case initial pressurization is achieved from a pump, the system then being switched over to the intensifier to produce and hold the final high pressure.

A diagrammatic representation of a typical air-hydraulic intensifier is given in Figure 14. Air pressure is used to drive the piston in a low pressure cylinder, the piston rod forming a plunger for a very much smaller diameter high pressure cylinder. Pressure intensification is directly related to the ratio of the cylinder bores, a typical intensifier ratio being of the order of 50:1. This would yield a hydraulic pressure of the order of 350 bar (5000 lbf/in$^2$) from a conventional air line supply at 7 bar (100 lbf/in$^2$). If necessary, very much higher pressures can be obtained from air hydraulic intensifiers. Normally a manufacturer would have a basic air cylinder and vary the diameter of the hydraulic cylinder to give the desired high pressure output. Output volume, of course, reduces with reducing diameter of the high pressure cylinder for a given stroke (available from the air cylinder), as well as the speed of reciprocation. Speed falls off as pressure builds up, until reciprocation finally ceases, when the loads on the low pressure piston and high pressure plunger are balanced. Reciprocation of the air cylinder is achieved by any suitable form of valve gear.

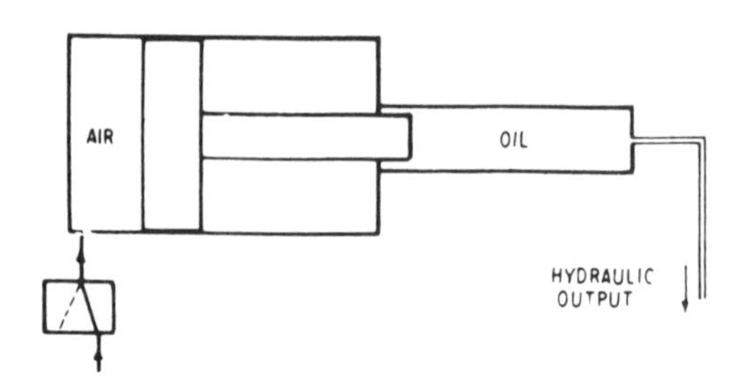


*Figure 14*

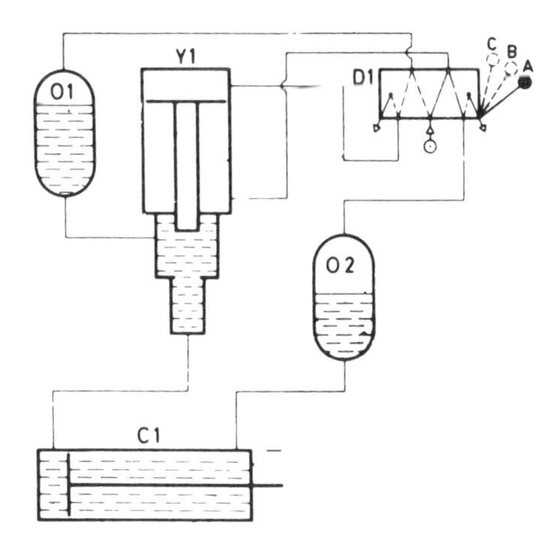


*Figure 15*

A typical circuit using a six-way directional control valve is shown in Figure 15. In the starting position D1 is in the position A. C1 and Y1 are in the withdrawn position. When D1 is moved to position B, reservoir D1 is pressurized and, simultaneously, reservoir 02 is vented. C1 moves outwards under low pressure. When D1 is moved to position C, Y1 moves downwards under air pressure and hydraulic pressure actuating the cylinder C1 is intensified. Return movement is accomplished by pressurizing 02 and venting 01.

*Figure 16*
*Miniature cylinders.*

**Miniature Cylinders**

Miniature pneumatic cylinders are divided into two basic types: *compact short-stroke* cylinders and *small bore pencil-type* cylinders. Both types are shown in Figure 16 and usually have a bore diameter under 25mm (1 in) to be considered in the miniature range.

The compact cylinders shown, more commonly known as *pancake* cylinders, are available in 15 mm (0.6 in) single-acting or both 15 mm (0.6 in) and 32 mm (1.25 in) double-acting bore diameters, with strokes of 5 mm (0.2 in) single-acting and 10 mm (0.4 in) double-acting. They are constructed with aluminium barrel and end covers, stainless steel piston and piston rod and are ideal for jig clamping operations.

Another range of compact short-stroke cylinders shown in Figure 17 have a bore range of 8 to 63 mm (0.3 to 2.5 in) diameter and a stroke range of 4 to 10 mm (0.15 to 0.4 in) and are available as either single-acting or double-acting through-rod cylinders. They are often used in robotic applications for gripping mechanisms on the robot's arm *etc*.

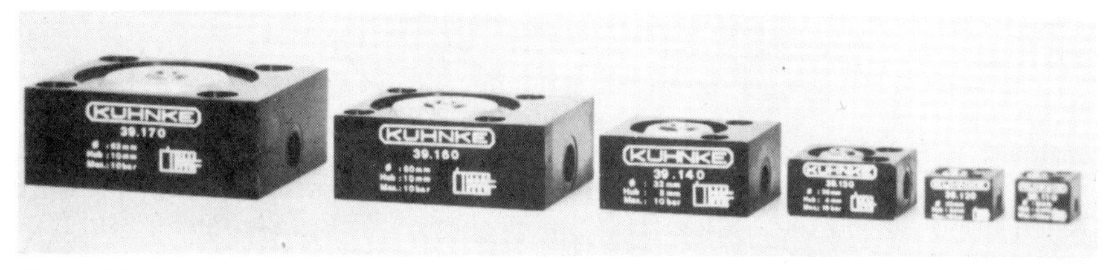


*Figure 17*
*Compact cylinders.*

The pencil-type cylinders (shown in Figure 16) are available as both double- and single-acting versions with bore sizes of 10 mm (0.4 in) and 15 mm (0.6 in) diameters. They are designed to operate with non-lubricated air and are constructed with stainless steel barrels and piston rods and magnetic pistons. The cylinders are shown fitted with hybrid IC hall effect non-contacting sensors with neon indicators for piston position switching and indication.

The dimensions of miniature cylinders up to 25 mm (1 in) bore have been standardized in BS 4862:Part 1 which specifies interchangeability dimensions for four mounting styles of cylinder with bores in the range 8 mm to 25 mm. An international standard ISO 6432 covers the same range of cylinders.

# Construction

### Materials

MATERIAL SELECTION for air cylinder construction must ensure adequate resistance to wear and corrosion, these being the two primary service hazards. Particular attention may have to be given to cylinders used in special environments, such as the use of non-sparking materials in explosive atmospheres, suitable chemical resistance in corrosive ambients, and so on.

Metal construction is more or less standard, although non-metallic tubes are now employed on light-duty cylinders. Suitable plastics include nylon, polyacetal, polycarbonate, rigid PVC and reinforced epoxy resin. One basic limitation applicable to all plastic materials is their relatively high coefficient of linear expansion which may present piston sealing problems if worked over a wide range of ambient temperatures.

Cylinder tubes are most easily, and most cheaply, made from hard drawn

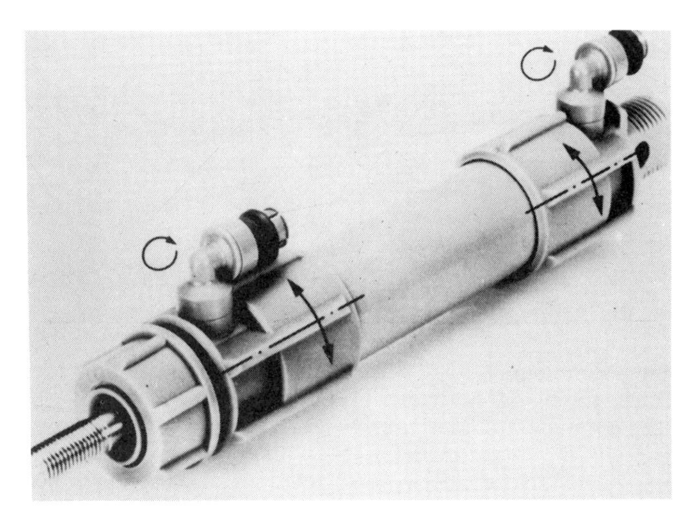

*Plastic cylinder.*

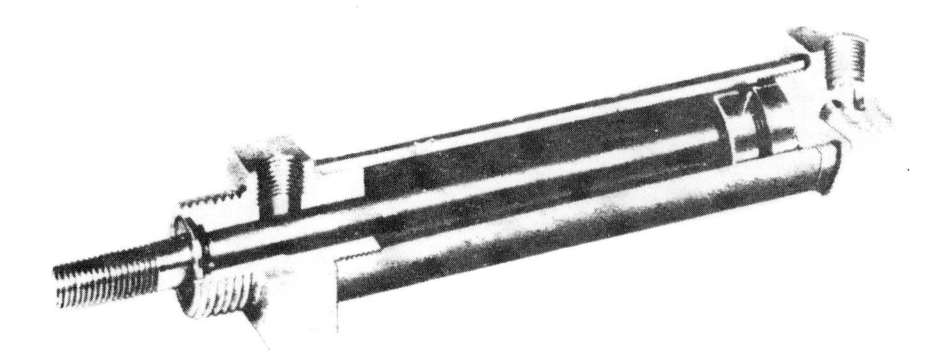

*Miniature cylinder.*

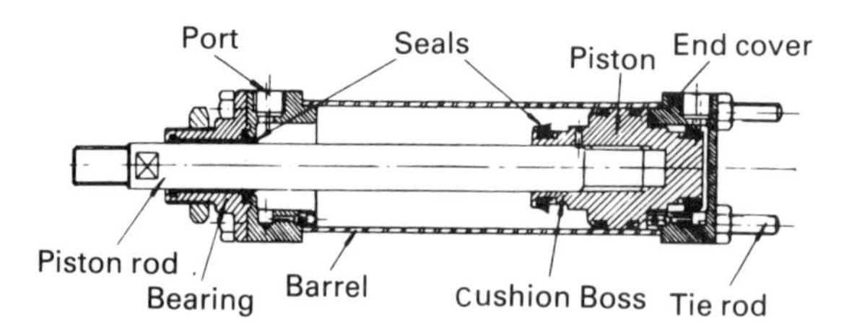


*Typical cylinder with cushion.*

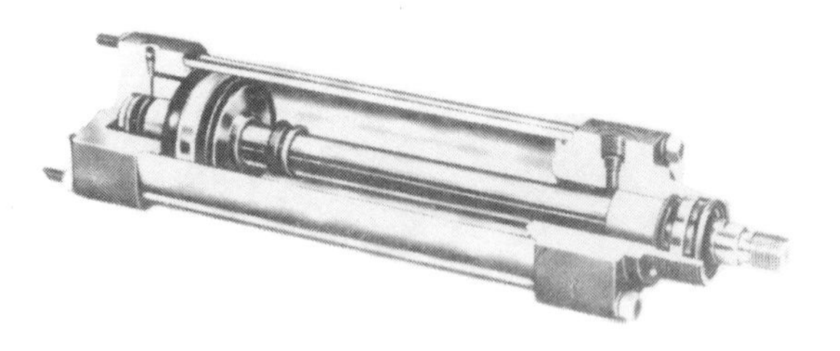

*Heavy-duty cylinder.*

metallic tubing. Aluminium (and to a lesser extent brass) is common for light-duty cylinders. Brass tubing is normally preferred for medium- and heavy-duty cylinders, although there are limitations to the bore sizes available in hard drawn brass tubing. Thus the larger cylinder tubes may be cast in aluminium, brass, bronze, iron or steel. Welded steel cylinder tubes are also used to some extent for large heavy-duty cylinders

For low friction and minimum seal wear, a fine bore finish is required, to

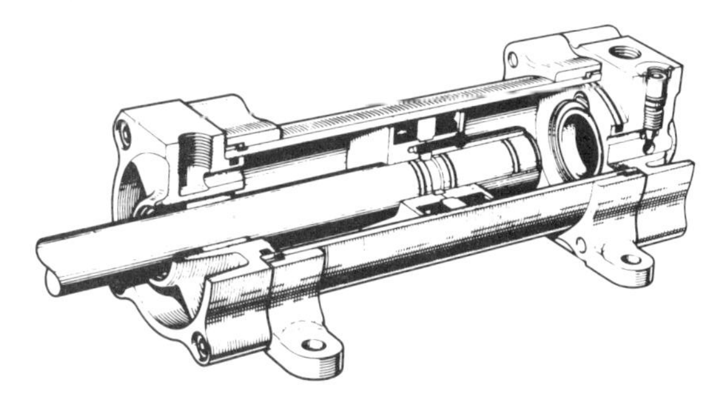

*Cutaway of air cylinder.*

600 $\mu$m (15 $\mu$n) or less where elastomeric seals are employed. Cylinders intended for very slow speed operation may need even finer finishing. Thus honing may be a necessary finishing operation, even with hard drawn seamless tubes. Hard chrome plating is also employed, honed and polished.

For working in corrosive ambients, cylinders may be fabricated from plastics, although this normally imposes low maximum service temperatures. The further development of plastic cylinders is a field in which considerable progress is being made.

**End Covers**

End covers may be cast or fabricated from solid stock, normally in the same material as the tube to avoid any possibility of electrolytic corrosion developing – particularly remembering that compressed air normally contains a considerable proportion of moisture. This same consideration favours the use of non-ferrous metals, for maximum resistance to corrosion. For heavy-duty cylinders, high tensile castings would be a normal choice.

Method of attachment of the end covers varies. With cast cylinders, one end is formed with the casting – the 'blind' end. Separated covers may be threaded (not usual), welded or mechanically secured in place.

A common construction in the case of medium- and heavy-duty cylinders is the use of tie-rods, enabling the assembly to be bolted up with the tie-rods in tension (Figures 1 and 2).This enables substantial covers to be employed, in easily fabricated shapes (usually square). Tie-rods are normally high tensile steel.

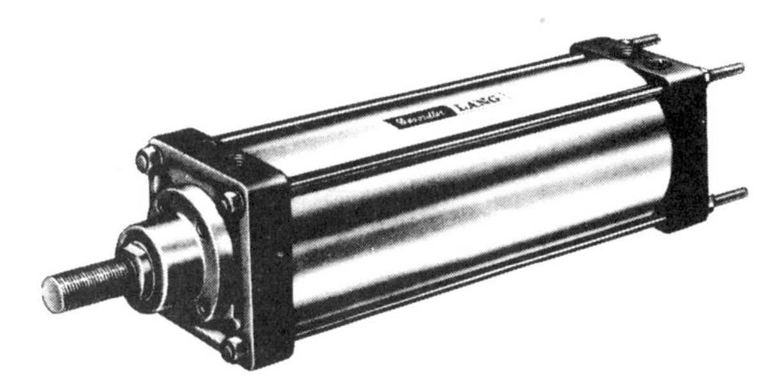

*Figure 1*
*Typical 'S' range cylinder.*

*Figure 2*
*Short-stroke cylinder with tie-rod assembly.*

**Tie-Rod Construction**

The main requirement of tie-rod construction is that the end covers cannot move when the cylinder is under pressure. This problem is exaggerated in the case of long stroke cylinders, when a tie-rod support ring may be added at the centre, into which the tie-rods are studded. This effectively halves any deflection or elongations of the tie-rods and makes for a far more rigid assembly, and is a generally more practical solution than using thicker tie-rods.

The end covers normally incorporate the inlet and outlet ports, and the cushion chamber(s) where fitted. One, or both, must also carry the rod bearing and rod seals, depending on whether the cylinder is of single-rod or through-rod design. In the case of a single-rod cylinder, the plain end is known as the *blind end*, and the other the *rod end*. A through-rod cylinder is symmetrical as regards construction, with two rod ends.

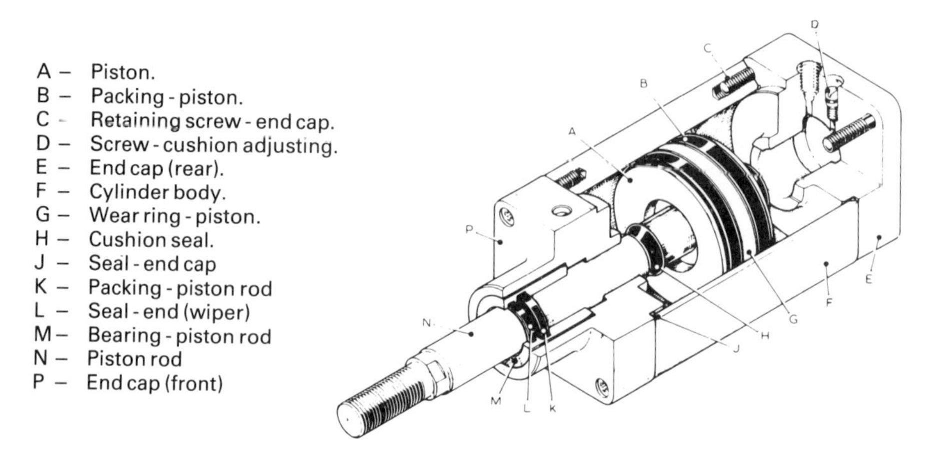


*'Square' cylinder construction.*

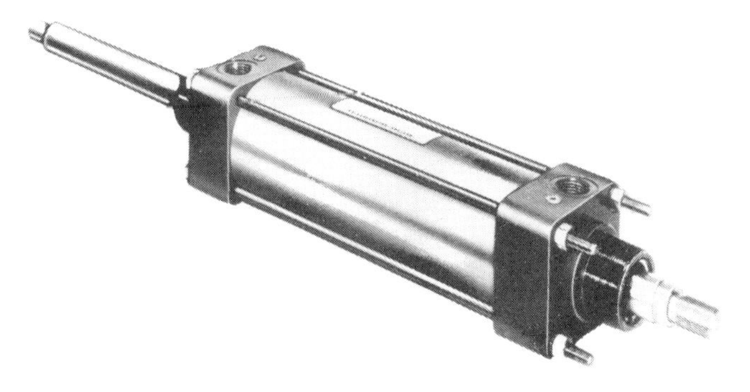

*Tie-rod construction.*

**Pistons and Rods**

Pistons are usually aluminium alloy castings for light- and medium-duty cylinders and aluminium forgings for heavy-duty cylinders. Other materials used for medium- and heavy-duty cylinders include brass, bronze and cast iron.

Pistons may be of one-piece, two-piece or three-piece construction, depending on the type of seals used. Cup seals demand the use of multiple construction. Other types of seals may be simply fitted in grooves machined in a one-piece piston.

An additional nylon wear ring (or two rings) may also be incorporated in the piston design to provide support for the whole assembly. In the case of a long stroke cylinder the piston length may also be increased for better support.

For operation at high service temperatures beyond the maximum rating of elastomeric or fabricated seals, automobile-type rings may be used as piston seals. In this case the requirements are basically the same as for internal combustion engines – a close-fitting piston in a material with a similar coefficient of expansion to the cylinder tube and a series of split rings assembled in grooves.

Leakage is invariably higher than with pressure-energized flexible seals and bore wear is also higher, so this form of piston seal would not be used for normal duties.

The usual choice for piston rods is either En8 or similar mild steel, ground and polished or chromium-plated or ground and polished stainless steel. In some cases hardened steel rods may be employed, although 0.012mm (0.0005 in) chrome plating usually provides suitable surface hardness. It is important that the rod has a high finish in order to minimise wear on the rod seals.

*Cut-away cylinder showing seals and ports.*

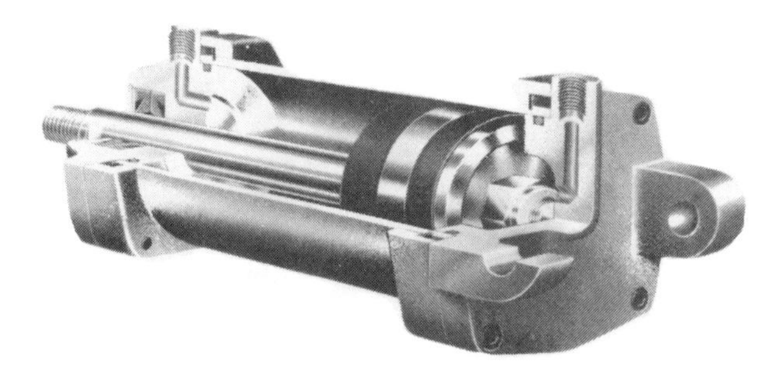

*Cut-away view of typical double-acting air cylinder, non-cushioned.*

Piston rod diameter is chosen with respect to the end loading and length – the latter governing whether the rod behaves as a rigid rod or as a column subject to buckling. Ideally, the rod diameter should be chosen so that the critical length is not exceeded.

Heavy-duty cylinders normally have larger rod diameters than other types, but choice of rod diameters is usually very restricted in the case of standard production cylinder sizes.

Long stroke cylinders again present a particular problem because of the length of rod involved. Besides introducing the possibility of buckling under compressive loading, the weight of a long rod will tend to make it sag in the case of horizontal cylinders.

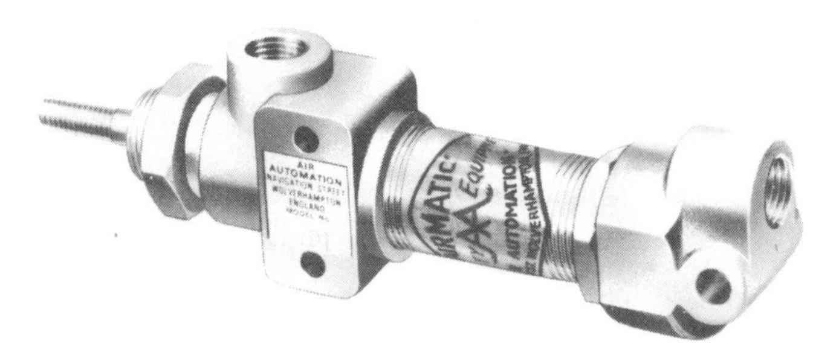


*Universal mounting cylinder.*

A solution which may be satisfactory in such cases is to increase the length of the rod bearing substantially and this is fairly common practice, *ie* the longer the stroke of the cylinder, the longer the length of rod bearing would be expected to be.

Piston rod bearings are commonly of sintered bronze, but may also be of other materials inserted in the form of a sleeve, or a simple plain bearing in the end cover material, *eg* in the case of a cast iron end cover. In addition this assembly also invariably incorporates a rod seal, choice again depending on the individual manufacturer's preference.

Simple elastomeric seals or fabricated rings are usually satisfactory on light-duty

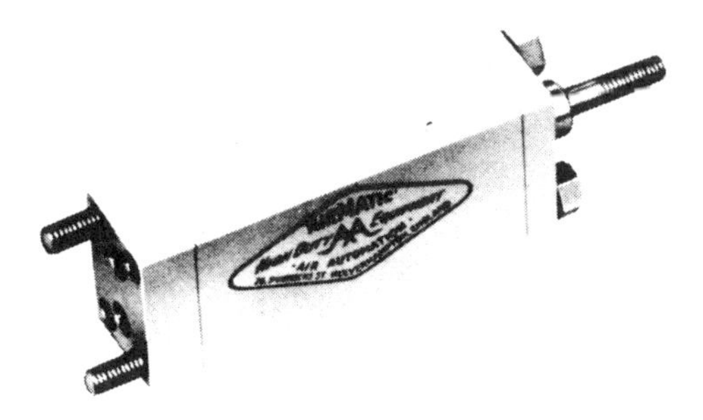

*Manifold mounting cylinder.*

cylinders. Medium- and heavy-duty cylinders normally employ multiple seals or seal sets. A second seal is also usually mounted on the outer end of the bearing, this being a wiper seal to prevent ingress of dirt, *etc*, into the bearing and main rod seal when the rod is retracting, *eg* see Figure 3.

While wiper seals are regarded as a standard form of protection for rod seals, their performance as wipers is essentially incomplete. Thus, if the cylinder is operating in a particularly dirty ambient, full protection can only be given by a protective cover enveloping the whole length of exposed rod.

Such covers are of bellows type to permit extension and retraction, the number and depth of corrugations depending primarily on the extended length of rod to be enclosed, *ie* a full stroke length. Their design is quite specialized because they must extend and retract without collapsing and must also be capable of 'breathing' to prevent collapse under internal 'pumping' action.

AA – Bearing Sealing Rod
BB – Wiper Seal Housing
CC – Retaining Ring
DD – Bearing Washer
EE – Wiper Seal Retaining Ring
W – Wiper Seal
V – Piston Rod Seal

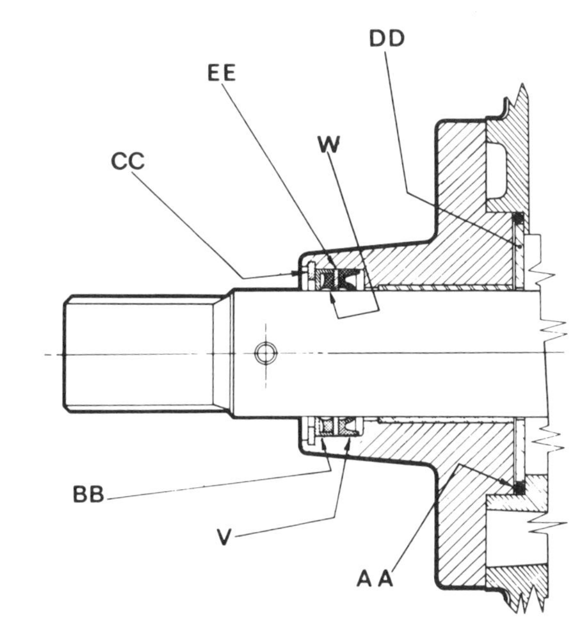


*Figure 3*
*Cylinder end assembly.*

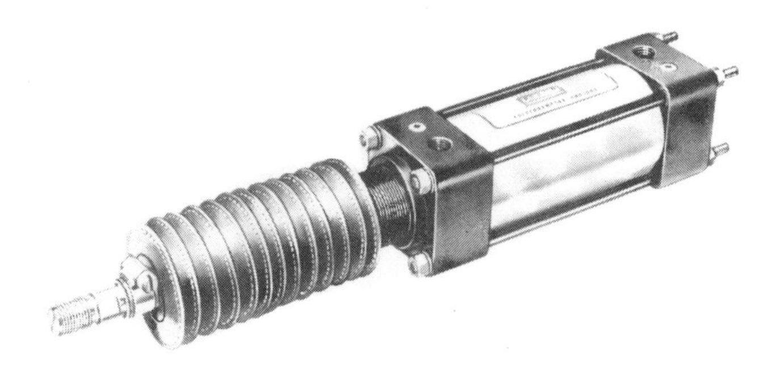

*Cylinder fitted with rod gaiter.*

Special seals are required in the case of rotating cylinders to provide a seal between the airshaft stem and the distributor. These can present a problem with large airshaft diameters and high rotational speeds and may even require cooling with a water jacket.

**Non-Lube Materials**

Cylinders have recently been introduced with special materials to enable them to work on a non-lubricated air supply in application areas such as the food industry and drugs manufacture. The use of bearing material with long-lasting lubrication properties and the design of the cylinder can make it suitable for operation without additional lubrication.

Lubrication free operation gives a better work environment, simpler installation and means that the cylinder does not require regular maintenance.

*Figure 4*
*Non-lubricated cylinders.*

The cylinders shown in Figure 4 are available in non-lube versions, which involves an internal finish coating of electroless nickel and PTFE, giving a very hard finish with good wear and self lubricating properties. This allows operation without the usual oil mist lubrication, where required, and the life of the cylinder is reduced by less than 20% of the life of a normally lubricated cylinder.

The non-lube version can be used with a lubricated air system if required. Additional benefits are obtained, by the very low friction coefficient of the material, in that the break out force of the piston occurs at very low pressures.

Other finishes are being developed for non-lube cylinders, involving anodised aluminium tubing and sprayed on coatings.

**Cylinder Mounts**

A wide variety of mounts are available for air cylinders and the choice of the most suitable mount is an important factor for optimum working. They can be grouped broadly as *floating* or *rigid mounts*, with numerous variations on individual attachments.

A floating mount anchors the cylinder at one point only, with freedom to move in one plane. This is often a preferred type because it allows the cylinder to compensate automatically for any mis-alignment of linkages without introducing side or bending loads on the rod.

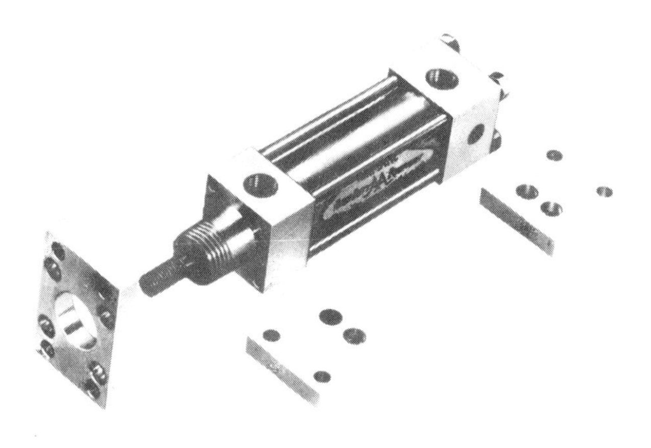

*Tie-bar type non-cushioned air cylinder with universal mountings, nose-rear trunnion and front face.*

For fully floating mounting, ball and socket or universal joint-type mounts can be used, fitted to the rear end cover. These are generally satisfactory, provided the stroke of the cylinder is comparatively short.

With longer strokes, fully floating end mounts are less satisfactory. A type of mounting which can be used in such cases is a *mid-position flexible mount*, trunnion pins locating in a ball joint bearing.

Such mounts must be designed to provide the most advantageous force distribution, relieving the mounting pin from shear loads.

The four main types of floating mounts are shown in Figure 5. The three end

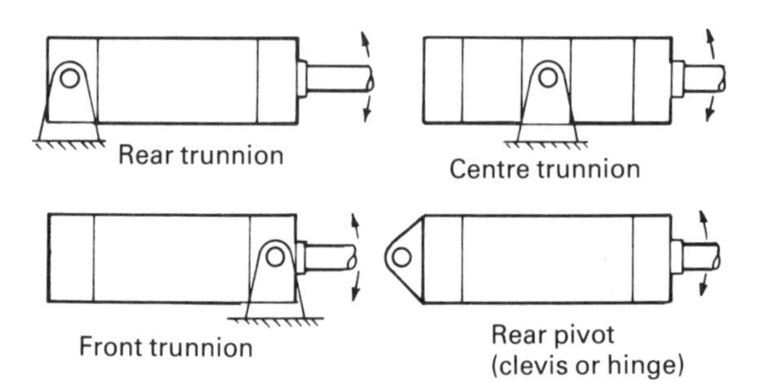


*Figure 5*

fixings are usually preferred to central trunnion mounting, but all have their specific applications.

Rigid mounting is used where the cylinder must have good support. These may be single or double mounts, the latter being particularly suitable for long-stroke cylinders. Basic forms are shown in Figure 6.

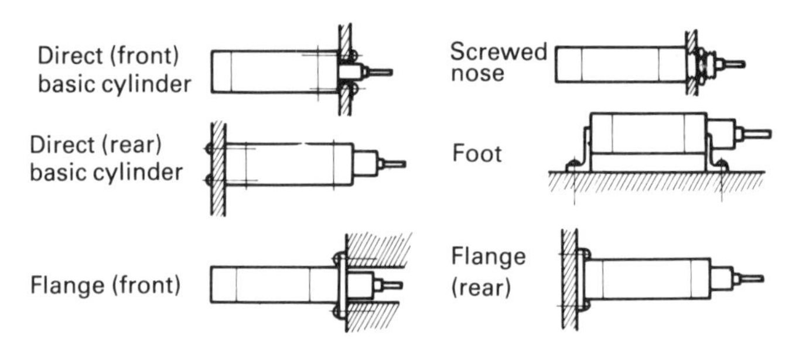


*Figure 6*

*Foot* and *centre-line* mounts are normally incorporated in the cylinder construction of end cover fabrication. Single foot mounts would normally only be used on light-duty cylinders. Pedestal and flange mounts are normally attachments to the end cover(s). The advantage of an 'attached' mount is that it enables a variety of different mounts to be offered to fit a standard design of cylinder.

Rigid mounts can either absorb the thrust force on the cylinder centre line or in a plane parallel to but removed from the centre line. The former are generally to be preferred where the forces to be absorbed are heavy, also for providing the most rigid form of mounting with a minimum of complication. The mounting bolts are only stressed in simple tension, provided there is no mis-alignment of the piston rod.

Examples of mounts which absorb the forces on the cylinder centre line are *centre line lugs*, mounting by extension of the cylinder tie-rods, *flange* mounting and *screwed-nose* mounting. Such mounts may provide some tolerance for mis-alignment at the end of a stroke, but little or none at the other end of the stroke.

Additional support may be necessary where a cylinder is end-mounted to provide support for the free end of the cylinder. Double mounts are normally

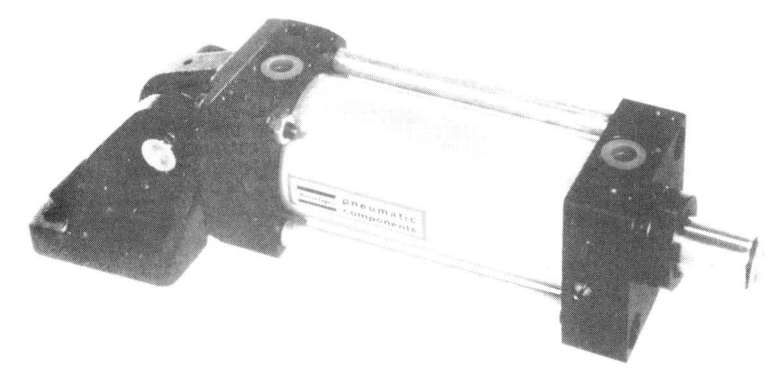


*Hinge-mounted cylinder.*

chosen where a long-stroke is completely overhung. Very long cylinders may be fitted with an additional central support.

Foot mounts which do not absorb the forces on the cylinder centre line may show distinct limitations where forces are heavy, or where heavy shock loads may be involved. The mounting bolts are subject to compound stresses, the cylinder itself is subject to bending or buckling stresses and the platform to which the mounts are attached may distort unless sufficiently rigid or strengthened.

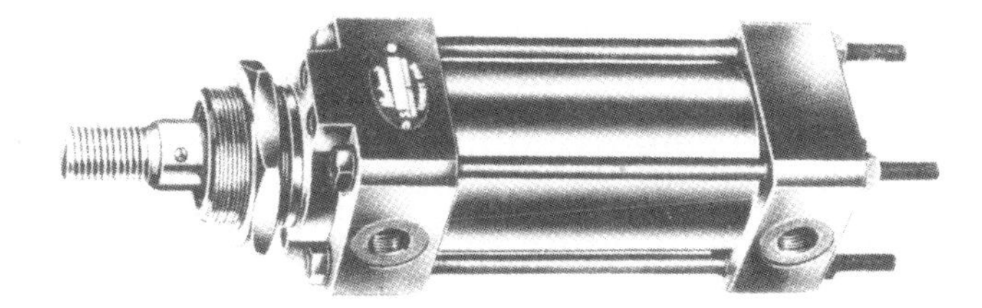

*Cylinder with screwed nose mount.*

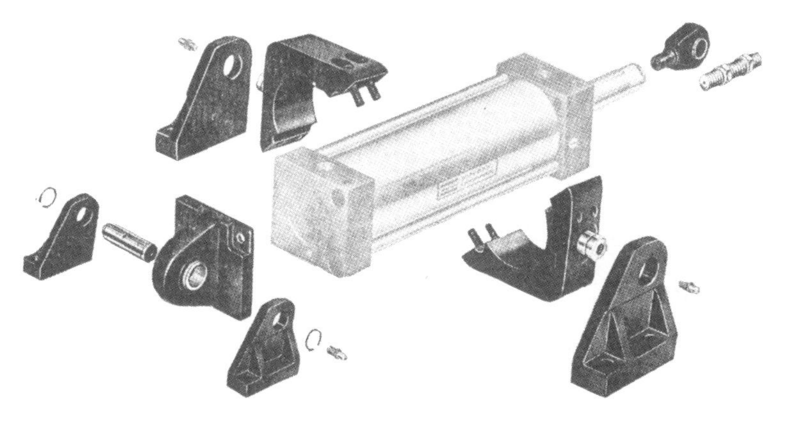

*Various mounts for medium-duty cylinders.*

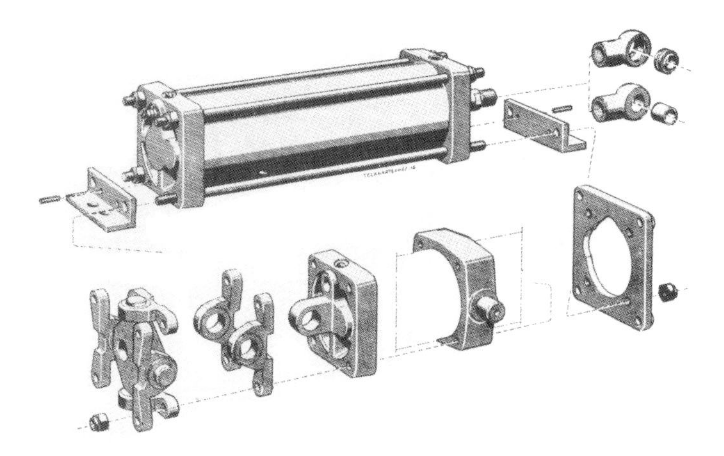

*Various mounts for heavy-duty cylinders*

A method of reducing the compound stresses in the mounting bolts is to fit keys or pins to take shear loads, leaving the mounting bolts in simple tension. Only one end of the cylinder should be keyed to provide movement for expansion and contraction of the cylinder due to changes in temperature and pressure. Sometimes a thrust key is incorporated on the cylinder itself, fitting directly into a keyway cut in the surface of the mounting member.

**Rod Ends**

Rod ends are finished in a variety of ways, the main alternatives being shown in Figure 7. Most or all of these alternatives are commonly offered on standard productions, although manufacturer's terminology may not always agree in the case of clevis and tongued ends. Essentially a clevis end can be recognised as a fork; and a tongue as a simple eye (which is associated with a bracket mount, or could accommodate another clevis fitting). Both the clevis and tongue provide pinned fittings which can substantially modify the column loading of the piston rod.

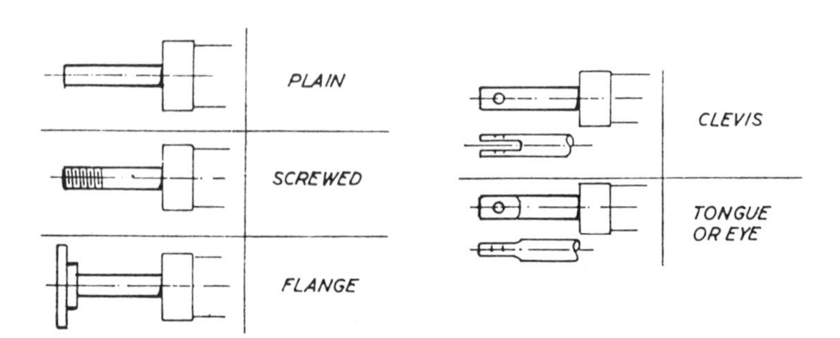


*Figure 7*

**TABLE 1 – CYLINDER TUBE MATERIALS**

| Light-Duty | Medium-Duty | Heavy-Duty |
|---|---|---|
| Plastic.<br>Hard drawn aluminium tube.<br>Hard drawn brass tube. | Hard drawn brass tube.<br>Aluminium castings. | Hard drawn brass tube.<br>Hard drawn steel tube.<br><br>Brass, bronze, iron or steel castings.<br>Welded steel tubes. |

**TABLE 2 – END COVER MATERIALS**

| Light-Duty | Medium-Duty | Heavy-Duty |
|---|---|---|
| Aluminium stock (fabricated).<br>Brass stock (fabricated).<br>Aluminium castings. | Aluminium stock (fabricated).<br>Brass stock (fabricated).<br>Bronze stock (fabricated).<br>Aluminium, brass, iron, or steel castings. | High tensile castings. |

**TABLE 3 – PISTON MATERIALS**

| Light-Duty | Medium-Duty | Heavy-Duty |
|---|---|---|
| Aluminium castings. | Aluminium castings.<br>Brass (fabricated).<br>Bronze (fabricated).<br>Brass, bronze, iron or steel castings. | Aluminium forgings.<br>Aluminium castings.<br>Brass (fabricated).<br>Bronze (fabricated).<br>Brass, bronze, iron or steel castings. |

**TABLE 4 – PISTON ROD MATERIALS**

| Material | Finish | Remarks |
|---|---|---|
| Mild steel. | Ground and polished.<br>Hardened, ground and polished.<br>Chrome-plated. | Generally preferred. |
| Stainless steel. | Ground and polished. | Less scratch resistant than chrome-plated rods. |

**TABLE 5 – MOUNT MATERIALS**

| Light-Duty | Medium-Duty | Heavy-Duty |
|---|---|---|
| Aluminium castings.<br><br>Light alloy fabrications. | Aluminium, brass and steel castings. | High tensile steel castings.<br>High tensile steel fabrications. |

# Performance and Selection

THE PNEUMATIC performance of a cylinder, regarding thrust and speed of operation, has been defined in Section 5. Due to the relatively low air pressures normally employed, cylinder walls are not highly stressed and wall thickness is normally determined by practical considerations rather than material stress requirements.

**Cylinder Strength**
The same formula as for hydraulic cylinders (see Section 3) can be used for stress calculations, *viz:*

$$\text{Material stress } S = \frac{PD}{2t}$$

Where t is the wall thickness, in the same units as diameter D.

In the case of plastic cylinders, stress determination can be based on the formula:

$$\text{Material Stress } S = \frac{P\ (D-t)}{2t}$$

With plastic materials the maximum permissible material stress is temperature dependent, decreasing appreciably with an increasing temperature. In the case of thermo-plastic materials there is also a maximum permissible temperature for pressure working, above which the material will start to flow.

Knowledge of the cylinder working temperature is, therefore, important for stress calculations with such materials, which have lower maximum permissible material stresses than metals to start with. In general, however, the design of plastic air cylinders is normally based on purely empirical lines.

**Rod Strength**
Cylinder rods can be stressed as rigid rods, provided the length of the rod does not exceed ten times its diameter. The stress formula in this case is:

$$\text{Material stress} = \frac{F}{A}$$

Where F = compressive or tensile force or load.
A = cross-sectional area of rod.

For general working the maximum permissible material stress may be based on the ultimate tensile stress of the material and a suitable factor of safety. This will then give adequate rod strength either in tension or compression.

If the length of the rod exceeds ten times its diameter, then it may be subject to buckling under compressive loading. Adequate strength in tension can then no longer be taken as an indication of adequate strength in compression.

The case of compressive loading must be analysed separately, considering the rod as a column. The material stress then depends on the method of end fixing which determines the equivalent strut length as shown in Section 3 for hydraulic cylinders.

For very long-stroke applications, rodless cylinders are now used where the transverse loading limitation depends on the design of the cylinder and is usually specified by the manufacturer.

Pneumatic cylinders are rarely used as struts or structural members of a machine, as hydraulic cylinders are, due to the compressibility of air and the lower working pressures. Stresses are more likely to be set up by dynamic loading due to the speed and inertia of the moving parts.

**Cylinder Cushioning**

Air cylinders are inherently fast-acting in the absence of throttling or inertia loading. If loads are light the piston will continue to accelerate rapidly until brought to a stop at the end of the stroke. The kinetic energy of the piston and rod is then dissipated in the form of a shock load on the cylinder end cover and end wall.

This can be avoided by incorporating some means of decelerating the piston in a uniform manner as it approaches the end of the stroke, known as 'cushioning'. Integral cushioning can be done pneumatically by providing a cushion chamber in the end of the cylinder into which a nose section on the piston enters progressively.

Air trapped in the cushion chamber escapes *via* a controlled bleed, thus providing adjustable damping or 'braking' effect over the latter part of the piston movement. Over the length of the cushion travel - which only needs to be an inch or so on most cylinders - the piston can be brought to rest with its kinetic energy dissipated in the form of heat or at least the majority of this energy can be dissipated with a minimum of shock when the piston is finally stopped (see Figure 1).

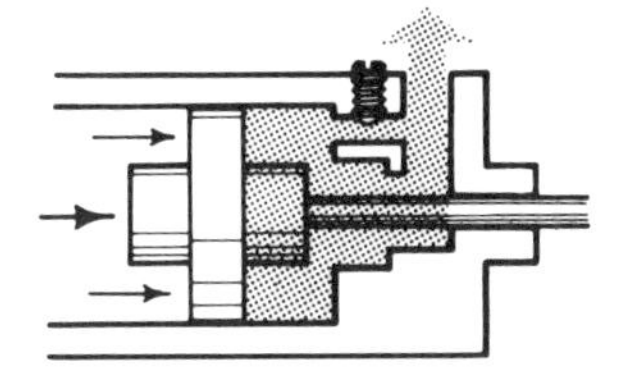

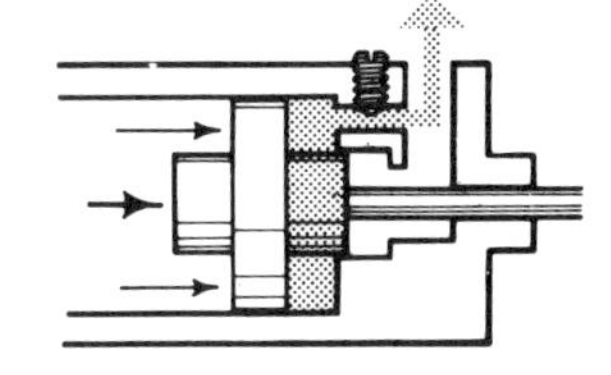

*Figure 1*
*Principle of air cylinder cushion.*

Cushions can be fitted to one or both ends of a cylinder. The cushion chamber may be incorporated in the end cover(s), or in the form of a cushion collar. The cushion boss or nose on the piston is usually tapered to provide smooth entry, but produces a plunger-type seal once it has fully entered the cushion chamber, trapping the air volume in the cushion chamber which is then compressed by further motion. The rate of bleed, and thus the cushioning effect, is normally adjustable *via* a needle valve.

Detail design may vary with different manufacturers. Originally cushions were of simple plunger type, with both the cushion chamber bore and piston nose in metal and this arrangement is still widely used. Metal-plastic combinations may, however, be preferred as being less affected by corrosion or dirt; also elastomeric ring seals may be used in place of plunger seals for similar reasons.

Still another type is based on a spring-loaded seal or an arrangement of *face* seals which forms its own cushion chamber automatically once the seal is brought into contact with the seating surface at the end of the cylinder. This eliminates the need for a separate cushion chamber, but requires a needle valve in both the blind end of the cylinder and the piston rod to bleed off trapped air.

The weakest point of all cushioning systems is usually the needle valve or bleed control which can be affected by dirt or corrosion with resulting modification of the cushioning action.

A cushioned cylinder also requires some minor modifications to the porting. On a double-acting cylinder provision must be made to supply air directly to the piston area and not through the cushion chamber, otherwise the intial thrust available would be very low. This is merely a matter of suitable geometry (see Figure 2).

*Figure 2*
*Typical cushioned end.*

With seal-type cushions the flexibility of the seal can be such that it will collapse when air pressure is applied from the reverse side, allowing flow to the full piston area. Flow to the full piston area may also be encouraged by distribution channels formed in the end cover or the face of the piston itself.

Cushioned cylinders are most suitable for low to moderate speed operation. At speeds above 460 mm/s (18 in/s) they can be relatively ineffective, the intitial shock as the piston reaches the cushion area being high unless a long cushion is used with progressive entry.

In such cases it may be necessary to arrange for some form of external cushioning, either by the use of a buffer cylinder or a form of speed control which provides retardation and cushioning over a suitable proportion of the stroke.

Two further limitations of integral cushioning are:

(1) They are effective only if the full stroke of the piston is used.

(2) They are relatively ineffective on pre-exhausted cylinders or those which have quick exhausting valves.

In the former case the piston never enters the cushion area, being brought to rest by external stops or some similar form of control. In the latter case the amount of air trapped in the cushion is very low and hence its cushioning effect is also low.

Proper attention should therefore be paid to the working requirements of cylinders fitted with integral cushioning in order to ensure that the cushioning is effective.

The actual amount of cushion movement is usually fixed by the design, and may vary from as little as 25 to 50 mm (1 to 2 in) on most cylinders, up to 150 mm (6 in) in the case of certain heavy-duty cylinders.

One design of sliding plate cushion valve, however, permits ready adjustment of the length of the stroke over which cushioning is provided by altering the length of the rods carrying the valves (see Figure 3).

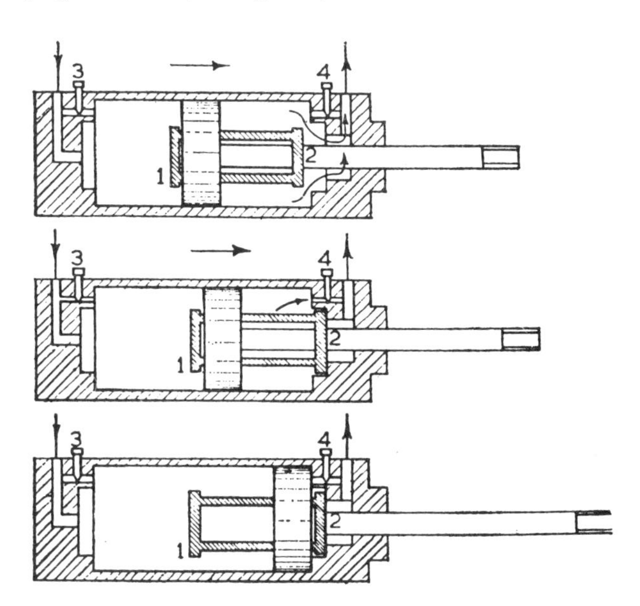


*As the piston approaches the end of its stroke the plate valve (2) seats against the end cap, sealing the main exhaust port. Exhaust air escapes through throttle valves (4).*

*Figure 3*
*Sliding plate valve cushion.*

**Installation**

The installation of standard pneumatic cylinders follows a similar procedure to that of hydraulic cylinders and depends on the mounting style chosen. This can be controlled to a considerable extent while the layout of the machine or plant is being arranged.

With special cylinders such as rodless, band and cable cylinders the machine or plant can be designed to take advantage of the special features of the cylinders and the manufacturers instructions should be followed.

Before fitting and actuating equipment, it is advisable to arrange a flexible air line complete with blow gun for test purposes. The air line should be water-free, well-filtered and lubricated. The cylinders should be installed and checked to see that they have been installed correctly and move freely.

It is important to ensure that where cylinders are operating guided slides, linkages, levers, *etc*, that the axis of 'normal' cylinder movement follows that of the associated guided component in all planes. The cylinder axis can be lined up by the use of dial gauges locating on the outside diameter of the cylinder tube.

When the cylinder has been lined up, it should be possible to manually move the load between the stroke extremities to make sure that the slide moves freely. If the mechanical arrangement is such that manual operation is not possible, due to the nature of the loading, then the blow gun can be used at appropriate ports to actuate movement. Only when the movement is smooth and steady, will the alignment be satisfactory.

For trunnion mountings it is necessary to ensure that the cylinders are free to oscillate on their mountings and the cylinder pipe connections must of course be flexible and clear to flex naturally during normal oscillation.

Trunnion pins should be lubricated before assembly with any standard type of bearing lubricant. These mounting styles give freedom of movement in one plane only, the alignment in the other 'rigid' plane must be carefully checked as for rigid mountings.

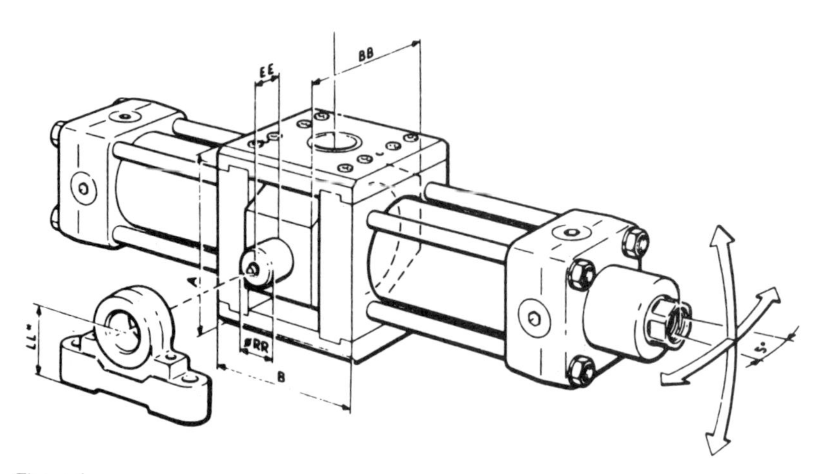


*Figure 4*
*Universal cylinder mount for flexible and shear-free installation.*

Pivot mountings incorporating universal joints or swivel joints are produced, however, and may be used where the path described by the rod transverses more than one plane as shown in Figure 4.

Where the piston rod thread is to be screwed to a mating part, it is better to connect the mating part first before fixing the cylinder in its mounting. Small cylinders may be rotated in order to screw the piston rod thread into initial engagement. For larger units, it is usually easier to rotate the connecting linkage or slide, *etc*, and this is to be preferred rather than rotating the piston rod within the cylinder.

**Standardization**

National and international standards have been drawn up for pneumatic cylinders, in conjuction with the standards for hydraulic cylinders, by bodies such as CETOP, BSI, DIN, ISO, *etc*.

The mounting dimensions of single rod double-acting pneumatic cylinders are covered by BS 4862. Part 1 of this standard specifies interchangeability dimensions for four mounting styles of miniature cylinder with bores in the range 8 to 25 mm, the associated ISO standard being ISO 6432. Part 2 specifies interchangeability dimension for seven mounting styles of cylinder with bores in the range 32 mm to 320 mm with detachable mountings, the associated ISO Standard being ISO 6341.

# Applications

### Mechanization

MECHANIZATION IS broadly defined as the replacement of manual effort by mechanical power. In all degrees of mechanization the operator remains an essential part of the system, although with changing demands on physical input as the degree of mechanization is increased.

Pneumatic cylinders are attractive actuators for low cost mechanization particularly for sequential or repetitive operations. Many factories and plants already have a compressed air system installed and manual machines can be mechanized with only moderate capital expenditure.

Additionally, compressed air is capable of providing both the power or energy requirements and the control system. The main advantages of an all pneumatic system is usually economy and simplicity, the latter reducing maintenance requirements to a low level. It can also have outstanding advantages in terms of safety.

The majority of air cylinders used for mechanization are 100 mm (4 in) bore or smaller, giving a maximum thrust of approximately 450 gf (1000 lbf) with a line pressure of 6 bar (90 lbf/in$^2$). Speed range usually allowed is from 0.5 to 2.5 m/s (1.6 to 8 ft/s), with speed control, if required, provided by throttling.

The cylinders are used for positioning movements and, as they are basically linear actuators of limited power, it is essential to have methods of converting linear motion into semi-rotary motion, or of magnifying the force for actuating clamps, cutters, *etc.*

### Limited Rotary Movements

Where rotary outputs are required to work for not more than one or two revolutions, the use of an air cylinder is again a logical choice. A simple method of producing semi-rotary motion for one or two revolutions is by rack and pinion. The rack can be of the cut gear-type or formed by stretching roller chain along a flat support.

The method adopted will normally depend on the local conditions. An angular movement of 90° or less can also be obtained by a crank mechanism, and this, by a ratchet and pawl, can be further extended to give an intermittently full rotary motion.

*Figure 1*
*Air cylinders fitted to a torque cylinder and index table.*

Manufacturers of pneumatic cylinders often also make semi-rotary units with the cylinders built in as a complete assembly as shown in Figure 1.

The rack and pinion torque cylinder shown is of sturdy tie-rod construction and applications include gate opening and closing, clamping and agitation of fluids and raw materials. The cylinders are available with or without adjustment to the angle of rotation at either or both ends.

The pneumatically-operated indexing table shown is complete with solenoid-operated valve and hydraulic check unit. A precision of ± 1 minute of arc can be obtained for an indexing table with four, six, eight, twelve or twenty-four equally-spaced divisions.

The hydraulic check unit acts as an hydraulic damper and enables a maximum applied force up to 7 KN on the piston to be retarded to closing speeds from 10 m/min down to 70 mm/min. Simple regulation enables fine manual tuning of the unit to achieve the required retardation speed.

**Force Magnification**

Force magnification is best used when the final large force has only to be applied over a short distance. Then simple mechanical methods will give a limited degree of magnification. Figure 2 shows a simple squeeze riveting machine with leverage of about 10:1.

A toggle mechanism has the advantage that a greater 'daylight' can be provided. The force exerted by a toggle mechanism at the end of the stroke is considerably influenced by the stiffness of the framework and links (see Figure 3).

**Industrial Robots**

Wide-spread use is made of pneumatics for the mechanization of industrial robots, taking second place after hydraulics. In terms of percentage of robots using the

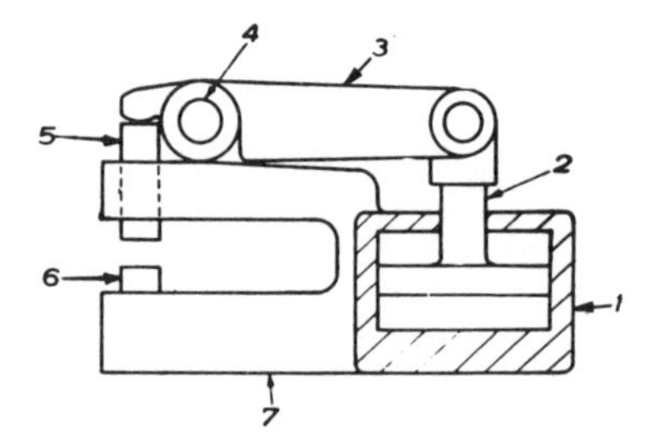


*1 – Cylinder and frame. 2 – Piston. 3 – Lever. 4 – Port. 5 – Plunger. 6 – Anvil. 7 – Frame.*

*Figure 2*
*Small riveting machine with lever to multiply the force exerted by the cylinder.*

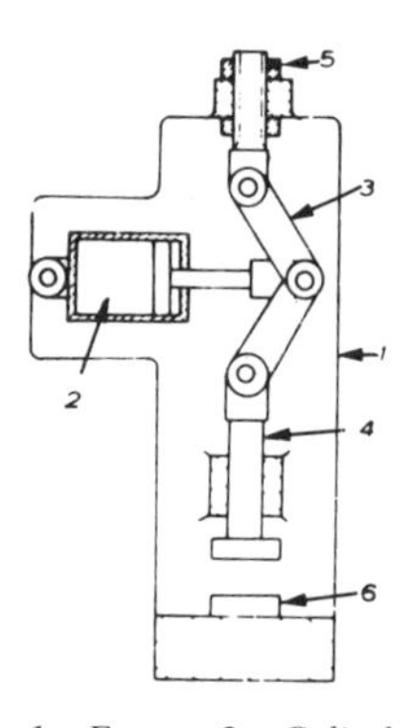


*1 – Frame. 2 – Cylinder. 3 – Toggle. 4 – Plunger. 5 – Adjusting Screw. 6 – Anvil.*

*Figure 3*
*Toggle press.*

different types of power systems in current use, hydraulics count for about 50%, pneumatics 30% and electrics 20%.

Power systems for industrial robots are simply advanced forms of mechanization rather than automation. The fact that industrial robots may be programmed, and thus work without attention from an operator, is reliant on a computer-type brain or microprocessor control unit taking the place of the human operator's brain.

The power system of the industrial robot remains, essentially, a mechanical system, which may however incorporate varying degrees of sophistication *via* closed-loop servo-systems with feedback and sensors. In this sense low technology robots can be described as coming under the heading of mechanization and high technology industrial robots under automation.

The point at which mechanization becomes automation is where the machine operator's role is reduced to that of a supervisory nature or even eliminated entirely. In other words, the essential feature of command is transferred to a timed operation of the system or semi- or fully-automatic control using a suitable detector generating feedback signals. True (full) automation starts when the operator is relieved of all responsibility.

Recent innovations such as rodless cylinders have made pneumatic power an effective and reliable power medium for mechanical robots where high speed linear movements combined with accurate positioning are demanded.

Robots for removing injection mouldings from a machine and for transporting them to, and positioning them on downstream finishing equipment are examples of the adoption of rodless cylinders shown in Figure 4.

*Figure 4*
*Rodless cylinders in robot handling devices.*

This robot employs two rodless cylinders. One provides 1.5m (4.9 ft) of horizontal movement to take the robot jaw assembly holding the finished mouldings to a position in front of the injection machine over a conveyor; the second, smaller diameter cylinder gives horizontal movement fore and aft within the 'daylight' of the machine to present the jaw unit to the moulding.

Some manufacturers, who supply air clinders and valves for the automation of engineering systems, have moved into the area of low cost automation and robotics as a natural extension of their traditional business. Harnessing the power of pneumatics and the flexibility of electronics has allowed companies to introduce ranges of low cost robotic devices. This has led to the development of peripherals such as *gripper devices* and *modular handling units*. Re-programmable robot manipulators were introduced as forerunners to today's range of programmable robot arms which exploit the flexibility of modular construction by using building block principles.

On most of the current models, pneumatic cylinders and adjustable hydraulically damped stops achieve the required point-to-point drive and there is a choice of dedicated or programmable pneumatic controls plus the option of electronic controls with programmable capability (see Figure 5).

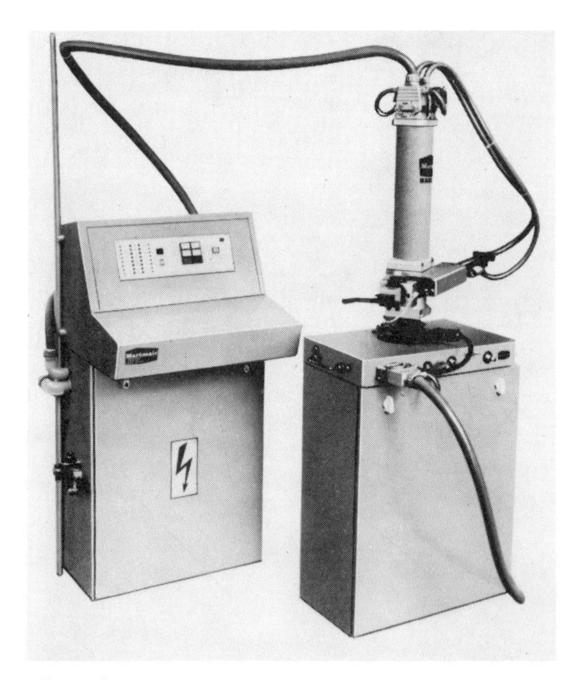

*Figure 5*
*Robot manipulators and controls.*

**Positioning Cylinders**
A positioning cylinder is one designed to provide intermediate movements or positioning of the piston. It is normally produced by adding an external control and/or indicating device known as a *position controller*.

Basically, in a positioning cylinder, the type of control provides positive movement of the piston rod to a pre-determined position when a signal is fed to it. Within limits, the positioning is independent of the load on the cylinder, with the position controller itself controlled by a variable pressure air supply, either varied manually or automatically.

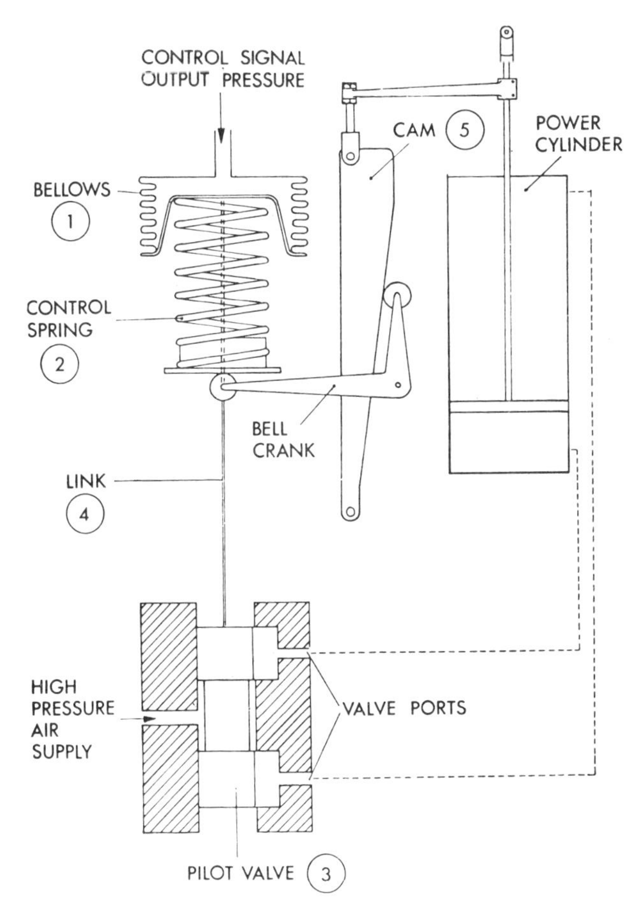


*Figure 6*
*Positioner unit.*

A positioner unit is shown in Figure 6. The principal components, as shown diagrammatically, comprise: a flexible brass bellows, control spring and pilot valve, which is connected to the bellows by a wire link. The control pressure signal is applied to the bellows which move against the control spring, the end of the spring being positioned by the power cylinder *via* the cam and bell crank.

The unit operates on the force/balance principle, the control signal output pressure being converted into force by the bellows and piston position by the control spring. These forces are arranged to oppose each other. While the forces are equal, the pilot valve remains in its neutral position but with a change in control signal pressure the bellows expand or contract resulting in movement of the pilot valve.

Air at high pressure, not exceeding 10 bar (145 lbf/in$^2$), enters the pilot valve as indicated in the illustration and is fed to one end of the power cylinder *via* the appropriate valve port and connecting pipe. At the same time the opposite end of the power cylinder is vented to atmosphere *via* the other valve port and movement of the piston results.

This movement, which is transmitted to the control spring by the cam and bell crank, continues until the force exerted by the spring exactly balances the force produced by the bellows.

Should the piston be subject to a restraining force (less than the design thrust of the power cylinder) its movement will not be prevented, as the pilot valve will remain open until the air pressure is sufficient to overcome the force. The piston will only come to rest when its position meets with the requirements of the control output pressure.

The most common application of positioning cylinders is where a quantity has to be varied continually, such as in *air-operated* valves, *speed variators* and similar devices.

**Linear Motors**

Cylinders designed to provide automatic reciprocation are commonly called *linear motors*, this description being particularly applicable where automatic reciprocation is achieved without the use of external valves or limit switches.

In this case, only a single connection is required to the air supply line, the reversal direction of the main air being provided automatically by internal valves. An example of this type of linear motor is shown in Figure 7. The control may be initiated by mechanical triggering or pneumatic or electronic signalling.

The reciprocating output of the cylinder (*ie* rod movement) may be used directly or to drive a crank for final rotary output. In the latter case, the crank throw must obviously be matched to the stroke available or rather the crank throw will govern the actual cylinder stroke achieved.

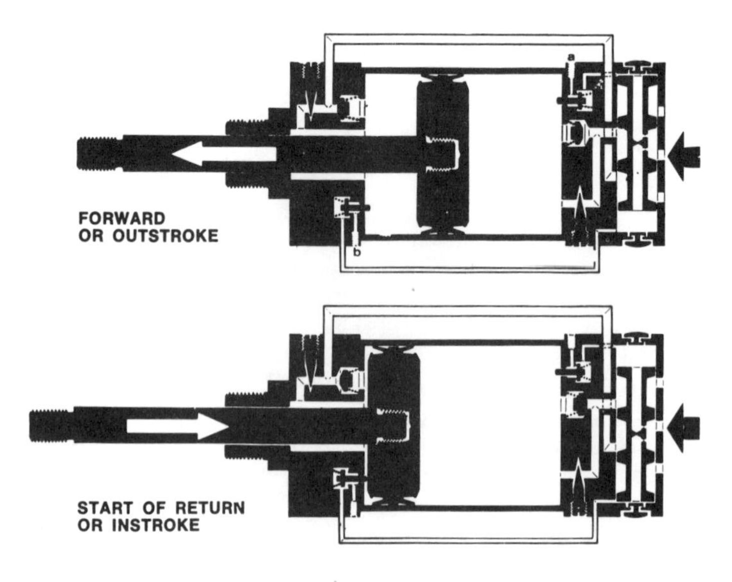


*Figure 7*
*Principle of operation of air motor.*

The versatility of the linear motor may be further improved by providing variable speed of stroking. This is usually done by fitting adjustable restrictors on the exhaust side controlling the rate of escape of pressurized air from the non-pressurized side of the piston and thus the back pressure, which effectively controls the speed.

**Feed Control Units**

All functions of movement performed with material being worked are covered by the definition *feeding*. The equipment required for such functions are *feeders* or *feed units*. Assemblies combining control and operating components are called *feed control* units.

These can be single- or double-acting air cylinders, with attached restrictor check valves for supply air throttling and electrically-actuated solenoid valves. The small standard assemblies can be installed quickly and easily into an electro-pneumatic control system from the simplest to the most intricate layout.

It is advantageous to have very short flow paths between the control valves and operating cylinder. Such assemblies are widely adopted for applications where the function of the air cylinder is as a controlling component, rather than an operating element, where relatively small control forces are normally adequate.

An advance on pure pneumatic feed control units is achieved by incorporating a hydraulic check cylinder to obtain an air-hydraulic feed unit. As a means of achieving uniform motion it is possible to install a through-rod air cylinder, so that the regulating function of the hydraulic cylinder will be performed as its piston rod is forced in.

Any kind of feed operation requiring constant rate of movement, is a potential application for an air-hydraulic feed control unit. Speed ranges are normally between about 20 and 3000 mm/min (0.8 and 118 in/min), thus extending into areas falling short of the scope of pure pneumatic feed control. Slow speed feed at a constant rate is essential in many metal-cutting processes, regardless of whether it is the work or the tool which is advanced.

As distinct from feed motions controlled by a single cylinder, whose stroke corresponds to the overall working stroke, certain material or work feeding operations require feed motion to be repeatable a certain number of times.

In these instances *progressive* feed control assemblies are used where one cylinder stroke represents a certain part of the overall working stroke. The overall motion is divided into progressive steps, the rate of progression being determined by the working machine. A typical application is the feeding of strip material from coils which are machined in separate indentical lengths.

With the progressive feeder, strip, rod or other continuous material can be supplied step-wise to a working machine, the signal initiating each feed motion being transmitted by the machine itself in accordance with its working cycle. Inside the body of the feeder is a double-acting air cylinder, the stroke of which is limited by adjustable stops.

In order to ensure that the material is indeed progressed by the set length on each stroke, the feed slide and the body of the unit each contain a diaphragm clamping cylinder, the two opening and closing alternately. The holding and slide clamps can be designed in a geometrical shape adapted to that of the material being handled, thus enabling sectional material to be progressed by the feeder.

**Bellows Actuators**

As linear actuators, bellows produce very high dynamic thrusts, ideal for numerous lifting and clamping applications. They have no sliding seals to create friction and wear, nor do they require lubrication, consequently they are almost maintenance-free.

Significantly, less space is needed for bellows that for their conventional cylinder counterparts – from a starting height of only 50 mm (2 in) they can provide stroke lengths of up to 250 mm (10 in) and more.

They are also lighter than comparable cylinders and their ability to flex and bend, further increases the bellows versatility, particularly in configurations where axial locations cannot always be rigidly maintained.

Another key property is its ability to absorb vibration. In this context it is considerably more efficient than the conventional spring or rubber mounting and much more versatile because it can be 'tuned', in situ, for optimum damping.

In direct contrast to its function as a vibration damper is its potential as a generator of vibration. Pulsing the air supply to the bellows causes the device to produce powerful low frequency oscillations that are ideal for certain types of shaker installations.

The ancillary equipment needed to control a bellows system is quite straightforward: a pressure regulator and suitable control valves which may be non-return, levelling, quick-exhaust, *etc*, depending on the application and mode of operation. Installation is usually simplicity itself because of the flexible nature of the devices.

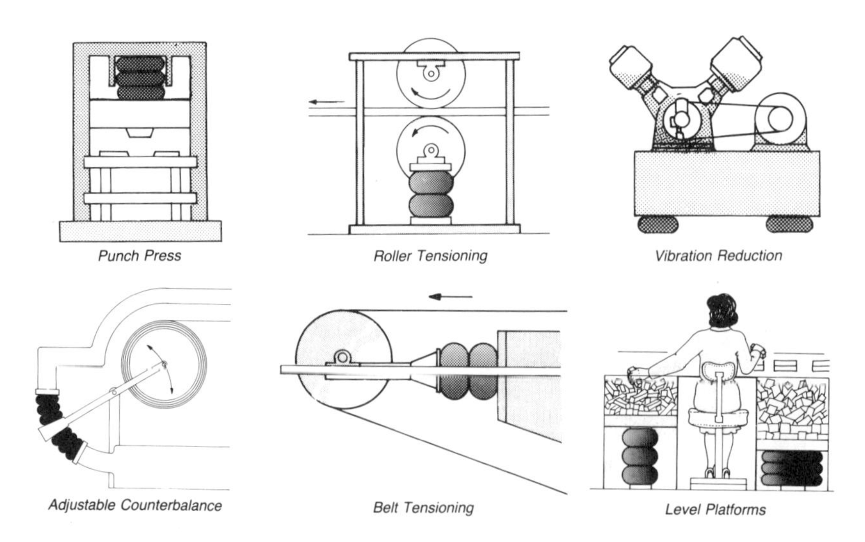


*Figure 8*
*Applications of bellows actuators.*

Typical applications using bellows: simple lifts on ramps, conveyors and tables, tensioning devices for conveyor belts, clamping mechanisms, punch presses – in fact any application that requires powerful, short-stroke actuators that are easily, non-critically installed and take up very little space, as shown in Figure 8.

# SECTION 8

## Pneumatic Valves

# Types of Valves

PNEUMATIC VALVES can be divided into three main categories defined by the function they perform in a similar manner to hydraulic valves, *ie*: directional control, flow control and pressure control.

However, they can be further sub-divided into valves which control the air flow to carry out a *power* function and valves which direct air flow to carry out a *control* function. The latter division includes a number of specialized valve types which do not fall into general categories, including valves employed as 'logic' elements.

Basically pneumatic valves can be divided into the same three classifications used for hydraulic valves, *ie* seated valves, sliding spool valves and variable orifice valves.

### Seated Valves

Seated valves can either be poppet, *ball* or *disc* valves with either hard metal-to-metal seats or soft metal to elastomer seats. The latter are generally preferred for pneumatic duties where good sealing is required to conserve the compressed air.

Poppet valves require minimum movements to achieve full opening, offer low resistance to flow when in the open position and have good sealing properties when closed. They can also be made to seal adequately for long periods but require relatively large operating forces.

A simple poppet valve designed for a pressure of 10.5 bar (150 lbf/in$^2$) may require an opening force as large as 10 kgf (22 lbf) because of its unbalance. On the other hand, the fact that full opening is achieved with only small travel, giving rapid full flow response to a signal, can be of particular advantage in high speed applications and for pilot circuits.

A limitation of poppet valves is restricted switching ability. The basic poppet valve is capable only of on–off switching function. A combination of poppet valves is needed to perform a three-way or four-way function. Other types of valves are thus normally used for such duties.

Multiple poppet valves base-mounted, that work with non-lubricated air, have been developed as self-cleaning poppet valves with a high tolerance to dirty air as shown in Figure 1.

The five-port two position valve, designed with a memory, has all internal parts except for seals made from durable metal.

*Figure 1*
*Self-cleaning poppet valve.*

*Floating exhaust* poppet seals automatically adjust to ensure proper sealing and to compensate for wear. The valve can be controlled by single- and double-solenoid pilot and single- and double-pilot controls.

Single control units use system pressure to return poppets when control pressure ceases. Double control units need only an impulse signal and use system air pressure to hold poppet position, with mechanical detents to assist during shut downs.

**Spool Valves**

Spool valves are attractive because they offer extreme flexibility in interconnection *via* a simple sliding mechanism. They are also relatively simple and economic to manufacture although, for adequate sealing, fine surface finish is required on both the spool and barrel bore with close tolerances to ensure practical minimum clearances. Glandless spool valves normally require a lapped fit between spool and body.

Spool valves with static seals offer simpler construction in this respect and also rather more flexibility in design with seals positioned between valve spaces so that a seal is situated between each subsequent port and one seal on the outside of each of the two outer ports (see Figure 2).

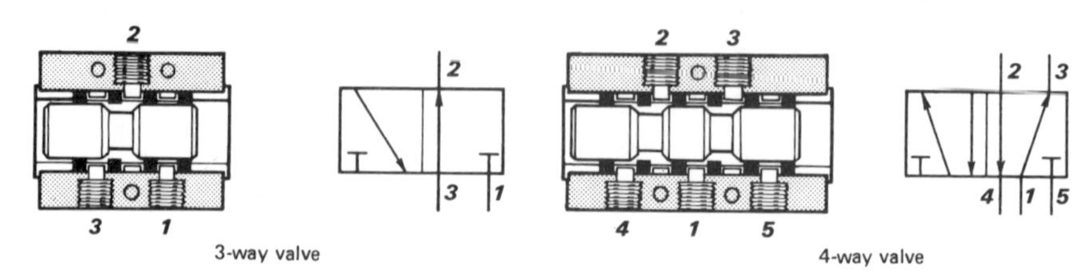


*Figure 2*

Spool valves have two basic elements: a cylindrical barrel in which slides a plunger. Port blocking is provided by lands or full diameter sections on the spool with intervening waist sections which provide port interconnection through the barrel. This makes it easy to provide multi-way and multi-position switching.

A possible limitation of glandless spool valves is that in certain applications and particularly with larger spool valves, the precision bore and fit may be distorted by externally applied loads. On the other hand, spool valves with static seals also have some limitations. A disadvantage in this case is that the sealing rings must slide over the ports where they may be subject to cutting action or distortion.

Variations, primarily again in the case of air valves, include the use of U-seals, bushed bodies and ports drilled with a number of fine holes where sealing rings pass and a linered construction with a liner floated on simple seal rings. All are aimed at reducing the possibility of damage or distortion of the sealing rings sliding over port openings.

A typical miniature solenoid pilot-operated spool valve is shown in Figure 3. The caged O-ring seals between each port are supported by glass-filled acetal spaces through which the hard anodized spool slides.

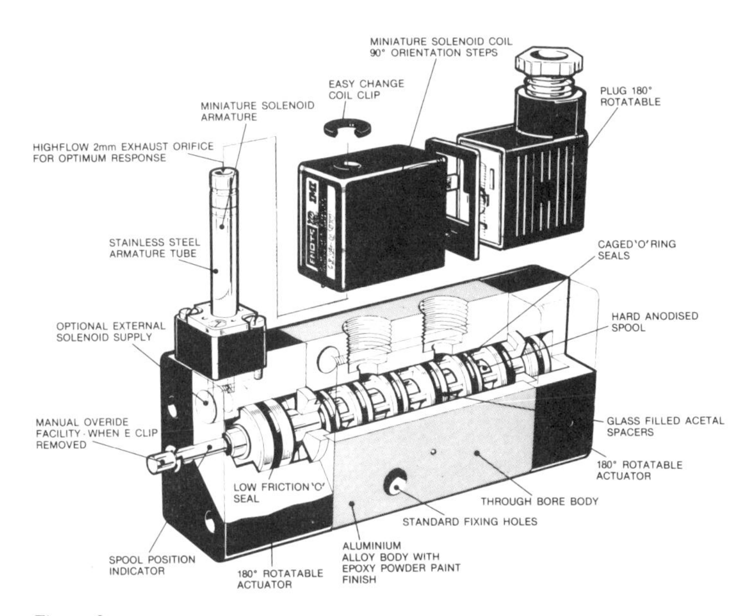


*Figure 3*
*Miniature solenoid-operated control valves.*

Compact spool valves have also been developed to operate on non-lubricated air as shown in Figure 4. The concept behind the valve is to provide an easily assembled system of basic control elements with flow rates covering most applications but occupying the minimum physical space. The valve uses a Teflon* spool and stainless steel sleeve assembly as shown.

**Slide Valves**

Slide valves offer an alternative approach equivalent in effect to an unwrapped spool. Ports are normally drilled in the fixed slide and matching recesses in the sliding member. The two members are then lapped to fit and held in contact by spring-loading. Various methods may be used to reduce friction forces. Thus, in

*Dupont Registered Trademark.

*Figure 4*
*Compact valve shown with its polytetrafluoroethylene non-lube spool and stainless steel sleeve.*

some valves, the sliding member is suspended on spring plates to prevent metallic contact.

In other cases, friction forces are reduced by incorporating hydrostatic pressure balance within the valve. The principal limitation of slide valves is that they demand extreme precision of manufacture in order to realize leakage rates comparable with those of other valve types. Slide valves are also sometimes known as *plate* valves or *sliding plate* valves.

**Rotary Plate Valves**

Rotary valves are another form of plate valve or sliding valve with the moving element rotating over a fixed plate to open or block ports drilled in that plate. Channels cut in the rotary plate can provide transverse flow. Reaction-forced compensation can be introduced by the use of deflector veins to reduce friction.

However, sliding plate valves, *ie* slide valves and rotary valves, are basically restricted to two-, three-, or four-way operation, see Figure 5.

Rotary valves have the particular advantage that they are readily adapted to multi-outlet working. For this reason they are often favoured as sequence valves in pneumatic circuits. Spool and plate valves are also commonly used in combination in two-stage valves.

Two-stage valves are intended to overcome the practical limitations of power response imposed by single-stage operation. A two-stage valve employs one valve as a first stage control and another at the second stage main valve, normally combined in one body or unit.

An example of a particularly versatile form of rotary valve is shown in Figure 6. This consists of a square or hexagonal section body with four or six peripheral ports communicating with a central chamber. An operating spindle passes through this chamber, operating spring-loaded poppets in individual ports by cam action. A number of valves of this type may be gauged together to form a multi-bank valve.

**Variable Orifice Valves**

Although sliding spool and plate valves create a variable orifice while they are changing positions, the term variable orifice valve is applied to valves where an orifice size is set to control the flow rate through the valve.

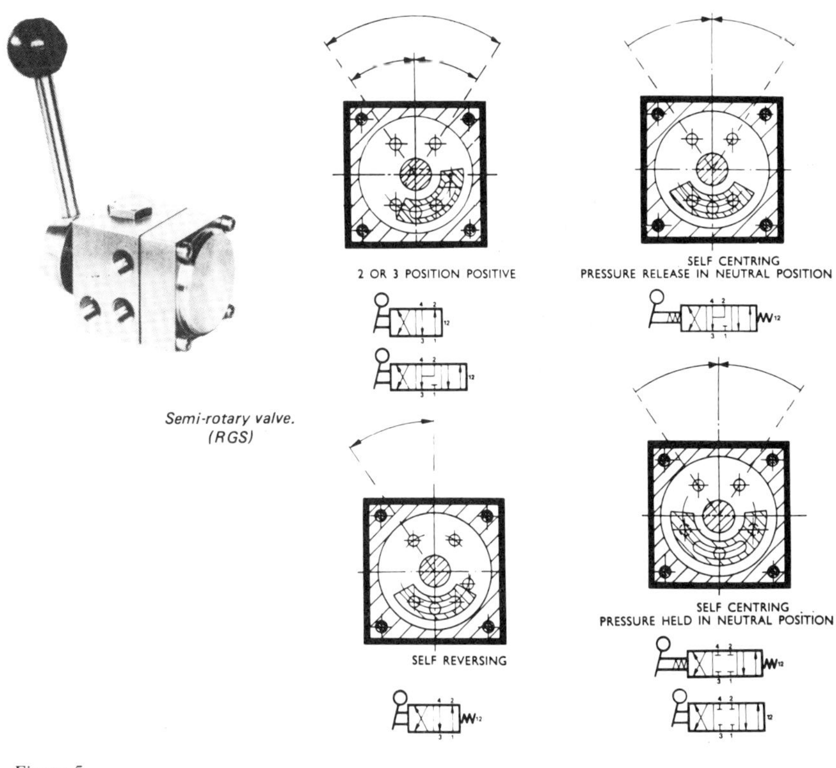


*Figure 5*
*Semi-rotary valve.*

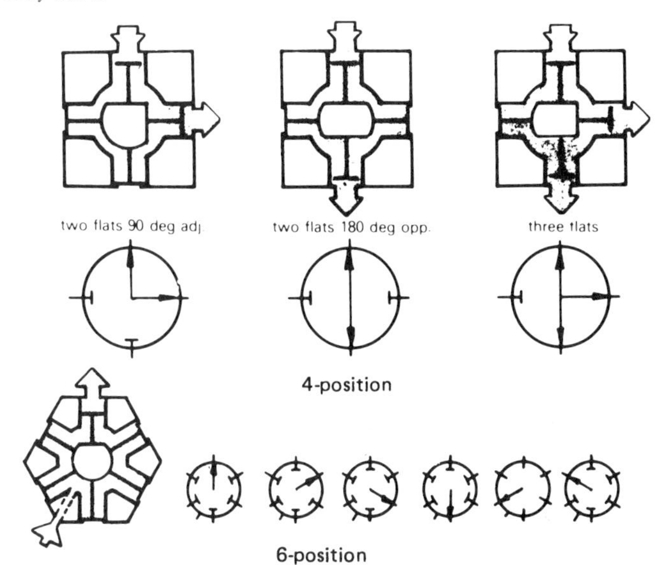


*Figure 6*
*Rotary switch valve.*

Needle valves are used as simple throttling devices to control flow rate, and also as screw down shut-off valves. The fineness of control depends on the angle of taper of the needle and the axial movement relative to the axis of a concentric orifice.

As previously mentioned needle valves used as speed control valves can be designed to be fitted directly into cylinder ports or valve exhaust ports and can be combined with a reverse-free flow check valve or an exhaust silencer.

Ball valves are used as shut-off valves or to divert air flow to, or isolate air flow from certain parts of a circuit. The basic geometry involves a spherical ball, located by two resilient sealing rings, with a hole through one axis which connects inlet to outlet with full bore flow when aligned with the axis of the valve.

The valve is completely closed, by rotating the ball through 90°, with positive sealing *via* the sealing rings – see Figures 7(a) and 7(b). This design has the advantages of a low pressure drop, due to the through bore offering minimum resistance to air flow in the open position, equal flow in either direction and a floating ball. The sealing rings or 'seats' on which the ball floats can be supplied in different resilient materials including PTFE.

*Figure 7(a)*
*Ball valve open.*

*Figure 7(b)*
*Ball valve closed.*

A design of ball valve known as an exhaust ball valve is available for pneumatic circuits as shown in Figure 8. This has a venting passage machined in the ball which opens the downstream side of the valve to an exhaust hole, when the valve is in the closed position. This enables any locked in residual pressure in downstream systems to be vented to atmosphere with the valve in the off position.

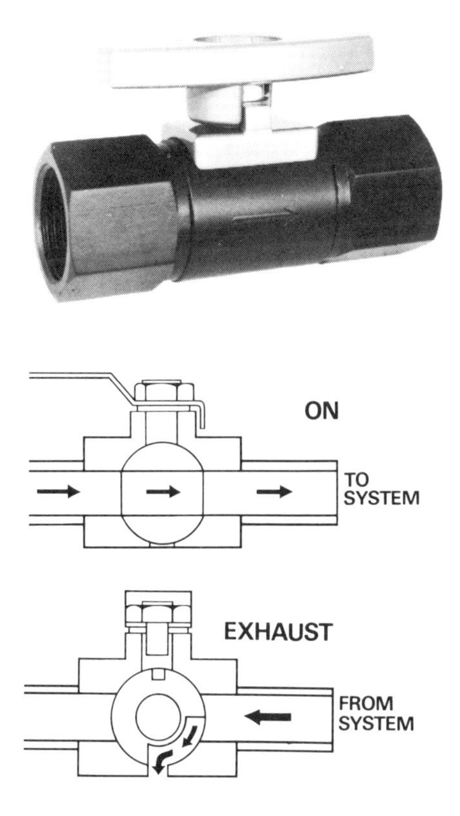


*Figure 8*
*Exhaust type ball valve.*

**Miniature Valves**

Miniature valves have been developed for pneumatic systems mainly to carry out control functions rather than power functions. These consist of the miniature spool valves which can be solenoid-, mechanically- or manually-operated as already mentioned.

In addition logic elements and sensing elements are miniaturized to enable them to be mounted in control cabinets as compactly as possible. These miniature valves can be considered as pneumatic switches rather than pneumatic power valves.

Figure 9 shows a group of pneumatic switches, microswitches and proximity switches for panel-mounting. This group of detectors includes four basic units with

*Figure 9*
*Miniature pneumatic switches and detectors.*

both pneumatic and electric output signals and provides facilities for controlling and sensing movements. Small dimensions, low actuation forces and stainless steel actuating devices give these detectors a broad range of applications.

The switches are basically three-way pneumatic and electric switches with two-position and three-position toggle switches and knobs with twist and return functions in the extensive range of manual actuating devices. The basic detectors can be selected for use with cylinders for built in pneumatic or electric end position signalling.

The pneumatic switches with mechanical push buttons are available as normally closed (NC) and normally open (NO) three-port valves. Using the valves in pairs with manual actuation provides a five-port function. Ports one and two, dimensioned G⅛ (BSP ⅛ in), are located on the underside for compact panel installation. The valves are vented *via* a built in silencer, thus reducing noise to a minimum, and can be used in systems without lubrication.

A time delay between subsequent functions of a pneumatic system can be achieved by either introducing an electrical timer to operate a solenoid valve or purely pneumatically, by metering the flow of air into a reservoir and at a predetermined pressure operate a valve.

The limitation of this method is the size of reservoir needed for long delays and the fact that it is not possible to have a graduated linear time scale. However, the pneumatic delay is useful where a fixed delay is required after the initial setting of the flow regulator. A group of miniature adjustable time delay elements are shown in Figure 10 giving time delays from between 0.1 to 30 s.

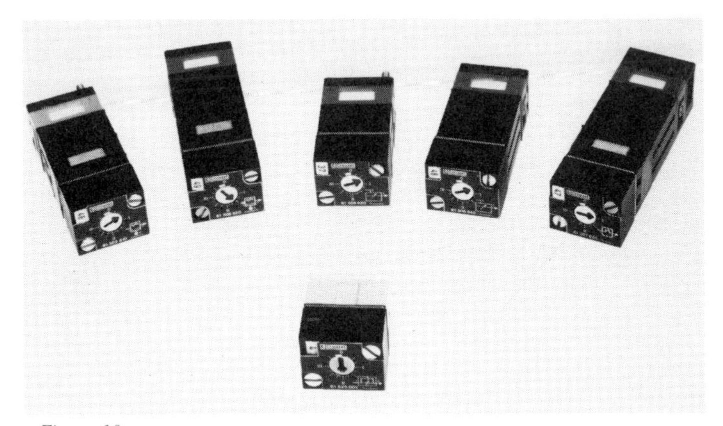

*Figure 10*
*Miniature time delay elements.*

**Pneumatic Sensors**

The most common form of pneumatic sensor is probably the *interruptible jet,* the principle of which is shown in Figure 11. Supply and receiver pipes are aligned axially, separated by a gap. The intrusion of any solid object into the gap, interrupting the jet, causes the pressure in the receiver to fall to atmospheric pressure. This change in pressure is used to operate a switching element controlling an appropriate circuit (*eg* a counter circuit). The switching element is normally a transducer giving an electric signal output.

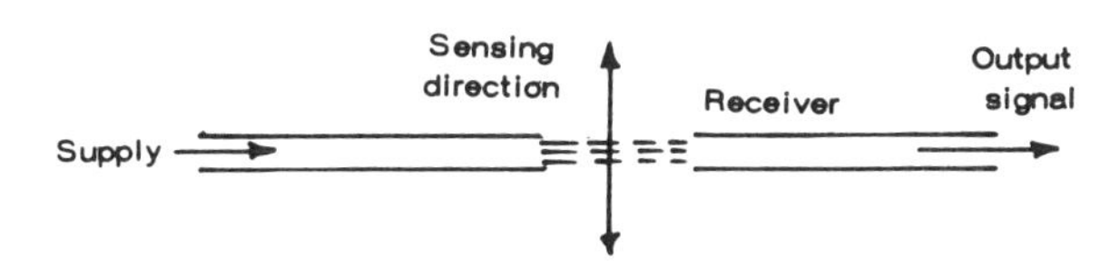


*Figure 11*

This type of sensor has the advantage that it is not critical on gap dimension, can change its state rapidly (*ie* count rapidly) and is not sensitive to object shape or texture. Its main disadvantage is that atmospheric air is entrained in the receiver and, if contaminated, can interfere with its performance.

This limitation can be overcome by lightly pressurizing the receiver in the opposite direction – Figure 12. The main jet then impinges on this secondary jet,

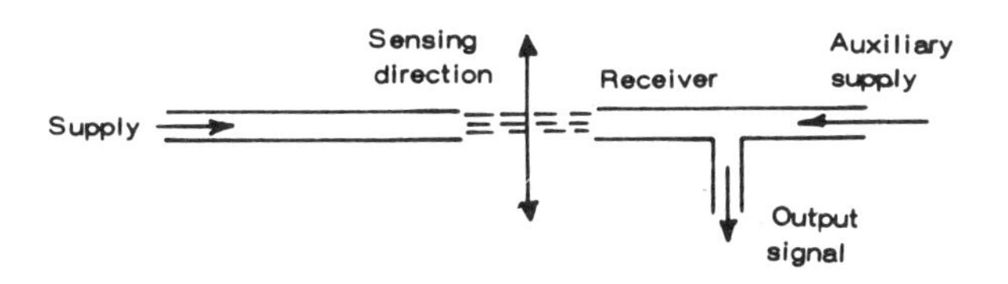


*Figure 12*

applying back pressure which appreciably increases the pressure in the receiver. An object interfering with the main jet cuts off this back pressure, when the receiver pressure falls to its normal pressure level. Under both conditions there is always outflow from the receiver, which thus cannot entrain ambient air.

Both of the aforementioned require small nozzle sizes in the supply (commonly provided by a restrictor) to achieve laminar flow. The effective gap length is then mainly the length over which laminar flow is maintained.

An alternative, shown in Figure 13, uses a third jet placed at right angles to the main (supply) jet. The receiver may or may not be pressurized. This works on the basis of a turbulence amplifier with the main jet flow (normally laminar) being rendered turbulent by impingement of the side jet but reverting to laminar flow when the side jet is interrupted. The practical measuring gap, which is between the side jet and the main jet, can be made much larger for the same pressure difference created in the receiver.

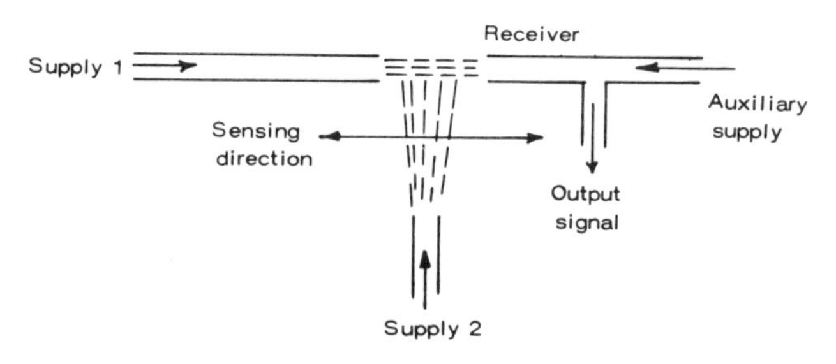


*Figure 13*

A more practical form of interrupted jet sensor with back flow from the receiver is shown in Figure 14, back flow being obtained from the same supply but with pressure reduced by a restrictor. Sensors of this type can be expected to have a maximum gap of about 20 mm working off a supply pressure of the order of 0.1 bar; but larger with higher supply pressures and larger nozzle sizes.

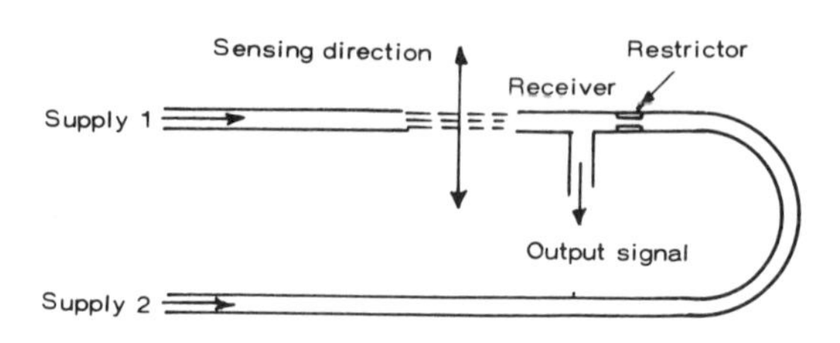


*Figure 14*

The system shown in Figure 15 is capable of working with larger gaps at the same pressure levels. Here a collector is incorporated to supply a second external gap (the *sensing* gap), with output signal derived from the same position as before. Specifically, this device is known as an *airstream detector*.

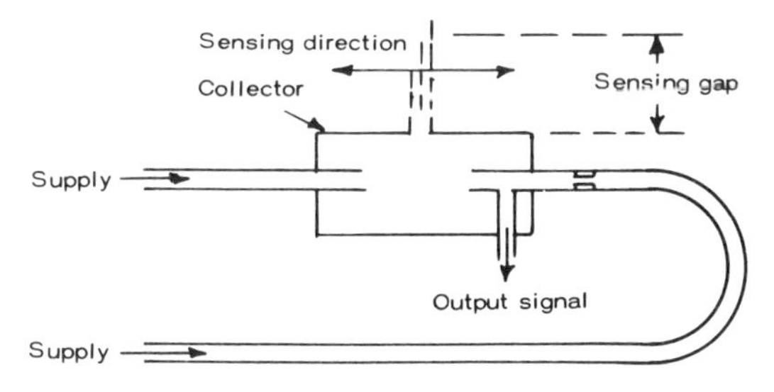


*Figure 15*

*Back pressure sensors* are an earlier type but are only effective when the object to be sensed can pass quite close to the jet. A single jet is used in this case, the presence of an object modulating the flow and causing a marked pressure change at the output (Figure 16). It can, however, be used to detect objects moving towards or away from it, as well as across the jet path, with suitable signal amplification.

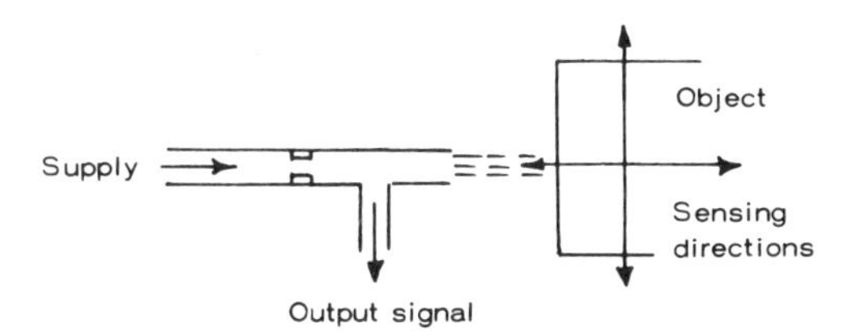


*Figure 16*

The *conical jet* sensor is a further type where the jet emerges from an annular nozzle and is thus of divergent conical form. Air inside the cone is entrained, resulting in a region of reduced pressure, sensed by the output. The degree of depression, and thus the output signal strength, is modified by any asymmetric distortion of the conical jet resulting from its impingement on a downstream object. It is, however, rather slower in response than other types of interrupted jet sensors and requires more power (higher air flows).

The reverse configuration or cone jet (Figure 17) employs an annular nozzle to

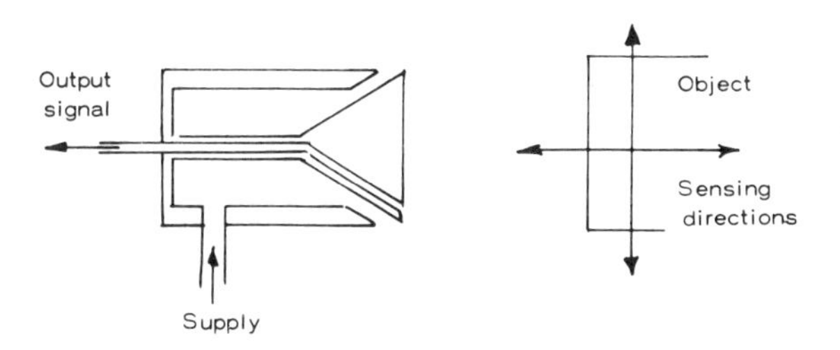


*Figure 17*

produce a convergent conical jet. When this impinges on an object pressure is increased within the cone, creating a back pressure through the outlet.

All single jet devices are, of course, free from contamination by dirty ambient air. Some, however – and particularly the cone jet – work with particularly small nozzle sizes and so require a supply of clean, uncontaminated air.

# Construction

**Valve Sizes**

VALVE SIZES are normally designated by the size of the port tappings. This offers a specific match between pipe size and valve *eg* a ¼ in valve would have ports tapped to take a ¼ in bore pipe threaded externally with BSP or NTP thread. The range of sizes most commonly covered is from 3 mm to 25 mm (⅛ in to 1 in) although main line valves are produced in larger sizes.

The latest ISO standard specifies a ¼ in BSP thread as G¼ for a parallel thread and R¼ for a BSPT taper pipe thread and this standard has been adopted throughout Europe. Although the British Standards have been in line with the ISO standard since 1973, the terms BSP and BSPT pipe threads are still used in British industry.

Most pneumatic valve ports are tapped with parallel 'G' thread sizes, due to the use of aluminium alloy and other light-weight body materials, which are not strong enough to seal on taper pipe threads especially if the pipe connection is continuously being broken and remade.

**Valve Configurations**

The majority of valves used in the UK are of *in-line* configuration, *ie* designed for direct connection of inlet, outlet and exhaust lines. Continental Europe and the USA favour the use of *sub-base* valves, *ie* with valve bodies in block form for mounting on a common baseplate or manifold. All lines are connected to the base.

With sub-base or manifold mounting the valve body normally has no tapped pipe connections. All internal ports are brought to the base of the valve and connection is completed by mounting on a matching sub-base or manifold carrying corresponding ports. Joints in such cases are normally sealed with gaskets or O-rings.

Sub-base mounts are designed to accommodate individual valves. Manifolds are designed to accommodate a number of valves on a common mount.

Port configurations are standardized to the ISO specification ISO 5599/1 in Europe, thus valves of different manufacture can readily replace ISO valves by another manufacturer on the same base and *vice versa*. It does not follow, however, that the actual overall dimensions of the valve or base will be identical under this standard, only the matching port configuration (see Figure 1).

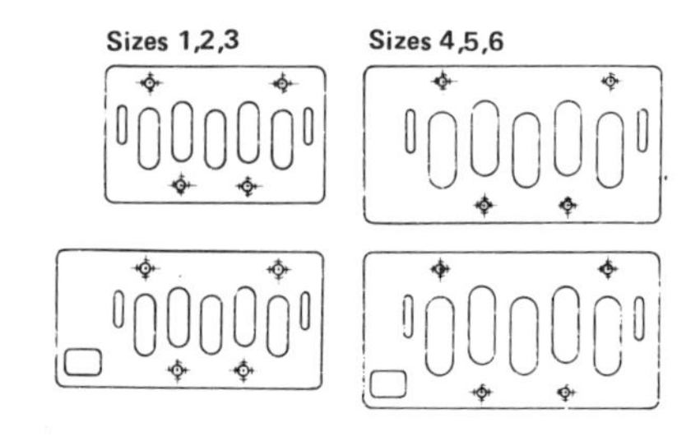

*Figure 1*
*ISO sub-base patterns.*

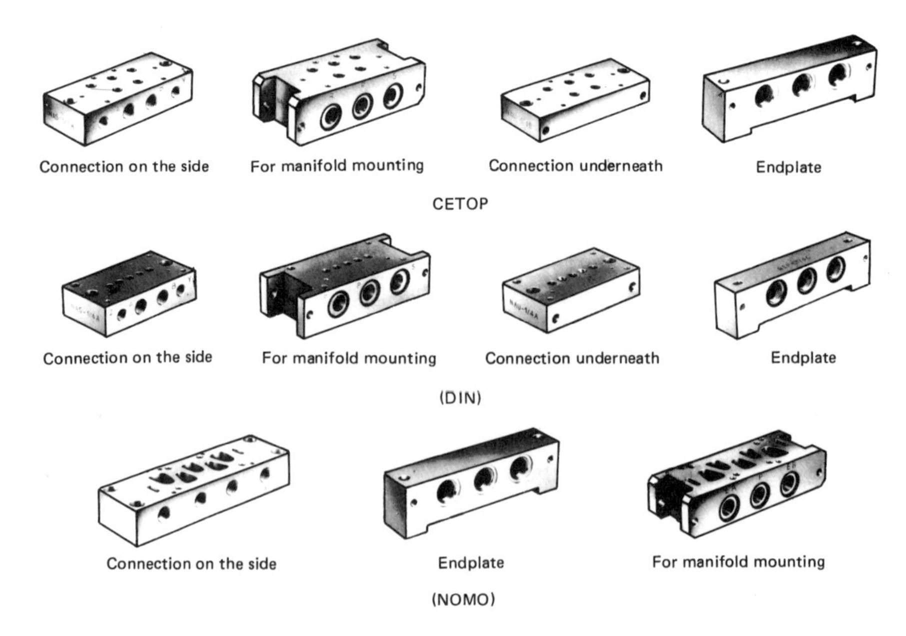


*Examples of sub-base patterns.*

*Figure 2*
*Examples of other sub-base patterns.*

Other standards for sub-base patterns exist, *eg:* DIN, CETOP, NOMO, ANSI, *etc,* as well as individual proprietary designs and may be found on existing valves (see Figure 2). They are not interchangeable with ISO or with each other.

A particular feature of the ISO pattern is the dimensioning of slots to allow for approximately 50% more flow than the corresponding pipe size so that the flow resistance in the sub-base can be ignored compared to the flow resistance in the valve itself (See Table 1).

**TABLE 1 – FLOW CAPACITIES COVERED BY ISO SUB-BASE SIZES**

| ISO Number | Pipe ports Recommended (inch) | $C_V$ – range | $k_V$ – range* | Flow Range ($Q_n$ = l/min) |
|---|---|---|---|---|
| 1 | 1/8, 1/4 | 0.5–1.0 | 0.4–0.9 | 400–1000 |
| 2 | 1/4, 3/8 | 1.0–2.0 | 0.9–1.8 | 1000–2000 |
| 3 | 3/8, 1/2 | 2.0–4.0 | 1.8–3.6 | 2000–4000 |
| 4 | 1/2, 3/4 | 4.0–7.0 | 3.6–6.3 | 4000–7000 |
| 5 | 3/4, 1 | 7.0–12.0 | 6.3–10.7 | 7000–12000 |
| 6 | 1, 1¼, 1½ | 12,0–17.0 | 10.7–15.2 | 12000–18000 |

$^*C_V = 1.12\ k_V$

**Manifold Mounting**

In-line valves can be mounted together for compact installations and one method is shown in Figure 3. The system consists of a range of manifold-mounting rail kits for quick and simple gang-mounting of groups of two to nine miniature solenoid valves. One rail kit can be used for manifolding common supply or common exhaust ports, two rail kits for a common supply and one exhaust port and three rail kits for a common supply and two exhaust ports.

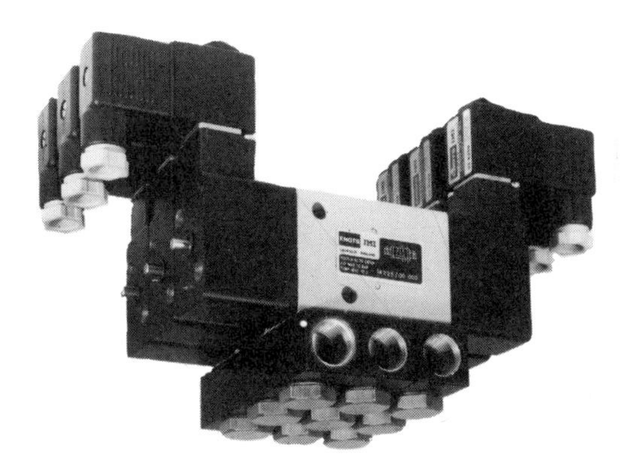

*Figure 3*
*Miniature solenoid valves gang-mounted on rail kits.*

The rails consist of square aluminium alloy sections with supply ports tapped at both ends of the rail. The valves are secured by brass *banjo bolts* with folded copper washers through the mounting holes drilled in the rails as shown in Figure 4. The rail kits are available with G1/8 and G1/4 threads on the banjo bolts.

*Figure 4*
*Six- and three-station rail kits.*

In order to facilitate the mounting of sub-base valves in the minimum space, some bases have provision for being mounted together on tie-rods. This system provides full access to individual valves for ease of assembly and replacement, and keeps all tubing together on one side to give compact pipe runs.

The manufacturer offers a complete range of ancillary components to enable the user to make up his own system (see Figure 5). Both three- and five-port valve bases can be mounted and mixed up in one system, with up to 20 valves being joined in one bank. Bases for several ancillary valves can also be incorporated into this system.

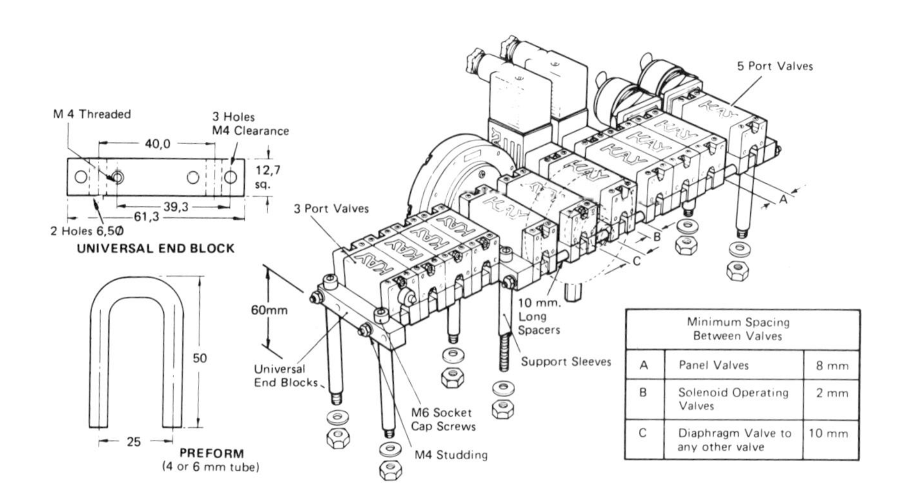


| Minimum Spacing Between Valves | | |
|---|---|---|
| A | Panel Valves | 8 mm |
| B | Solenoid Operating Valves | 2 mm |
| C | Diaphragm Valve to any other valve | 10 mm |

*Figure 5*
*ISO Gang-mounting system.*

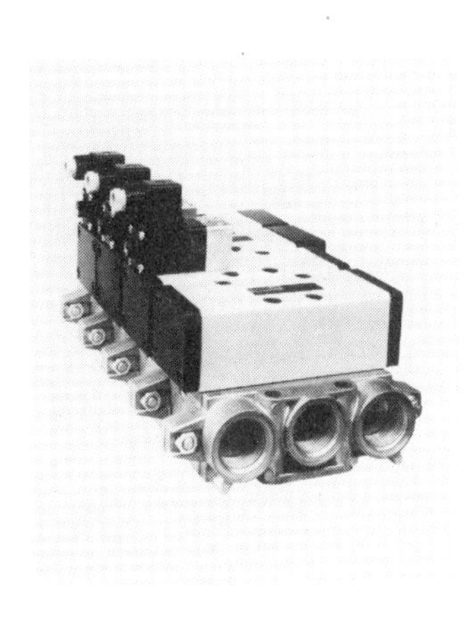

*Figure 6*
*ISO Sub-base valves.*

Where the inter-valve tubing has to be bent into a tight radius 'pre-forms' are available as shown to avoid tube flattening. The piping of such compact systems can often be made by using push-in fittings.

ISO sub-bases can also be ganged together to form a compact manifold assembly as shown in Figure 6. These are designed to ISO specification ISO 5599/1 and are

*Figure 7*
*ISO Sub-base manifolds with adaptor plates.*

die-cast aluminium sub-bases which provide the user with a choice of installation method.

There are also single-station side entry bases as well as the gang mounting types with manifold supply and exhausts. The tapped port sizes for the single station bases are G¼ for ISO 1, G⅜ for ISO 2 and G½ for ISO 3.

Gang-mounting of ISO sub-bases is used where a number of valves have to be built into a system and the problem of connecting all the main air supplies has to be overcome. This problem is repeated if it is required to pipe up the exhausts for reasons of silencing, cleanliness within the cabinet or contamination on the machine or product.

Gang-mounting bases connect the air supply and exhausts from each valve into a common passage. Thus a single air supply pipe is all that is required and there are only two exhaust ports to be piped as required. Not only does this simplify assembly by reducing the cost of fittings and T-pieces, it reduces the time taken to build a system. There is no need to attach each base to a panel. The gang may be self-supported with only the bases at each end fastened to the machine.

Where a system combines valves of two or three different sizes, an adaptor plate may be fitted to reduce to the smaller size as shown in Figure 7. The supply and two exhaust passages are continued through the manifold. The use of an adaptor plate often simplifies cabinet assembly and piping.

There may be a need to provide a lower pressure supply to some of the valves in a gang or to switch off a part of the system. The use of an intermediate supply to some of the valves in a gang could even be used to connect a vacuum.

# Valve Operation

MOVEMENT OF a pneumatic valve element from one position to another may be performed by any of the following methods:

(1) *Manually* – by pushbutton, lever or pedal, *etc.*
(2) *Mechanically* – by spring, plunger or roller, *etc.*
(3) *Electrically* – by solenoid (or in some cases by electric motor).
(4) *By fluid pressure* – air-actuation.
(5) *Self-actuation.*

Configurations of these methods may also be used. Standard symbols defining these methods of actuation are given in Figure 1. (See also *Data Section* for other symbols).

From the practical point of view manual and mechanical operation can be considered basically similar in that a mechanical force is applied directly to the moving member of the valve. With lever movements the mechanical advantage can readily be adjusted. However, the particular application may demand careful

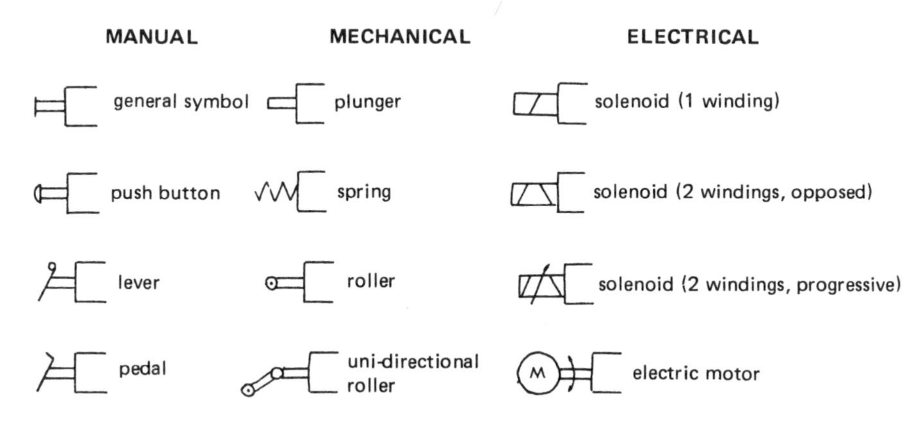


*Figure 1*
*Standard symbols for valve control methods.*

consideration of the working conditions and the possible necessity of incorporating spring detents to hold a valve in a particular position until reversed by the next movement on the part of the operator.

**Manual and Mechanical**
Generally these valves fall into one of two categories: direct or indirect operation of the spool or poppet. In the case of direct operation, the spool is linked or is actuated directly by the mechanism which can be mechanically- or manually-operated. Typical arrangements of this are shown in Figure 2.

*Figure 2*
*Manually- and mechanically-operated miniature pilot valves.*

Indirectly operated valves are where the spool is moved by the application or removal of an air supply acting either directly on the end of the valve spool or on a pilot piston. The thrust exerted by the air pressure acting over the surface area of the pilot piston overcomes the friction in the valve and return spring, where fitted, and causes the spool to move. Valves can be either dead movement or spring return and could combine a direct mechanism with an indirect pilot air reset.

*Pedal* valves are available as light-weight foot-operated valves, suitable for use with both light and heavy machinery. Standard direct-acting spring return plunger valves can be mounted inside a light-weight die-cast pedal mechanism of compact overall dimensions.

The valve can be operated by a cam mechanism that prevents overtravel of the valve spool and the unit can be left free-standing on rubber feet or secured to a base or floor. A compact unit with small movement gives less operator fatigue during repeated operations because great movement of the lower leg is not necessary.

*Two-handed control units* are intended for use whenever it is essential that both hands of an operator are engaged before an operation commences. The control of a press or guillotine is among the most obvious examples of their application.

A control unit can be designed to pilot operate a pressure-operated spring return valve, which in turn controls an air cylinder which may be used directly as a press, or to actuate a mechanical or hydraulic press.

The control unit protects the person who is actually operating the buttons. If there is a danger of persons being injured other than the operator, then surrounding guards or interlocks are necessary.

A detailed and simplified circuit symbol of a two-handed control unit is shown in Figure 3 which has push-buttons built into the end covers. The positioning of the

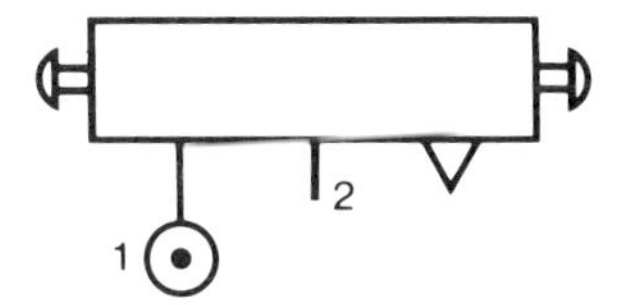


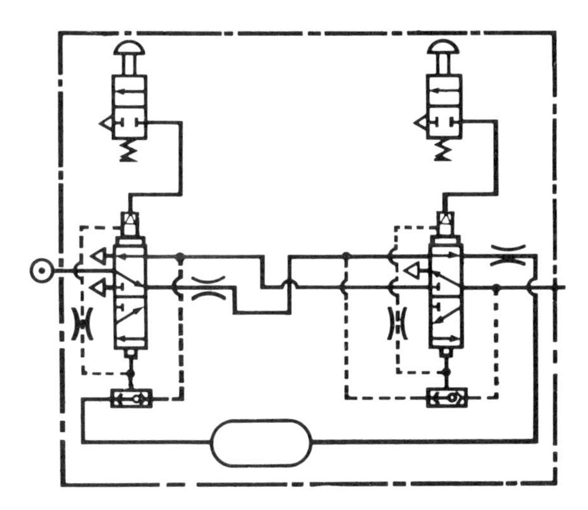

*Figure 3*
*Two-handed control unit, simplified and detailed circuit symbols.*

push-buttons is such that they cannot be spanned with one hand or easily bridged with a tool. The unit should be so installed so as to prevent it being operated by one hand and another part of the body.

In addition the push-buttons are shrouded to guard against accidental operation while it is still possible for the operator to wear industrial gloves. The control unit will supply pilot pressure to the operating valve only if the two push-buttons are pressed within one second of each other. No signal will be given if one button is continually depressed or under any circumstances other than both buttons being pressed together.

Manual- and mechanically-operated miniature valves can be supplied with a variety of actuating mechanisms, examples of which are shown in Figure 4. The *rocker arm* valve has a maximum angular movement of 90°, the actuation angle being 40° either side of centre. The actuation distance in mm (or in) depends on the length of the arm which is adjustable.

The *whisker-operated* valve is completely actuated after movement of 30° with a maximum angular movement of 60°. The whisker can be actuated from any direction but the actuating force should be applied against the plastic knob at the end of the whisker.

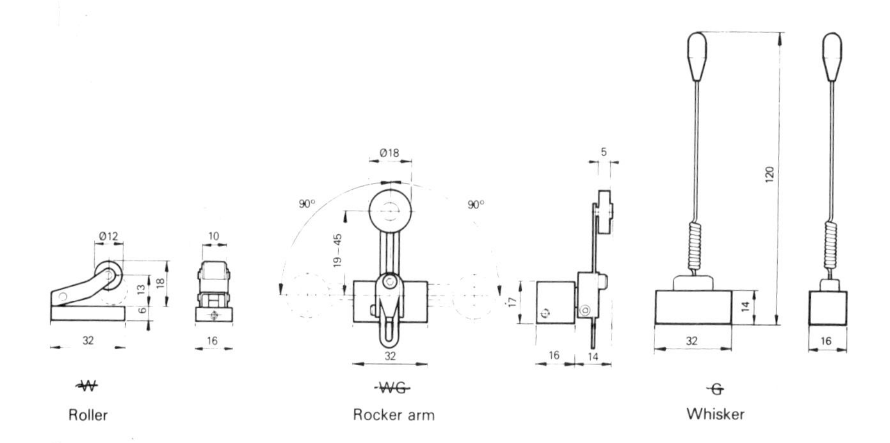


*Figure 4*
*Mechanically-actuated miniature valves.*

**Air Operation**
In the case of air operation this may be direct, indirect or pilot-operated. With a direct air-operated valve, signal pressure is applied directly to a piston formed by an end cap or full-bore section on the end(s) of the valve spindle.

Air-operated movement can be in one direction (with spring return) or in both directions. Either method is suitable for establishing two-position movements.

For three-position movements a central position can be given by applying signal pressure to both ends simultaneously, releasing pressure on alternate sides, as required.

The signal pressure required is generally substantially less than the line pressure carried by the valve. Thus lower signalling pressures can be used in many cases. However, if the signal pressure available is too low, the valve may be operated indirectly by a low pressure air signal through a suitable actuator, such as a diaphragm device.

One-direction or two-direction actuation is possible by feeding signal air pressure to one or both sides of the diaphragm, respectively, with a central position obtained by pressurizing both sides of the diaphragm simultaneously.

Basically pilot-operation or air-pilot operation, as it is more correctly described, provides the main valve with a pneumatic actuator capable of moving the valve when the low pressure pilot signal is applied.

The pilot valves used are normally three-way or four-way configuration, connecting the two chambers alternately to pressure and exhaust for pressurized operation.

Alternatively, exhaust operation would require the use of two two-way pilot valves which allow the chambers to exhaust when opened. In this configuration the chambers are permanently pressurized through the main inlet port.

*Air-actuated* valves can be designed to provide a delay in shifting the valve after receipt of pilot air ('timed-in') or cessation of the pilot supply ('timed-out'). For timed-in operation, air is admitted slowly so that a few seconds are required for enough air to be accumulated in the head to shift the valve. For timed-out operation the process is reversed, causing a delay in the exhausting of the pilot air from the head.

The valve thus remains actuated for a few seconds after the pilot air supply is turned off. The delay mechanism can be adjusted to provide delays of varying duration. The possibility of having an automatic sequence of operations without the need for operator control is obvious with this type of valve. Pilot air can be supplied automatically by some other operation of the machine and the subsequent step can be delayed for a pre-set period of time.

**Solenoid Valves**

When control is to be initiated by an electrical signal from a switch, some form of solenoid operator is required in order to generate the linear force necessary to operate the valve. A simple solenoid valve comprises an integrally-mounted solenoid with the armature directly linked to the valve movement.

Alternatively, to minimize the electrical power required, the solenoid may only act as a pilot control, the valve itself then being operated by fluid pressure available.

By arranging the solenoid armature to work in a sealed tube with the solenoid coil enveloping it, the sealing glands can be dispensed with, so simplifying the construction and eliminating one possible point of leakage. This principle has been applied extensively to the smaller valves. A typical type is shown in Figure 5.

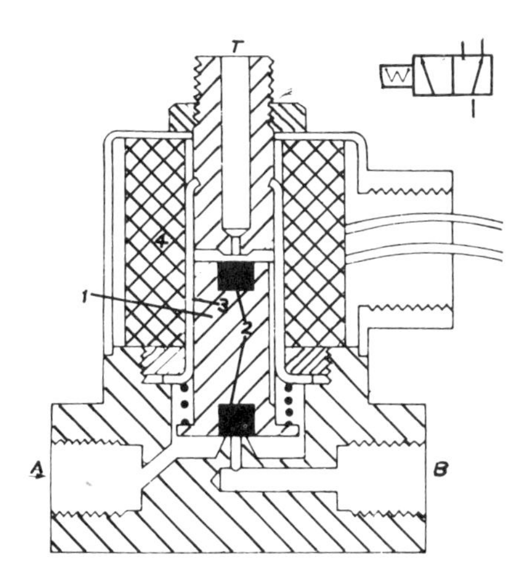


*1 – Plunger. 2 – Synthetic seats. 3 – Sleeve.*
*4 – Coil. A – Cylinder. B – Pressure. T – Exhaust.*

*Figure 5*
*Glandless solenoid valve.*

This valve is T-shaped with two ports opposite each other, while the third is at right angles to them. The plunger, usually of a corrosion-resistant ferrous material, is spring-biased so that when unenergized it closes the lower orifice, while leaving the other open.

When energized, the plunger is pulled up so that the lower orifice is opened and the upper closed. If desired, the spring can be arranged to bias the plunger in the other direction.

The plunger is provided with plastic valve discs, *eg* nylon or synthetic rubber. Because the plunger is unbalanced, the force due to the pressure must be limited and the size of orifice and, therefore, the flow and pressure drop, is usually related to the pressure.

Sealing is normally 'bubble-tight' but this is to some extent dependent on the cleanliness of the air. Lubrication is not essential but the valve life is increased by air-line lubrication.

Glandless valves can be installed in any position and will withstand appreciable shock loads. Response time is extremely short; 5 ms on a.c. and 10 to 15 ms on d.c. electrical supply and it is claimed that speeds of up to several hundred cycles per min are possible.

Most of the makers of these valves also supply pilot-operated valves incorporating the basic glandless valve; four-way valves and two- and three-way valves for larger flows are made in this way.

Figure 6 shows miniature solenoid valves of the direct control poppet-type with G⅛ size ports. These valves can be obtained in two versions; one for normal power coils and one for 2 W low power coils. One feature is a manual override lever which can be finger or screwdriver set and is designed to be snapped off to render the unit tamperproof if required.

*Figure 6*
*Direct control poppet type mini-solenoid valves.*

The valves can be supplied in either single form or bank-mounted as shown with up to six valves per manifold. Manifolds can also be fitted with a special blanking plate if not all valve connections are in use.

The valve body is made of reinforced plastic with seals of oil-resistant rubber. Either connector plugs, with or without integral lamps, or simple push-on connectors can be used in conjunction with the encapsulated spade connections of the coil.

**Solenoid Pilot Valves**

An example of this type of solenoid valve is shown in Figure 7. This valve controls a pilot supply for the indirect operation of the main valve spool. When the solenoid

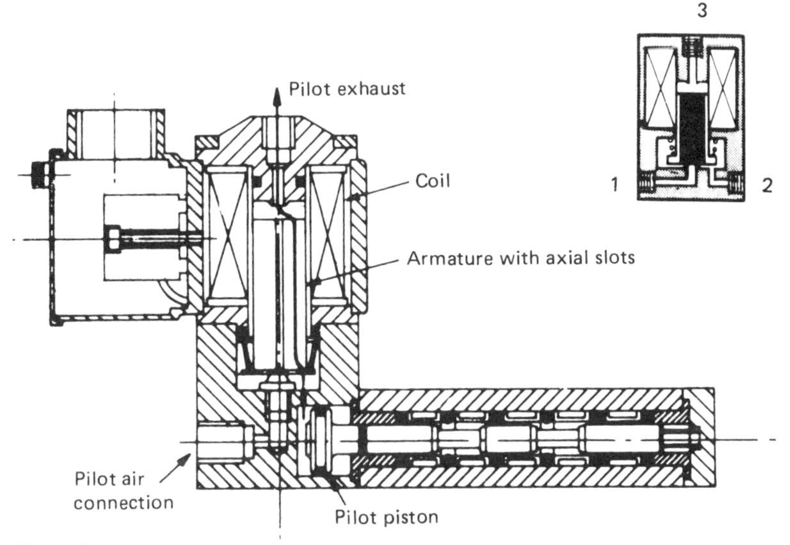


*Figure 7*
*Three-way solenoid valve.*

is de-energized the pilot air is blocked off and the pilot chamber is vented to exhaust by way of axial slots in the armature of the solenoid.

Energizing the solenoid applies the pilot air to the pilot valve. Thus a small pilot solenoid valve operates a spool valve that would otherwise require a large solenoid with its armature directly linked to the spool. The single solenoid-operated spool valve shown is normally supplied with spring return, although any other mechanism could be fitted as a re-set. In the same way a further solenoid could be fitted for this purpose, which would give a latching function from two independent pulses.

By fitting low power solenoid valves to a larger pilot-operated valve, special low power/high flow valves can be produced for use with microprocessor-controlled systems. Figure 8 shows such a valve which combines low power with high flow and fast response.

*Figure 8*
*Low power solenoid valve.*

An electric signal of only 1.9 W is required to actuate the valve to give an air flow of 52 l/s (110 $ft^3$/min) from a five-port version valve. The life expectancy of the valve is only slightly reduced when connected to a non-lubricated air supply.

Single-solenoid, double-solenoid and pilot-operated valves are shown mounted on ISO 5599/1 standard manifolds in Figure 9 and are part of a non-lubricated range of pneumatic valves. Two types of manifolds are shown, one with ports at the bottom face and the other with ports at the side.

*Figure 9*
*ISO Manifold-mounted valves.*

On the bottom-ported-type, each valve has its own manifold base and each complete assembly of manifolds requires an end plate kit consisting of two ends. To assemble the manifolds, diagonally positioned screws are tightened down on to the adjacent manifold or base, which automatically pulls the two together, compressing and effecting the seal.

*Figure 10*
*ISO Size 1 non-lubricated solenoid valve.*

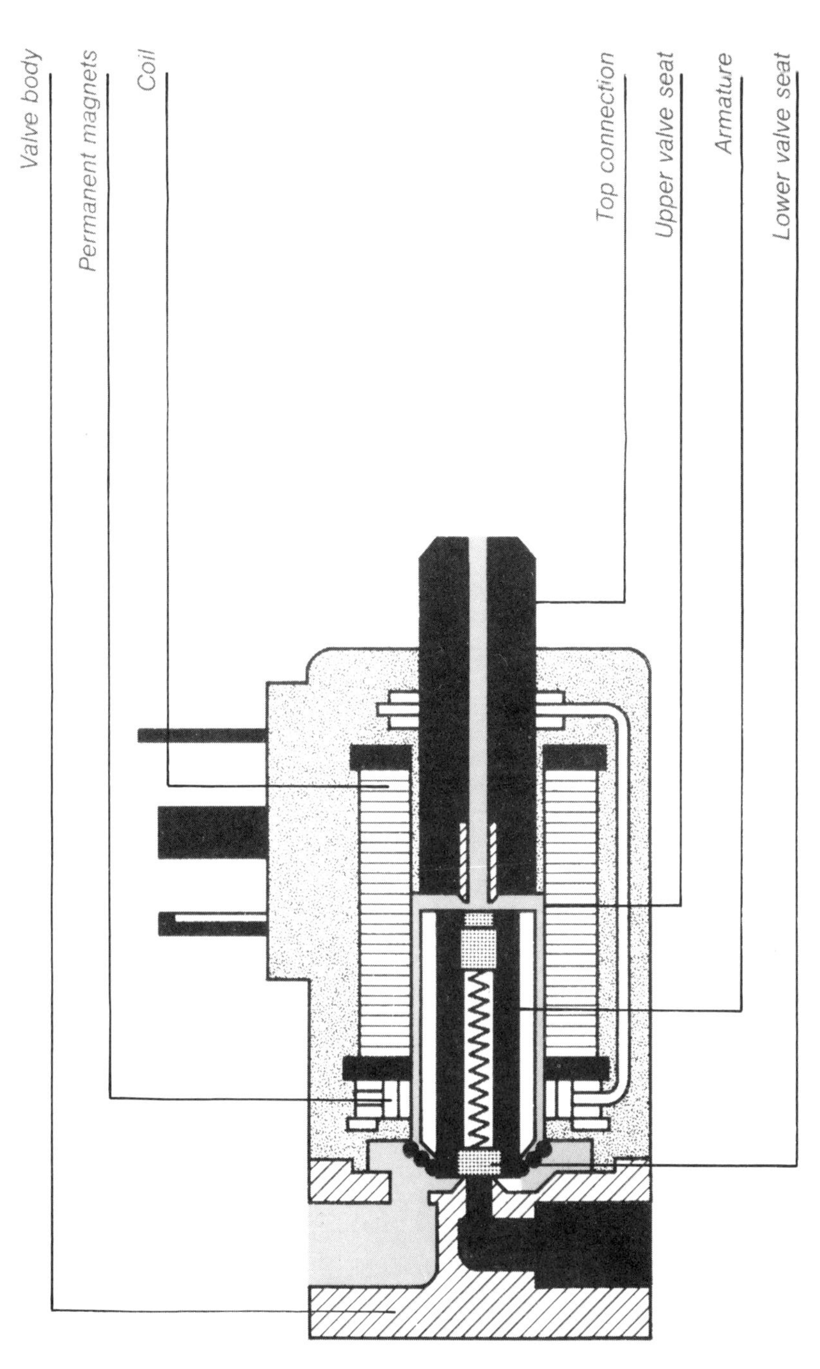


*Figure 11*
*Impulse solenoid valve.*

On the side-ported manifold, one of five selector plates is fitted between each base, allowing various options, including blanking of all internal connections and combinations of common internal paths, to permit the use of an external air supply to the manifold. This is advantageous when the main operating pressure is insufficient to switch the valve or when a number of valves require switching simultaneously.

A single *side-ported* valve is shown in Figure 10 to ISO 5599/1 size 1 standard. The main feature of this valve is an aluminium spool assembly with a composite nitrile-based seal, which gives a non-lubricated life expectancy equal to that of a standard valve. The tell-tale spool position indicator/manual override can be seen at the end of the main valve and this is recessed to prevent damage or accidental operation.

**Impulse Solenoid Valves**

Solenoid valves can be supplied which switch on impulse power signals only and thus consume no power in the steady state. The advantages of these 'zero-watt' coils are; no continuous power consumption, negligible heating effects and safety, because the valve is bi-stable and will remain in its last position in the event of a power failure.

A section through a typical valve is shown in Figure 11. The actuator consists of an epoxy-encapsulated coil containing windings for pull and throw functions. A system of permanent magnets incorporated in the outer iron circuit is completely separated from the fluid medium.

The impulse coil for plunger-armature systems works in such a way that switching is achieved by means of a short power impulse on the electro-magnet. The permanent magnets enable the valve to retain this position without the need for continuous application of electrical energy. Only when a second impulse is applied does the valve switch back. No power is therefore required to maintain the operated position.

Consequently, the heating effect of the coil is insignificant and the seal materials of the valve are not subjected to any thermal loading. The advantages of this impulse coil can be appreciated in applications involving the control of multi-way pneumatic valves.

These valves have usually required either an electrical control unit for switching pneumatic cylinders, which had to provide a continuous electrical or pneumatic signal during the entire period of solenoid valve operation. The impulse coil simplifies matters considerably when used in conjunction with reed-contacts for the control of pneumatic cylinders.

# Valve Selection

ONCE THE valve function and mode of operation has been decided upon, the main parameters which determine the size of the valve are the flow rate and corresponding pressure drop through the valve. These are not so easy to define with compressible air, as with an incompressible liquid for hydraulic valves.

Also difficulty can be experienced when trying to relate different manufacturers methods of measuring and stating these parameters. A provisional recommendation RP50 P has been drawn up by CETOP in 1973 to define the flow capacity value of pneumatic components and this has been adopted by many manufacturers in Europe.

**Flow Capacity Value**

The CETOP standard defines the characteristic flow parameters of a pneumatic component and gives the method for determining them. It can be applied to all kinds of pneumatic components, which have a fixed flow path and provision for connections at the inlet and outlet ports. These include directional control, shut-off, shuttle, check and restrictor valves as well as pipe and hose assemblies, couplings, connectors and manifolds.

The two parameters which are defined by the standard are the *conductance* (C) and the *critical pressure ratio* (b). The conductance (C) of a component is the ratio between the flow rate (Q) and the inlet pressure ($P_1$) when the flow is choked and the air temperature is + 20° C, *ie*:

$$C = \frac{Q}{P_1} \text{ in units dm}^3\text{/s. bar.}$$

Choked flow occurs when the primary pressure ($P_1$) is so high in relation to the secondary pressure ($P_2$) that the flow is proportional to the primary pressure ($P_1$) and independent of $P_2$. This situation arises when the velocity at some part of the component becomes sonic.

The critical pressure ratio (b) is the pressure ratio $\frac{P_2}{P_1}$ at which flow becomes choked. Where $P_1$ is the absolute pressure upstream of the component and $P_2$ is the absolute pressure downstream of the component, measured in bar.

For any specified combinations of flow path and direction of flow the performance of a pneumatic component can be stated in terms of conductance (C), which is a measure of flow capacity, and critical pressure ratio (b), which is representative of the flow characteristic and necessary for predicting total performance. These values alone are sufficient to define the flow performance of a pneumatic component over all of its range.

Manufacturers can also give further graphical illustrations of the performance, as shown in Figure 1 which gives flow graphs for three sizes of valves meeting ISO 5599/1 standard.

CETOP is studying a further recommendation which permits the measurement of flow, not only through individual devices but through a complete circuit. It aims to achieve this by the principle of adding the flow ratings of individual sections of the circuit.

**Pressure Limitation**

Details of maximum operating pressure for valves are given by manufacturers and this is usally set by a pressure regulating valve and safety valve at the compressor and air receiver. As the air is compressible and is rarely stored at pressures much higher than 10 bar (145 lbf/in$^2$), high peak pressures are seldom encountered with pneumatic valves, as they are with hydraulic valves.

The development of higher working pressures requires an increasingly disproportionate expenditure of power for compression, as well as increased component stresses and explosion hazard in the event of failure. The marked difference between pneumatics and hydraulics in these respects is due to the compressible nature or 'elasticity' of air.

For general applications, the only advantage offered by higher compression ratios is in storage systems where high pressure air can be stored in a suitable pressure vessel and 'let down' to conventional delivery pressures. Storage at high pressure substantially reduces the volume required for the storage vessel.

For other more specialized applications there are definite requirements for higher pressure compressed air, which may extend to pressures of around 1000 bar for air blasting and metal forming. High pressure air systems can, in fact, be categorized in four pressure ranges, with conventional industrial pneumatics rated as low pressure systems:

(1) *Moderate pressure* (17.5 to 35 bar) normally derived directly from a compressor or reservoir charged by a compressor.

(2) *High pressure* (35 to 210 bar) normally derived from a charged air bottle.

(3) *Very high pressure* (210 to 450 bar) either derived from high pressure reservoirs charged by multi-stage compressors for large volume applications or from intensifiers for 'single-shot' or small volume application.

(4) *Ultra – high pressure* (700 to 1400 bar) for highly specialized applications with supply from multi-stage reciprocating compressors.

**Temperature**

In operation, the temperature rise of pneumatic equipment is negligible. It may therefore, be more readily installed in relatively high temperature installations, *ie* near furnaces and similar equipment, than equivalent electrical and hydraulic

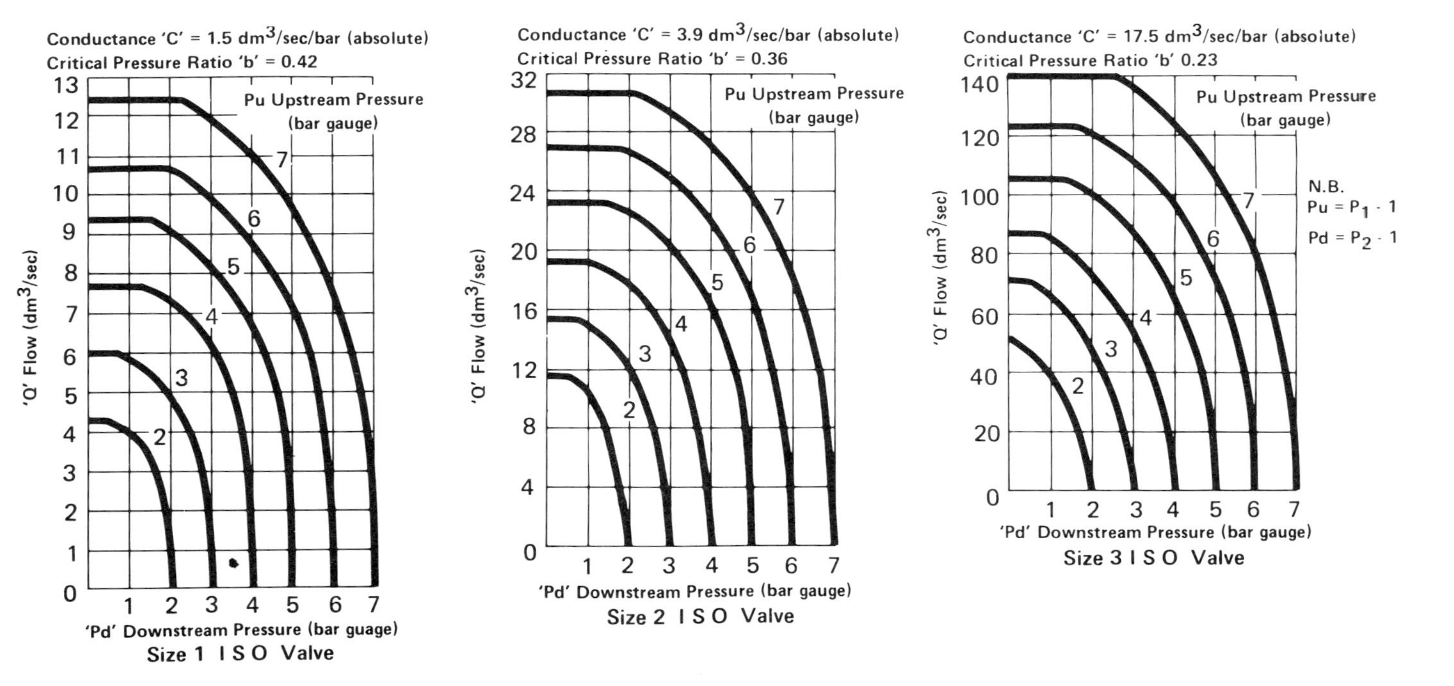


*Figure 1*
*Flow capacity graphs for ISO valves.*

equipment. Care must be taken, however, with pneumatic equipment to ensure that the temperature limitations imposed by the manufacturer are not exceeded.

The limitation of temperature in pneumatic valves is largely due to materials used and the types of seals used. Most units are so constructed that they are perfectly satisfactory for use in temperatures of between 10 and 90° C(50 and 194° F).

Where specific precautions are taken to ensure that the air supply is 100% dry, then the lower limit is frequently extended to below 0° C (32° F). If, under these circumstances, wet air were to be employed, then considerable ice formation would result, causing malfunction of the valves. Special seals are available for temperatures outside the ranges quoted but the valve manufacturer should be consulted before attempting to use such units.

**Fluid Compatibility**

The ideal fluid for pneumatic systems is clean, 100% dry, atmospheric air, but it is not always possible to achieve this ideal in practice. Many valves and cylinders are being made in materials which need no lubrication during the life time of the units and can be operated with 'non-lube' air.

However, it is not always easy to remove all moisture from the air and lubrication will reduce corrosion, which can occur in pneumatic equipment due to the presence of moisture. Lubrication will also reduce the friction and wear of working parts.

Most valves made of materials which are suitable for non-lube air will operate satisfactorally with lubrication. In addition to the normal greasing carried out during servicing and maintenance, additional lubricating oil is usually required to be fed to the equipment during operation.

Most pneumatic equipment is lubricated by injecting oil into the compressed air supply by means of a suitable airline lubricator. The oil is then carried through by the air flow to be deposited in valves and cylinders as required. The design is usually such that lubricant is only supplied while the air supply is flowing and automatically stops when the air flow ceases.

The introduction of non-lube air is being called for in many applications such as the food industry and for drug manufacture. A lubrication-free operation gives a better work environment, simpler installation and less maintenance.

*Filters* are needed to clean the air at the intake to the compressor and also in the line from the receiver to the valves. Atmospheric air, especially in industrial areas, is extremely dirty and to prevent harmful matter from entering compressors, filters are fitted on the suction side.

Filters on the output side are also fitted largely to remove compressor products, carbonized oil and condensate, *etc*. Most airline filters are constructed to remove both solid particles and water from the airline.

Rust, pipe-sealing compound, grit particles, *etc*, must be prevented from reaching the equipment being operated. Abrasive particles and impurities of this type can cause damage to both valves and cylinders, some of which employ fine metallic clearances.

Such particles also lead to the cutting of synthetic rubber pistons, valve seals and seatings, *etc*, as well as the clogging of fine bleed holes in pneumatic instruments.

Valves should be selected to be compatible with the grade of filtration and air condition of the system.

# Installation and Commissioning

THE INSTALLATION of pneumatic equipment requires semi-skilled mechanical fitting services only and the additional need of separate electrical services where required. It does not need as much fitting skill and care as high pressure hydraulic equipment which is necessarily of precision construction, involving units with fine mechanical clearances.

**Installation**

Valves and equipment are usually supplied ready lubricated and it is not necessary to add grease or other lubricants and equipment can be fitted immediately. All portings should be covered with plastic dust caps which should only be removed as and when pipe connections are made.

If valves are left with open portings for any length of time then the protective plugs should be loosely replaced. Valve mechanisms are often covered with plastic peel for protection during transit and this should be removed prior to assembly.

During fitting, no dirt, swarf or dust, *etc*, must be allowed to enter the valves and when checking cylinder alignment by manual or blow-gun operation no dust, *etc*, should be be allowed to enter any open ports.

Valves which directly control cylinders, *ie* double-air-operatated, double-solenoid-actuated, air-operated and similar free spool types should be fitted in the horizontal postition. If mounted vertically, there is a tendency for the spool in this type of valve to move downwards when air is completely off, particularly if the valve unit is subjected to mechanical vibration.

Circuits are usually drawn and should be piped up in the commencement *ie* 'at normal rest' position and valves are usually supplied in the 'as delivered' condition. If the circuit indicates reversed porting conditions, then the position of the valve spool should be altered.

Valves when first fitted to a circuit should, therefore, be placed in the position to give the connections required and by applying an air supply, *ie* a blow-gun, to the free spool type, the spool can be blown over to give the required initial circuit starting connections.

The valve spool of some double-air-operated valves can be actuated by inserting a small screwdriver. The valve spool should move over easily and undue force should not be used. Air can be fed to the pilot exhaust port to obtain actuation of the valve spool of pilot solenoid actuated valves.

Manual valves with two- or three-position mechanisms should also be operated to give the required circuit at normal rest connections.

Valves which directly control the cylinders should be fitted close to the cylinder, keeping the cylinder connecting pipes as short as possible. This will give maximum speed of cylinder operation and is more effective than overlong cylinder connecting pipes. The valves should be firmly and squarely bolted down.

Mechanically-tripped valves, such as roller-operated or plunger-operated types should be firmly and squarely bolted in position and either the valve or tripping mechanism adjusted until the valve is being fully actuated. Valves having overtravel can easily be checked, but valves which do not incorporate overtravel require careful setting to ensure that the mechanism does not go 'solid' during normal tripping. It should be possible to fit a 0.12 to 0.25 mm (0.005 to 0.010 in) feeler between the valve mechanism and mechanical trips before the solid condition is reached.

Connection of pipes should be made in an orderly fashion, commencing with the valve operated by the first movement in the cycle and piping the valves in turn as they would normally be actuated during the cycle. Each connection, should be checked and double checked and the remainder of the circuit inter-connecting pipework connected to the main air supply.

**Commissioning**

Where flow regulators are fitted as part of a time delay in circuits, they should be initially adjusted to be one turn from the fully in closed position. Final adjustment is made later when the whole circuit is operating. All dust plugs should have been removed from valve ports and silencers screwed in where they are to be fitted.

A final check should be made through the entire circuit of all pipes and connections to ensure that these have been correctly made before the pilot supply is switched on. The pressure should be checked first to see that it is satisfactory.

On actuating the 'start' valve in the circuit, the circuit should cycle immediately and correctly. If this is not so, the air supply should be shut off and pipes and valves be re-checked to make sure that all valves are in the at normal rest positions. The main supply to the cylinder main control valves and the pilot main supply can be switched on again and another try made.

When the circuit is sequencing correctly, all the flow regulators connected to cylinders should be adjusted to obtain the desired speeds. Where flow regulators are used in pneumatic time delay circuits, these should now be adjusted for the appropriate time delay required. Where these are used for simple sequence signalling, they should be adjusted to give a time delay slightly longer than the associated cylinder movements.

All cylinder cushion screws should be adjusted to give the required cushioning. For this purpose and also during cylinder speed adjustments, the machine should run fully loaded.

Comments made concerning the connection of valves also apply to electrical connections made to solenoids and electrical switches and sensing devices but these should be made by a qualified electrician.

In order to trace faults, it is necessary to fully understand the normal functioning of a particular circuit or pneumatically-operated machine and by a process of deduction and elimination, the fault can be located and corrected.

Although not absolutely essential, the theoretical circuit diagram, showing circuit functions and pipe connections is invaluable and should always be obtained. Faults can arise due to a variety of causes, both mechanical and pneumatic and many circuits will have individual peculiarities, but by following a logical and orderly proceedure even unusual faults can be deduced.

**System Maintenance**
Scheduled periods for the inspection of drains, moisture traps and separators should be based on experience with the particular components involved (manufacturer's recommendations should be used as an initial guide).

Filters and filter/separators fitted at individual tool stations must receive regular attention, particularly if they are of the manual drain type. The level of contaminants will normally be visible in the filter bowl or an excessively high level indicated by a tell-tale. Both types – manual drain and automatic – will also require periodic cleaning of the filter element; thus a regular period must be set down, depending largely on the 'dirtiness' of the system, for cleaning or replacement of the filter elements as appropriate.

Some types of filter have an element which can be cleaned very simply by blowing out; others have an element which can only be cleaned by special techniques (and which thus need replacement while the element is returned to the manufacturers for cleaning); still others have elements which are discarded and replaced after a specified period of use. A filter element accidently damaged during cleaning should always be replaced by a new element.

Clogged filters are the most common source of pressure drop. Air leaks or constrictions in a flexible line are other obvious sources of pressure drop at tool points. The former can normally be heard with the tool connected but switched off. Kinks or damage to hose that has not produced puncturing may show up in the form of variable performance as the tool position is changed. In this case end fittings should first be checked for air tightness, then the hose length itself checked for damage.

Less obvious sources of pressure loss may occur farther back in the system. With adequate line sizing and everything in good condition, pressure drop between the air receiver and the tool station take-off point should not normally exceed 0.2 bar to 0.35 bar (3 to 5 lbf/in$^2$) or in some installations 10% of the nominal system pressure. Additional losses at the tool point itself should not normally exceed 0.35 bar (5 lbf/in$^2$), although again a maximum figure of 15% total pressure drop to the tool is sometimes accepted.

Regardless of the actual system pressure drop as installed and originally worked, any further loss of pressure at take-off points is a subject for immediate attention so that the fault can be rectified. Loss of pressure not only reduces working efficiency, but may also be expensive in terms of compressed air wastage if due to a leak. Leakage may develop at joints, connections, *etc* or at traps where the valve has been jammed open by contaminants, apart from such obvious sources as mechanical damage to the lines.

It is not commonly appreciated that an excessive line pressure can be as bad as lack of line pressure. While the latter reduces the working efficiency of the tool or cylinder – a tool designed to operate on 5.6 to 7 bar (80 to 100 lbf/in$^2$), for example, may suffer a loss of efficiency of 25% operating on 5 bar (70 lbf/in$^2$) – excessive pressure results in accelerated wear and possible over-stressing of parts, which could lead to early failure.

Lubricators at tool points also need regular attention to ensure that an adequate supply of oil is maintained and the rate of oil injection can also be checked if necessary. It is particularly important that only the specified oil be used to re-fill the lubricators as an oil of different viscosity may not only upset the lubrication rate but lack the qualities necessary for a pneumatic tool lubricant. Light oils of approximately SAE 10 viscosity are normally specified, specially compounded for pneumatic system use (*eg* normally containing emulsifying agents and rust inhibitors).

# SECTION 9

## Data

STANDARDS AND PUBLICATIONS
GRAPHICAL SYMBOLS FOR
PNEUMATIC SYSTEMS AND COMPONENTS

# Standards and Publications

| Standard | Subject | Cross Reference(s) |
|---|---|---|
| BS21 : 1973 | Pipe threads for tubes and fittings where pressure-tight joints are made on the threads. | ISO 7 |
| BS1123 : 1976 | Specification for safety valves, guages and other safety fittings for air receivers and compressed air installations. | |
| BS1387 : 1967 | Steel tubes and tubulars suitable for screwing to BS21 pipe threads. | ISO 65 |
| BS1701 : 1970 | Air filters for air supply to internal combustion engines and compressors other than for aircraft. | |
| BS1710 : 1975 | Identification of pipelines. | ISO/R 508 |
| BS1780 : | Bourdon tube pressure and vacuum guages. | |
| Part 1 : 1960 | Imperial units. | |
| Part 2 : 1971 | Metric units. | |
| BS2051 : | Tube and pipe fittings for engineering purposes. | |
| BS2779 : 1973 | Pipe threads where pressure-tight joints are not made on the threads. | ISO 228 |
| BS2917 : 1977 | Specification for graphical symbols used on diagrams for fluid power systems and components. | ISO 1219<br>CETOP R3 |
| BS3600 : 1976 | Specification for dimensions and masses per unit length of welded and seamless steel pipes and tubes for pressure purposes. | ISO 336<br>ISO 64<br>ISO 1179 |
| BS4062 : | Valves for hydraulic fluid power systems. | |
| Part 1 : 1982 | Methods for determining differential pressure/flow characteristics. | |
| Part 2 : 1982 | Methods for determining performance. | |
| BS4231 : 1982 | Classification of viscosity grades of industrial liquid lubricants. | ISO 3448 |
| BS4742 : | Hydraulic equipment for agricultural machinery. | |
| Part 1 : 1971 | Cylinders. | ISO 5669 |
| Part 2 : 1983 | Hydraulic couplers for general purposes. | ISO 5675 |

| Standard | Subject | Cross Reference(s) |
|---|---|---|
| BS4862 : 1983 | Mounting dimensions of single rod double-acting pneumatic cylinder, 10 bar. | CETOP RP 43P, RP 51P, RP 62P, RP 53P, ISO 6341/6432. |
| BS5118 : 1980 | Rubber air hose. | ISO 2398 |
| BS5169 : 1975 | Fusion-welded steel air receivers. | |
| BS5200 : 1975 | Hydraulic connectors and adaptors. | |
| BS5242 : | Tubes for fluid power cylinder barrels. | CETOP RP 79P |
| Part 1 : 1975 | Steel tubes with specially finished bores. | |
| Part 2 : 1983 | Non-ferrous tubes with specially finished bores. | |
| BS5276 : | Pressure vessel details (dimensions). | |
| Part 4 : 1977 | Standardized pressure vessels. | |
| BS5319 : 1976 | Specification for quick-release vacuum couplings (screwed and clamp type). | ISO 2986/1 |
| BS5380 : 1976 | Hydraulic port and stud couplings using 'O'-ring sealing and 'G' series threads. | |
| BS5409 : | Specification for nylon tubing. | |
| Part 1 : 1976 | Fully plastized nylon tubing types 11 and 12 for use primarily in pneumatic installations. | |
| BS5500 : 1976 | Unfired fusion welded pressure vessels. | |
| BS5543 : 1978 | Vacuum technology – graphical symbols. | ISO 3753 |
| BS5555 : 1976 | SI units and recommendations for the use of their multiples and of certain other units. | ISO 1000 |
| BS6331 : | Mounting dimensions of single rod double-acting hydraulic cylinders. | |
| Part 1 : 1983 | Specification for 160 bar medium series. | ISO 6020/1 |
| Part 2 : 1983 | Specification for 160 bar compact series. | ISO 6020/2 |
| Part 3 : 1983 | Specification for 250 bar series. | ISO 6022 |

**CETOP Publications**

The following documents are printed in English, French and German; CETOP recommendation; documents without the suffix 'P' apply equally to hydraulics, those with suffix 'H' apply to hydraulics only.

| | |
|---|---|
| R.2 | Classification of fluid power terms and documents. |
| RP.4P | Pneumatic cylinders, suggested data for inclusion as a minimum in manufacturers' technical sales literature. |
| RP.5P | Specification for pneumatic cylinders. |
| RP.6P | Pipe couplings for pneumatic piping – coupling thread. |
| RP.7P | Pneumatic cylinders, recommended minimum relation of port size (thread) to cylinder bore. (In part superseded by ISO R.1939.) |
| R.10H | Hydraulic cylinders – dimensions. |
| R.11H to 17H | Starting, servicing and maintenance instructions for: |

| | |
|---|---|
| R.11H | Complete hydraulic systems. |
| RP.15H | Hydraulic valves. |
| RP.16H | Hydraulic cylinders |
| RP.17H | Hydraulic intensifiers. |
| RP.19 to RP.30P | Recommended data for inclusion in manufacturers' technical sales literature. |
| RP.19P | Pneumatic directional control valves. |
| RP.20P | Pneumatic flow control valves. |
| RP.21P | Pneumatic pressure control valves. |
| RP.22P | Pneumatic shuttle, non-return and quick exhaust valves. |
| RP.23P | Pneumatic pressure intensifiers. |
| RP.24P | Pneumatic rectilinear piston type cylinders. |
| RP.25P | Pneumatic filters and water traps. |
| RP.26P | Lubricators. |
| RP.27P | Air dryers. |
| RP.28P | Connections. |
| RP.29P | Pneumatic quick-action couplings. |
| RP.30P | Pneumatic rotating and telescopic joints. |
| RP.32P | Subplates for pneumatic directional control valves. |
| RP.33 | Graphical symbols and definitions for operations of logic and related functions in fluid logic circuits. |
| RP.34P | Couplings for industrial air hoses – 10 bar. |
| R.35H | Mounting surfaces for hydraulic directional valves. |
| RP.37 | Recommended diameters for pneumatic tubes and hoses. (Under revision.) |
| RP.38P | Guidance on relation between port threads and pipe hose diameters. (Under revision.) |
| R.39H | Schedule of required data for hydraulic fluids. |
| RP.40P and amendment | Hose couplings, claw type. |
| RP.41 | Hydraulic and pneumatic circuits, circuit diagram. |
| RP.43P | Pneumatic cylinders 10 bar, mounting dimensions (bores 32 to 100mm). |
| RP.49P | Technological symbols for fluid logic and related devices with and without moving parts. |
| RP.48H | Evaluation of the anti-corrosive qualities of water-based fire-resistant fluids. |
| RP.50P | Flow capacity value of pneumatic components. (Under revision.) |
| RP.51P | Pneumatic cylinders; basic data. |
| RP.52P | Pneumatic cylinders, operating conditions and dimensions (bores 8 to 25mm). |
| RP.53P | Pneumatic cylinders, operating conditions and dimensions (bores 125 to 320mm). |
| RP.54P | Specification for polyamide tubing 11 and 12 bar for pneumatic transmissions. |

| | |
|---|---|
| RP.55H | Schedule of fire-resistant tests for fire-resistant fluids. |
| RP.56H | Test method to determine the fire resistance of FR fluids. |
| RP.57P | Pressure relief valves – recommended data for inclusion in manufacturers' technical sales literature. |
| RP.58H | Hydraulic cylinders – 160 bar medium series – mounting dimensions. |
| RP.59P | Quick action couplings – plug dimensions – 10 bar. Supplements Nos 1 & 2 (1975). |
| RP.60H | Specification for oil hydraulic piston rods. |
| RP.68 | Identification code of ports and operators of pneumatic and hydraulic control valves and other components. (Under review.) |
| RP.69H | Mounting surfaces for hydraulic flow control, pressure control and check valves. |
| RP.71 | Quantities, symbols and units of the International System (SI) to be used for fluid power. |
| RP.73H | Hydraulic cylinders – 250 bar – mounting dimensions. |
| RP.76 | Outside diameters for tubes in fluid power applications. |
| RP.78H | Hydraulic cylinders – tolerances. |
| RP.79P | Specification for non-ferrous pneumatic cylinder tube (ready to use). |
| RP.80 | Cone type connection – 24° – for fluid power tubes and hoses. |
| RP.81H | Compatibility of hydraulic fluids with elastomeric materials. |
| RP.82P | Response time characteristics of pneumatic directional control valves. |
| RP.83P | Characteristics of the pressure medium to be supplied to pneumatic, fluid logic and fluidic devices and systems. |
| RP.84P | Flow coefficients of pneumatic components. |
| RP.85P | Characteristics of pneumatic components. |
| RP.86H | Guidelines for the use of fire-resistant fluids in hydraulic systems. |
| RP.87H | Hydraulic cylinders – mounting dimensions for rod end plain eye. |
| RP.88H | Hydraulic cylinders – mounting dimensions for rod end spherical eye. |
| RP.89P | Clevis for pneumatic cylinders. |
| RP.90P | Rod end attachment (spherical) eye. |
| RP.92H | Statement of requirements for filters in hydraulic systems. |
| RP.96H | Cavities for two-port slip-in hydraulic cartridge valves. |
| RP.99H 1982 | Modular stack and directional control hydraulic valves – sizes 03 and 05 – clamping dimensions. |
| RP.100 1976 | Hydraulic and pneumatic fluid power glossary. |
| RP.101 | Lexicon of terms from hydraulic and pneumatic fluid power glossary. |
| RP.102P | Rod end clevis for pneumatic cylinders. |
| RP.103P | Rod end bearing for pneumatic cylinders. |
| RP.104H 1983 | Acceptance test for hydraulic cylinders. |
| RP.106H | Hydraulic quick-action couplings. |
| RP.109H | Hydraulic filter elements – low temperature integrity test. |

**BCAS Publications**

Guide to the selection and installation of compressed air services.

Buyers' guide to compressed air plant and equipment.

BCAS Brochure.

The principal SI units to be used by the compressed air industry.

A guide to compressor noise reduction.

# Graphical Symbols For Pneumatic Systems and Components

THE FOLLOWING symbols are specified in ISO 1219 : 1976 (E/F). They are virtually identical to those specified in BS2917. The other standard source of reference for Europe, CETOP RP3, is superseded by ISO 1219.

**General (Basic and Functional Symbols)**
The symbols for hydraulic and pneumatic equipment and accessories are *functional* and consist of one or more *basic symbols* and in general of one or more *functional symbols*. The symbols are neither to scale nor in general orientated in any particular direction.

## BASIC SYMBOLS

| Description | Application | Symbol |
|---|---|---|
| Line:<br>– continuous | Flow lines. | |
| – long dashes | Flow lines. | $L > 10E$ |
| – short dashes | Flow lines. | $L < 5E$ |
| – double | Mechanical connections (shafts, levers, piston-rods). | $D < 5E$ |
| – long chain thin (optional use). | Enclosure for several components assembled in one unit. | $L$ = length of dash<br>$E$ = thickness of line<br>$D$ = space between lines |
| Circle, semi-circle. | As a rule, energy conversion units (pump, compressor, motor). | |

cont...

## BASIC SYMBOLS (cont'd)

| Description | Application | Symbol |
|---|---|---|
| | Measuring instruments. | |
| | Non-return valve, rotary connection, *etc*. | |
| | Mechanical link, roller, *etc*. | |
| | Semi-rotary actuator. | |
| Square, rectangle. | As a rule, control valve(s), except for non-return valves. | |
| Diamond. | Conditioning apparatus (filter, separator, lubricator, heat exchanger). | |
| | Flow line connection. | $d$ $d \approx 5E$ |
| | E = thickness of line | $E$ = thickness of line |
| | Spring. | |
| | Restriction:<br>– affected by viscosity | |
| | – unaffected by viscosity. | |

## FUNCTIONAL SYMBOLS

| Description | Application | Symbol |
|---|---|---|
| Triangle. | The direction of flow and the nature of the fluid. | |
| – solid | Hydraulic flow. | ▼ |
| – in outline only. | Pneumatic flow or exhaust to atmosphere. | ▽ |

cont...

## FUNCTIONAL SYMBOLS (cont'd)

| Descripton | Application | Symbol |
|---|---|---|
| Arrow | Indication of:<br>– direction | |
| | – direction of rotation | |
| | – path and direction of flow through valves. | |
| | For regulating apparatus as in *Pressure Control Valves* both representations with or without a tail to the end of the arrow are used without distinction. | |
| | As a general rule the line perpendicular to the head of the arrow indicates that when the arrow moves the interior path always remains connected to the corresponding exterior path. | |
| Sloping arrow. | Indication of the possibility of a regulation or of a progressive variability. | |

## PUMPS AND COMPRESSORS

| Description | Remarks | Symbol |
|---|---|---|
| Fixed capacity compressor. | | |
| Fixed capacity pneumatic motor:<br>– with one direction of flow | | |
| – with two directions of flow. | | |

cont...

## PUMPS AND COMPRESSORS (cont'd)

| Descripton | Remarks | Symbol |
|---|---|---|
| Variable capacity pneumatic motor:<br>– with one direction of flow<br>– with two directions of flow. | | |
| Oscillating motor:<br>– hydraulic<br>– pneumatic. | | |
| Pump/motor units | Unit with two functions, either as pump or as rotary motor. | |
| Fixed capacity pump/ motor unit:<br>– with reversal of the direction of flow | Functioning as pump or motor according to direction of flow. |  |
| – with one single direction of flow | Functioning as pump or motor without change of direction of flow. | 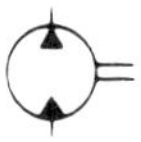 |
| – with two directions of flow. | Functioning as pump or motor with either direction of flow. |  |
| Variable capacity pump/motor unit:<br>– with reversal of the direction of flow<br>– with one single direction of flow<br>– with two directions of flow. | | |

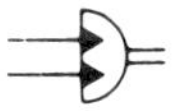

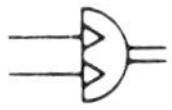

# CYLINDERS

| Description | Remarks | Symbol | |
|---|---|---|---|
| | | Detailed | Simplified |
| Single-acting cylinder: | Cylinder in which the fluid pressure always acts in one and the same direction (on the forward stroke). | | |
| – returned by an unspecified force | General symbol when the method of return is not specified. | 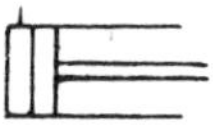 | 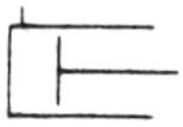 |
| – returned by spring. | | 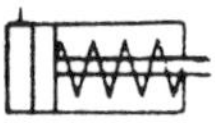 | 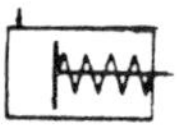 |
| Double-acting cylinder. | Cylinder in which the fluid pressure operates alternately in both directions (forward and backward strokes). | | |
| – with single piston rod | | 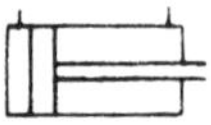 | 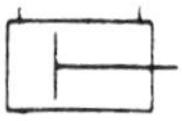 |
| – with double-ended piston rod. | | 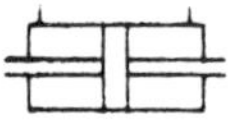 | 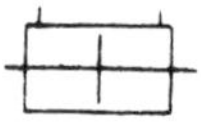 |
| Differential cylinder. | The action is dependent on the difference between the effective areas on each side of the piston. | 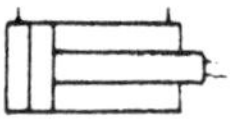 | 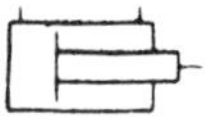 |
| Cylinder with cushion: | | | |
| – with single fixed cushion | Cylinder incorporating fixed cushion acting in one direction only. | 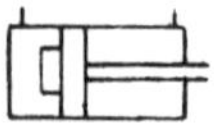 | |
| – with double fixed cushion | Cylinder with fixed cushion acting in both dierections. | 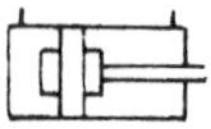 | |
| – with single adjustable cushion | | 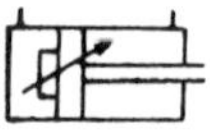 | |
| – with double adjustable cushion. | | 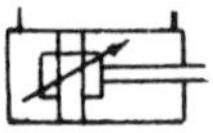 | |
| Telescopic cylinder: | | | |
| – single-acting | The fluid pressure always acts in one and the same direction (on the forward stroke). | | |
| – double-acting | The fluid pressure operates alternately in both directions (forward and backward strokes) | 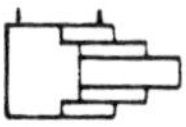 | |

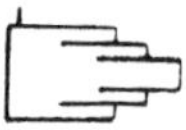

## PRESSURE INTENSIFIERS

| Description | Remarks | Symbol | |
|---|---|---|---|
| | | Detailed | Simplified |
| For one type of fluid. | *eg* a pneumatic pressure *x* is transformed into a higher pneumatic pressure *y*. |  | 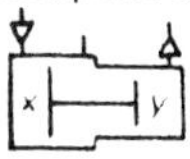 |
| For two types of fluid. | *eg* a pneumatic pressure *x* is transformed into a higher hydraulic pressure *y*. |  | 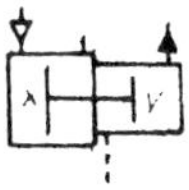 |
| Air-oil actuator. | Equipment transforming a pneumatic pressure into a substantially equal hydraulic pressure or vice versa. | 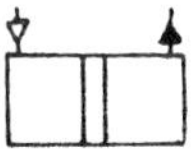 | |

## CONTROL VALVES

| Description | Remarks | Symbol |
|---|---|---|
| Method of representation of valves. | Made up of one or more squares and arrows. *(In circuit diagrams hydraulic and pneumatic units are normally shown in the unoperated condition).* | |
| One single square. | Indicates unit for controlling flow or pressure, having in operation an infinite number of possible positions between its end positions so as to vary the conditions of flow across one or more of its ports, thus ensuring the chosen pressure and/or flow with regard to the operating conditions of the circuit. | |
| Two or more squares. | Indicate a directional control valve having as many distinct positions as there are squares. The pipe connections are normally represented as connected to the box representing the unoperated condition. The operating positions are deduced by imagining the boxes to be displaced so that the pipe connections correspond with the ports of the box in question. | 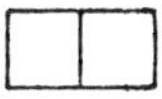 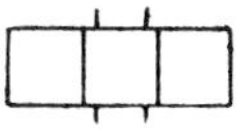 |

cont...

## CONTROL VALVES (cont'd)

| Descripton | Remarks | Symbol |
|---|---|---|
| Simplified symbol for valves in cases of multiple repetition. | The number refers to a note on the diagram in which the symbol for the valve is given in full. |  |

## DIRECTIONAL CONTROL VALVES

| Description | Remarks | Symbol |
|---|---|---|
| Flow paths:<br>– one flow path<br>– two closed ports<br>– two flow paths<br>– two flow paths and one closed port<br>– two flow paths with cross connection<br>– one flow path in a by-pass position, two closed ports. | Square containing interior lines. | <br><br> <br><br><br> |
| Non-throttling directional control valve. | The unit provides distinct circuit conditions each depicted by a square.<br><br>Basic symbol for two-postion directional control valve.<br><br>Basic symbol for three-position directional control valve.<br><br>A transitory but significant condition between two distinct positions is optionally represented by a square with dashed ends.<br><br>A basic symbol for a directional control valve with two distinct positions and one transitory intermediate condition. | 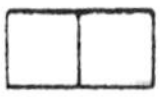<br>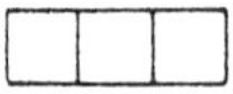<br> |

cont...

## DIRECTIONAL CONTROL VALVES (cont'd)

| Description | Remarks | Symbol |
|---|---|---|
| Designation:<br>The first figure in the *designation* shows the number of ports (excluding pilot ports) and the second figure the number of distinct positions. | | |
| Directional control valve 2/2:<br>– with manual control<br>– controlled by pressure operating against a return spring (*eg* on air unloading valve). | Directional control valve with two ports and two distinct positions. | |
| Directional control valve 3/2:<br>– controlled by pressure in both directions<br>– controlled by solenoid with return spring. | Directional control valve with three ports and two distinct postions. | |
| Directional control valve 4/2:<br>– controlled by pressure in both directions by means of a pilot valve (with a single solenoid and spring return). | Directional control valve with four ports and two distinct positions. | Detailed 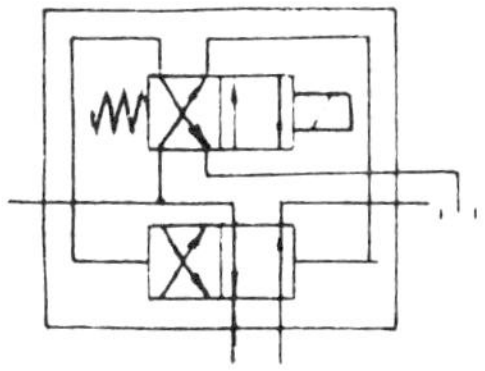<br>Simplified  |

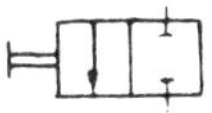

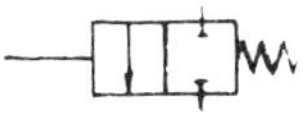

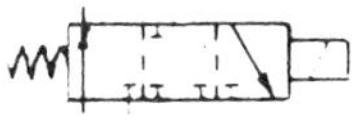

cont...

# DIRECTIONAL CONTROL VALVES (cont'd)

| Description | Remarks | Symbol |
|---|---|---|
| Directional control valve 5/2: | Directional control valve with five ports and two distinct positions. | |
| – controlled by pressure in both directions. | |  |
| Throttling directional control. | The unit has two extreme positions and an infinite number intermediate conditions with varying degrees of throttling. | |
| | All the symbols have parallel lines along the length of the boxes. | |
| | Showing the extreme positions. | 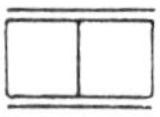 |
| | Showing the extreme positions and a central (neutral) position. | 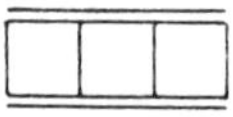 |
| – with two ports (one throttling orifice) | *eg* tracer valve plunger operated against a return spring. | 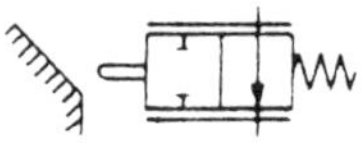 |
| – with three ports (two throttling orifices) | *eg* directional control valve controlled by pressure against a return spring. | 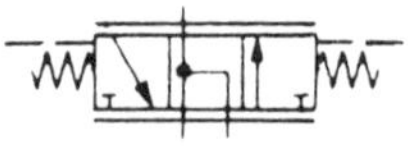 |
| – with four ports (four throttling orifices) | *eg* tracer valve, plunger operated against a return spring. |  |
| Electro-hydraulic servo-valve: Electro-pneumatic servo-valve: | A unit which accepts an analogue electrical signal and provides a similar analogue fluid power output. | |
| – single-stage | – with direct operation. | 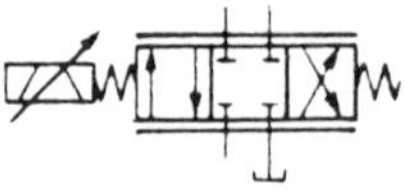 |
| – two-stage with mechanical feedback | – with indirect pilot-operation. | 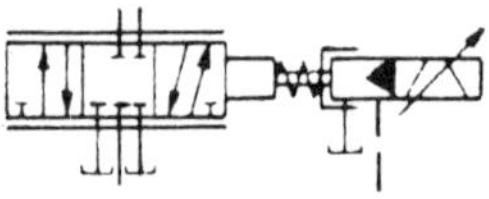 |

cont...

## DIRECTIONAL CONTROL VALVES (cont'd)

| Descrlpton | Remarks | Symbol |
|---|---|---|
| – two-stage with hydraulic feedback. | – with indirect pilot-operation. | 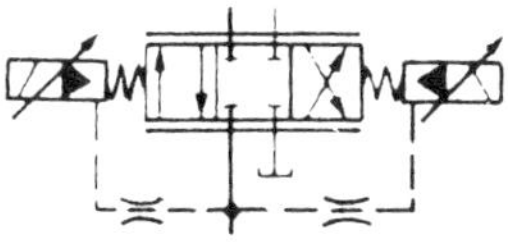 |

## NON-RETURN VALVES, SHUTTLE VALVE, RAPID EXHAUST VALVE

| Description | Remarks | Symbol |
|---|---|---|
| Non-return valve: | | |
| – free | Opens if the inlet pressure is higher than the outlet pressure. |  |
| – spring-loaded | Opens if the inlet pressure is greater than the outlet pressure plus the spring pressure. |  |
| – pilot-controlled | with pilot control it is possible to prevent:<br>– closing of the valve | 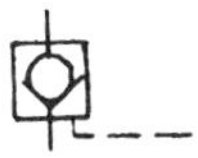 |
| | – opening of the valve. | 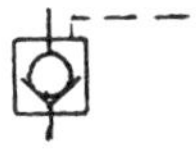 |
| – with restriction | unit allowing free flow in one direction but restricted flow in the other. | 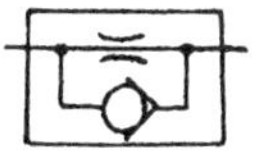 |
| Shuttle valve. | The inlet port connected to the higher pressure is automatically connected to the outlet port while the other inlet port is closed. | 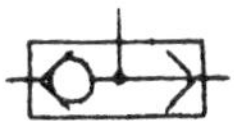 |
| Rapid exhaust valve. | When the inlet port is unloaded the outlet port is freely exhausted. | 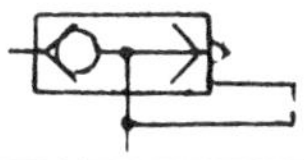 |

# PRESSURE CONTROL VALVES

| Description | Remarks | Symbol |
|---|---|---|
| Pressure control valve:<br>– one throttling orifice normally closed<br>– one throttling orifice normally open<br>– two throttling orifices, normally closed. | | 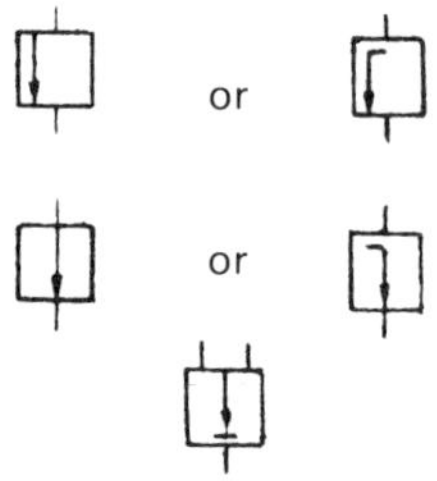 |
| Pressure relief valve (safety valve) | Inlet pressure is controlled by opening the exhaust port to the reservoir or to atmosphere against an opposing force (for example a spring). | 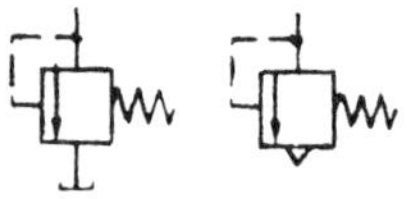 |
| – with remote pilot control. | The pressure at the inlet port is limited, or to that corresponding to the setting of a pilot control. | 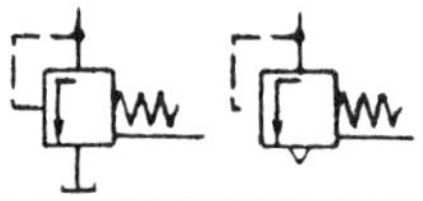 |
| Proportional pressure relief. | Inlet pressure is limited to a value proportional to the pilot pressure. | 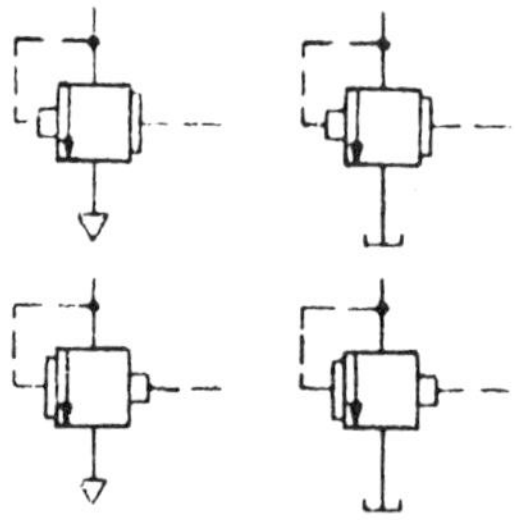 |
| Sequence valve. | When the inlet pressure overcomes the opposing force of the spring, the valve opens permitting flow from the outlet port. | 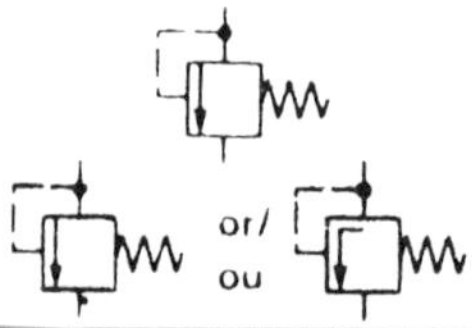 |
| Pressure regulator or reducing valve (reducer of pressure). | A unit which, with a variable inlet pressure, gives substantially constant output pressure provided that the inlet pressure remains higher than the required outlet pressure. | |
| – without relief port | | 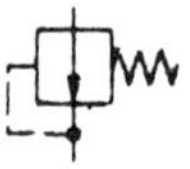 |
| – without relief port with remote control | Outlet pressure is dependent on the control pressure. | 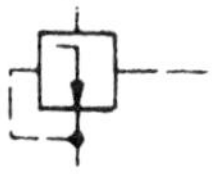 |

cont...

## PRESSURE CONTROL VALVES (cont'd)

| Description | Remarks | Symbol |
|---|---|---|
| – with relief port | | 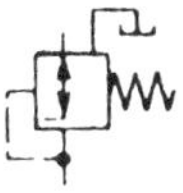 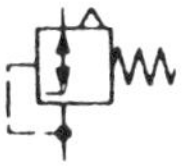 |
| – with relief port, with remote control. | Outlet pressure is dependent on the control pressure. | 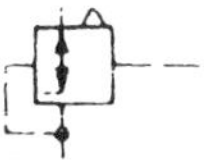 |
| Differential pressure regulator | The outlet pressure is reduced by a fixed amount with respect to the inlet pressure. | 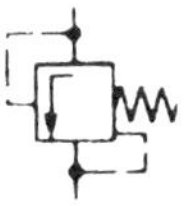 |
| Proportional pressure regulator. | The outlet pressure is reduced by a fixed ratio with respect to the inlet pressure. | 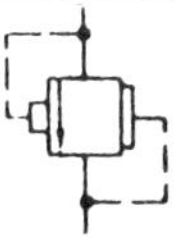 |

## FLOW CONTROL VALVES

| Description | Remarks | Symbol | |
|---|---|---|---|
| Throttle valve: | Simplified symbol (does not indicate the control method or the state of the valve). | | |
| – with manual control | Detailed symbol (indicates the control method or the state of the valve). | | |
| – with mechanical control against a return spring (braking valve). | | | |
| Flow control valve: | Variations in inlet pressure do not affect the rate of flow. | Detailed | Simplified |
| – with fixed output | | | |
| – with fixed output and relief port to reservoir | with relief for excess flow. | | |
| – with variable output | | | |

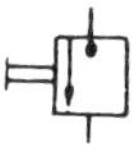

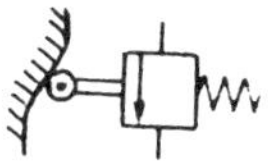

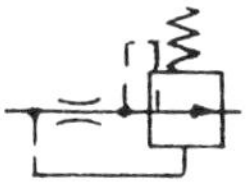

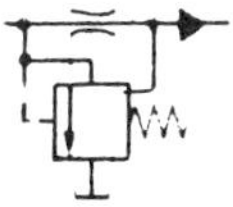

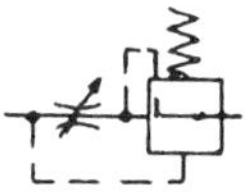

cont...

## FLOW CONTROL VALVES (cont'd)

| Descripton | Remarks | Symbol | |
|---|---|---|---|
| | | Detailed | Simplified |
| – with variable output and relief port to reservoir. | with relief for excess flow. | 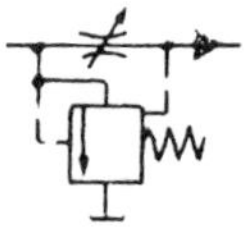 | 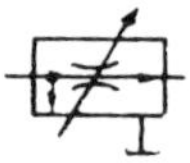 |
| Flow dividing valve. | The flow is divided into two flows in a fixed ratio substantially independent of pressure variations. | 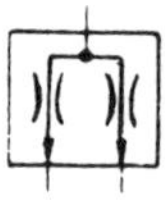 | |
| Shut-off valve. | Simplified symbol. |  | |

## SOURCES OF ENERGY

| Description | Remarks | Symbol |
|---|---|---|
| Pressure source. | Simplified general symbol. | 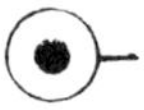 |
| Hydraulic pressure source.<br><br>Pneumatic pressure source. | Symbols to be used when the nature of the source should be indicated. | 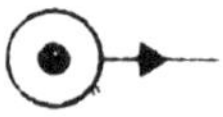 |
| Electric motor | Symbol 113 in IEC publication 117.2 | 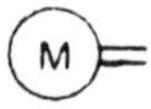 M |
| Heat engine. | | 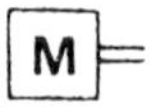 M |
| Flow lines and connections<br>Flow line:<br>– working line, return line and feed line<br>– pilot control line<br>– drain or bleed line | | 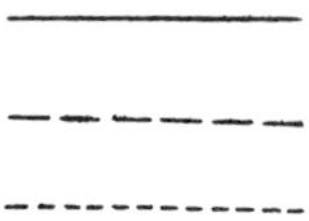 |

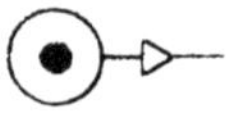

cont...

## SOURCES OF ENERGY (cont'd)

| Description | Remarks | Symbol |
|---|---|---|
| – flexible pipe | Flexible hose, usually connecting moving parts. |  |
| – electric line. | |  |
| Pipeline junction. | |  |
| Crossed pipelines. | Not connected. |  |
| Air bleed. | |  |
| Exhaust port: | | |
| – plain with no provision for connection | |  |
| – threaded for connection. | |  |
| Power take-off: | On equipment or lines, for energy take-off or measurement. | |
| – plugged | |  |
| – with take-off line. | |  |
| Quick-acting coupling: | | |
| – connected, without mechanically opened non-return valve | |  |
| – connected, with mechanically opened non-return valves | |  |
| – uncoupled, with open end | |  |
| – uncoupled, closed by free non-return valve. | | 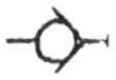 |
| Rotary connection: | Line junction allowing angular movement in service. | |
| – one-way | |  |
| – three-way | | 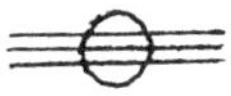 |
| Silencer. | | 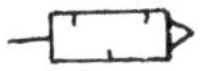 |

# RESERVOIRS

| Description | Remarks | Symbol |
|---|---|---|
| Reservoir open to atmosphere: | | |
| – with inlet pipe above fluid level | | 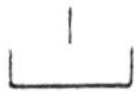 |
| – with inlet pipe below fluid level | |  |
| – with a header line. | | 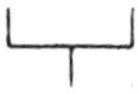 |
| Pressurized reservoir. | | 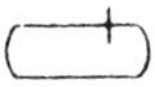 |
| Accumulators | The fluid is maintained under pressure by a spring, weight or compressed gas (air, nitrogen, *etc*). |  |
| Filter or strainer. | | 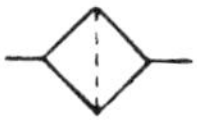 |
| Water trap:<br>– with manual control | | 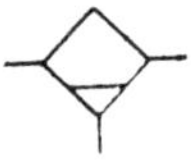 |
| – automatically drained. | |  |
| Filter with water trap:<br>– with manual control | |  |
| – automatically drained. | |  |
| Air dryer. | A unit drying air (*eg* by chemical means). | 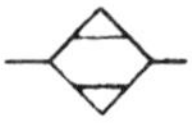 |
| Lubricator. | Small quantities of oil are added to the air passing through the unit, in order to lubricate equipment receiving the air. | 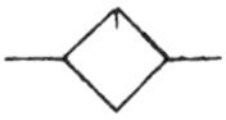 |

cont...

## RESERVOIRS (cont'd)

| Description | Remarks | Symbol |
|---|---|---|
| Conditioning unit. | Consisting of filter, pressure regulator, pressure guage and lubricator. | Detailed symbol 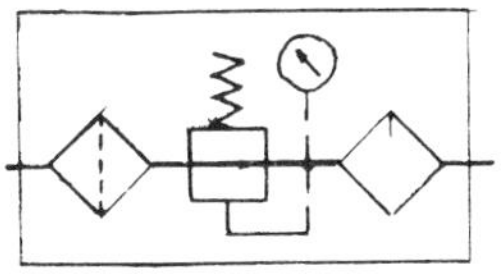<br>Simplified symbol  |
| Heat exchangers. | Apparatus for heating or cooling the circulating fluid. | |
| Temperature controller. | The fluid temperature is maintained between two-pre-determined values. The arrows indicate that heat may be either introduced or dissipated. |  |
| Cooler. | The arrows in the diamond indicate the extraction of heat.<br>– without representation of the flow lines of the coolant<br>– indicating the flow lines of the coolant. |  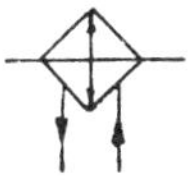 |
| Heater. | The arrows in the diamond indicate the introduction of heat. |  |

## CONTROL MECHANISMS

| Description | Remarks | Symbol |
|---|---|---|
| Rotating shaft:<br>– in one direction<br>– in either direction. | The arrow indicates rotation. |   |
| Detent. | A device for maintaining a given position. |  |
| Locking device. | The symbol* for unlocking control is inserted in the square. | 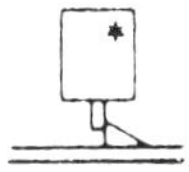 |

cont...

## CONTROL MECHANISMS (cont'd)

| Descripton | Remarks | Symbol |
|---|---|---|
| Over-centre device. | Prevents the mechanism stopping in a dead centre position. |  |
| Pivoting devices:<br>– simple | |  |
| – with traversing lever | | 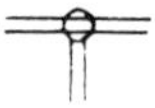 |
| – with fixed fulcrum. | | 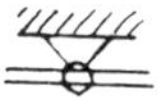 |

## CONTROL METHODS

| Description | Remarks | Symbol |
|---|---|---|
| Muscular control: | General symbol (without indication of control type). | 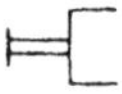 |
| – by push button | | 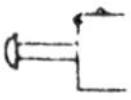 |
| – by lever | |  |
| – by pedal. | |  |
| Mechanical control:<br>– by plunger or tracer | | 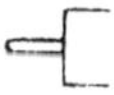 |
| – by spring | |  |
| – by roller | | 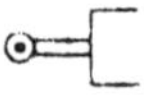 |
| – by roller, operating in one direction only. | | 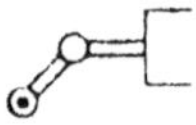 |
| Electrical control:<br>– by solenoid | – with one winding | 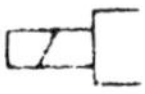 |

cont...

## CONTROL METHODS (cont'd)

| Description | Remarks | Symbol |
|---|---|---|
| | – with two windings operating in opposite directions | 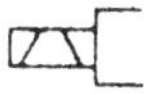 |
| | – with two windings operating in a variable way progressively operating in opposite directions. |  |
| – by electric motor. | | 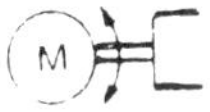 M |
| Control by application or release of pressure<br>Direct-acting control:<br>– by application of pressure. | | 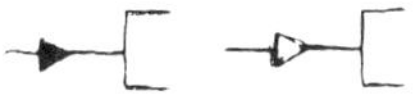 |
| – by release of pressure | | 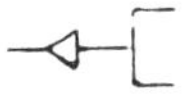 |
| – by different control areas. | In the symbol the larger rectangle represents the larger control area, *ie* the priority phase. | |
| Indirect control, pilot actuated:<br>– by application of pressure | General symbol for pilot directional control valve. | 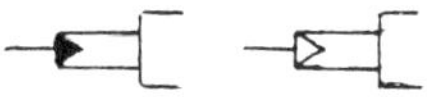 |
| – by release of pressure. | | 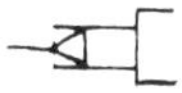 |
| Interior control paths. | The control paths are inside the unit. |  |
| Combined control:<br>– by solenoid and pilot directional valve | The pilot directional valve is actuated by the solenoid. |  |

cont...

## CONTROL METHODS (cont'd)

| Descripton | Remarks | Symbol |
|---|---|---|
| – by solenoid or pilot-directional valve. | Either may actuate the control independently. | |
| Mechanical feedback | The mechanical connection of a control apparatus moving part to a controlled apparatus moving part is represented by the symbol which joins the two parts connected. | 1)<br>2) 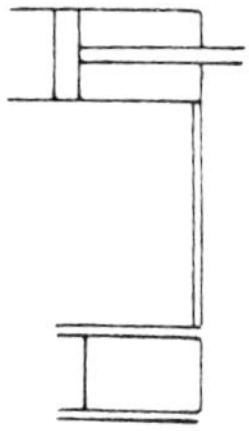<br>1) Controlled apparatus.<br>2) Control apparatus. |

## SUPPLEMENTARY EQUIPMENT

| Description | Remarks | Symbol |
|---|---|---|
| Pressure measurement:<br>– pressure guage. | The point on the circle at which the connection joins the symbol is immaterial. |  |
| Temperature measurement:<br>– thermometer. | The point on the circle at which the connection joins the symbol is immaterial. |  |
| Measurement of flow:<br>– flow meter<br>– integrating flow meter. | | 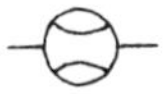  |
| Other apparatus<br>Pressure electric switch. | | 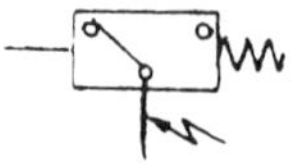 |

**CETOP Recommendations for Circuit Diagrams**

The following rules and recommendations are extracted from CETOP provisional recommendations RP41 – *Hydraulic and Pneumatic Systems Circuit Diagrams* – published by the Comite Europeen des Transmissions Oleohydrauliques et Pneumatiques.

The purpose of RP41 is to facillitate the design, construction, description and maintenance of fluid power systems by aiding communication through uniform presentation and by preventing confusion. The recommendation is based on the applicable ISO standards and CETOP recommendations. It is not necessary for a circuit diagram to incorporate all the requirements listed but anything which is shown must be in accordance with the appropriate sections of the document. The recommendation is generally applicable to the representation of all fluid power systems.

**Design of Circuit Diagrams**

The circuit diagram must show fluid circulation for all the control and motion functions. All hydraulic and pneumatic components and flow line junctions in the plant will be represented functionally. In addition all control components that affect the operational processes of the fluid power system will be shown.

The circuit diagram need not be constructed with reference to the physical arrangement of the equipment in the system. Interdependent circuits should be shown on one diagram. For electro-hydraulic or electro-pneumatic controls the circuit diagram will, however, be divided into one hydraulic or pneumatic circuit diagram and, if necessary, a separate electrical circuit.

The following circuit diagram formats should be used:

ISO-A-Format (ISO recommendation R216)
Length: Preferably 297 mm
exceptionally 420 or 594 mm.
Width: up to 1189 mm.
Diagrams will be folded to ISO A4 format with binding margin (except for diagrams for reproduction).

The layout of the circuit diagram must be clear. Care should be taken to see that the circuit paths are easily followed. Pneumatic circuits should be drawn if possible in the sequence of the operational process of functions. The components of the individual control chains and groups should be drawn wherever possible in the direction of energy flow.

Cylinders and directional control valves should preferably be shown horizontally. Fluid lines should be represented wherever possible by straight lines without intersections.

The location of control components such as sequence switch cams and simple cams, *etc*, should be indicated. If a signalling element is to be actuated by a one-way trip, then the actuating direction will be indicated by an arrow. In the case of electro-hydraulic and electro-pneumatic control systems, signalling and final control elements such as limit switches and electro-magnetic valves will be shown in both circuit diagrams.

Individual sub-circuits may be identified. The function of each operating element should be indicated beside it (*eg* clamping,lifting, *etc*). Where necessary for better understanding a function chart can be drawn up.

Components should be represented by symbols in accordance with the applicable standards, *eg* ISO R1219 for fluid power equipment. Where no symbols exist, the operation of the part should be clearly shown, *eg* by means of a simplified sectional representation or the like.

Symbols that are repeated in the diagram can be represented by a numbered rectangle and the corresponding symbols shown separately.

**Data to go on Circuit Diagram**

Every component must be clearly identified,*eg* by item number. Components in a given sub-circuit may be given a common prime identification plus a running sub-number, *eg* sub-circuit 2 component 3 = 2.3.

**Component Operating Positions**

In the case of valves with discrete switching positions these positions may be identified by letters, the letter 'O' being used only for the spring returned positions (*eg* spring-loaded valves).

**Identification of Ports**

Ports will be identified on the circuit diagram by the characters marked on the components or the connection plate. This applies to individual components and sub-assemblies.

**Identification of Oil or Air Lines**

The pipelines are to be drawn according to their function in line with ISO recommendations ISO R1219 (5.2.1).

In cases where hydraulic and pneumatic circuits are represented on one diagram they should by differentiated in accordance with ISO R1219.

Should it be necessary to identify the pipelines more precisely than in ISO R1219, the following code applies:

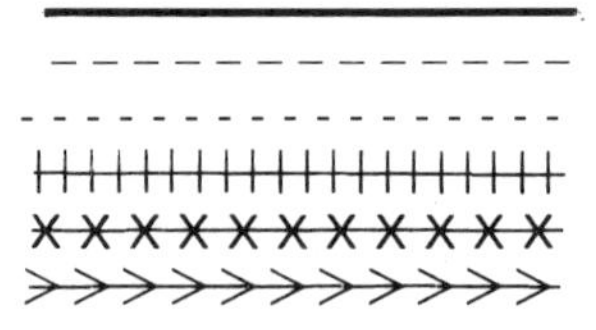

| | |
|---|---|
| full line | pressure line. |
| dashed line | pilot control line. |
| short dotted line | drain or bleed line. |
| full line with short cross lines | return or exhaust line. |
| full line with crosses | replenishing line. |
| full line with arrows | pump inlet line. |

If pipelines are identified by colour, the following colour code should be used:

| | |
|---|---|
| red line | pressure line. |
| red dashed line | pilot control line. |
| blue line | return or exhaust line. |
| blue dashed line | drain or bleed line. |
| green line | replenishing line. |
| yellow line | pump inlet line. |

Any deviations from the above or additional colours should be explained on the circuit diagram.

The following technical data should be given beside each individual component or on a component list.

For reservoirs, give the volume represented by the maximum quantity of fluid in the system and, if required, the minimum quantity, the latter figure indicating the point at which replenishment is necessary. Show also the type and viscosity grade of the fluid. For pneumatic reservoirs indicate the volume.

For constant delivery pumps, indicate nominal delivery rate for the application and, for variable delivery pumps, minimum and maximum delivery rate. In both cases show speed of rotation and nominal power of the prime movers. For prime movers indicate nominal power at stated speed and speed range if applicable. For pressure valves and pressure switches, show pressure setting or permissible pressure range for the system.

For cylinders, the internal diameter of cylinder, the diameter of piston rod, maximum stroke in mm, *eg* 100/50 x 500, and if necessary speed and force should be given. For each cylinder describe the functions, *eg* clamping, lifting, transverse feed *etc.* For telescopic cylinders this information should be given for each stage. Where necessary the circuit should be supplemented by a stroke-time diagram showing the speed.

For semi rotary actuators show torque in relation to pressure, if necessary in

**STANDARD ELECTRICAL SYMBOLS**
**(Based on IEC publication 117.2)**

Conduit, multi-core line.

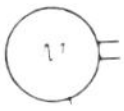

Direct current (d.c.).

Alternating current (a.c.).

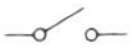

Earth connection.

Limiting line.

Branching.

Crossing point without electric connection.

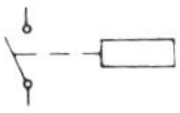

Fuse, the side under tension is marked.

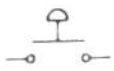

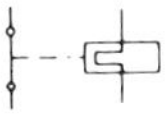

Socket-contact.

Contact-pin.

Pilot lamp.

Diode.

Three-phase delta connection (D-connection).

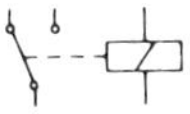

Three-phase star connection (Y-connection).

Electric motor.

Three-phase transformer.

Switch

Two-way switch with interruption.

Push button switch.

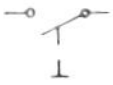

Switch, manually operated.

Switch, electromagnetically-operated.

Switch, thermally-operated.

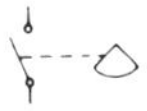

Pressure switch with contact closing when the pressure exceeds a pre-set value.

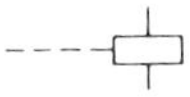

Electromagnetic operation.

Thermal operation.

One-coil relay.

One-coil relay with two-way switch with interruption.

each direction, the angular movement, the speed and the function.

For fixed displacement motors, the nominal capacity and, for variable displacement motors, the maximum capacity per revolution should be indicated. For both give the torque in relation to pressure for the application, speed range, direction of rotation and function.

For gas loaded hydraulic accumulators indicate the precharge pressure, type of gas, effective gas volume, working pressure range and corresponding draw off capacity. See RP62 H when available.

For pipes, give the external diameter in millimetres, wall thickness in millimetres and pipe material if necessary. For flexible hose, indicate the nominal diameter and type.

For delaying devices the time delay in seconds or range of adjustment should be shown. Any temporary connection ports such as bleed and test points should be identified by size, type and as marked on component.

For filters give the nominal filtration rating, flow rate and nominal pressure.

# Editorial Index

**K**



**L**



**M**



**N**



**O**



**P**

**R**

**W**